Foundations of

PHOTONIC CRYSTAL FIBRES

Foundations of
PHOTONIC CRYSTAL FIBRES

Frédéric Zolla

Gilles Renversez

André Nicolet

(Institut Fresnel, University of Aix-Marseille I & III, France)

Boris Kuhlmey

(Centre for Ultrahigh-bandwidth Devices for Optical Systems (CUDOS), The University of Sydney, Australia)

Sébastien Guenneau

(Imperial College, London, & Liverpool University, United Kingdom)

Didier Felbacq

(Groupe d'Etude des Semiconducteurs (GES), University of Montpellier II, France)

NEW JERSEY • LONDON • SINGAPORE • BEIJING • SHANGHAI • HONG KONG • TAIPEI • CHENNAI

Published by

Imperial College Press
57 Shelton Street
Covent Garden
London WC2H 9HE

Distributed by

World Scientific Publishing Co. Pte. Ltd.

5 Toh Tuck Link, Singapore 596224

USA office: 27 Warren Street, Suite 401-402, Hackensack, NJ 07601

UK office: 57 Shelton Street, Covent Garden, London WC2H 9HE

British Library Cataloguing-in-Publication Data
A catalogue record for this book is available from the British Library.

FOUNDATIONS OF PHOTONIC CRYSTAL FIBRES

ISBN-13 978-1-86094-507-6
ISBN-10 1-86094-507-4

Printed in Singapore

Foreword

by
Professor Ross C. McPhedran
CUDOS ARC Centre of Excellence
The University of Sydney

One influential view of the progress of science is that it proceeds by gradual evolution, punctuated by abrupt paradigm shifts and periods of rapid development. We can identify clearly such revolutions when we look back, say in optics, to the period when the laser was developed, and, after a period of being "a solution in search of a problem", rapidly provoked a renaissance of the whole subject. It is my contention that we are living through another such period in optics and photonics, in connection with the related topics of photonic band gap technology, and photonic crystal fibres. The former field started with groundbreaking research by Eli Yablonovitch and Sajeev John, and promises fair to deliver us compact, ultra-fast devices with unparalleled power to manipulate photons for a whole range of applications. The latter field had its origin in the work of Philip Russell and his group at the University of Bath, and has opened up new vistas for optical fibres in communication, sensing, generation of light, etc. Both fields have progressed amazingly rapidly, through a combination of improvements in fabrication technology, sophisticated computational modeling and radical thinking "outside the square". The study of photonic crystal fibres (PCFs) then represents a new frontier for those interested in the science or technology of optical fibres. For those familiar with conventional fibres, it gives conceptual difficulties, because the guidance method in PCFs is completely different from that in conventional fibres, and the old ways of analyzing them simply do not work. Thus, researchers interested in this new field need expert guidance, whether they are experienced in classical fibre technology or new to the field. They will appreciate the range of perspectives on

PCFs afforded by this book. What they will find is a thorough treatment of the theoretical, physical and mathematical foundations of the optics of PCFs. The spread of expertise of the authors is reflected in the depth and strength of their coverage, which will benefit those approaching the subject with a diversity of motives. PCFs may be studied in order to understand how to optimize their application in communication or sensing, as devices confining light by new mechanisms (such as photonic band gap effects), or as physically-important structures which require sophisticated mathematical analysis bearing on questions like the definition of effective refractive index, and the link between large but finite systems, and infinite, periodic systems. All these topics, and more, are treated here. In conclusion, one of the exciting features of this book and the field from which it stems is that it cannot be regarded as the "last word" on its subject. Instead, it offers much of the essential information and concepts for those who wish to join in the exploration of an emerging and important branch of optics and photonics.

Foreword

by
Professor Roy J. Taylor
Femtosecond Optics Group
Physics Department, Imperial College London

The most familiar guise of the optical fibre is that of a hair-like single strand of solid glass, which over the past thirty years has underpinned telecommunications and has been widely utilised in numerous applications from cheap decoration to medical diagnostics. Guiding, a result of total internal reflection, could be controlled via the core-cladding refractive index interface, however, variations have been somewhat limited. Based on concepts of the photonic bandgap, the now enviously simple schemes of the 2-dimensional microstructured optical fibre, introduced in the 1990's, have changed basic ideas regarding optical fibre. Light can now be guided and controlled in ways that were simply not possible, nor even envisaged, in conventional fibre and all because of that array of fine air holes running down their length. The microstructured optical fibre can guide single mode in an air core through photonic bandgap effects, while the air core minimizes optical non-linearity. Alternatively, fibres can be made with exceedingly small glass cores to permit optical non-linearities to be dramatically enhanced and observed over unprecedented short lengths. The dispersion can also be tailored through modifying the waveguide structure such that the zero dispersion is in the visible region, allowing anomalous dispersion, with solitons and associated effects to be observed at wavelength unconceivable with conventional fibres. Another unique property is the endlessly single mode fibre, while large mode area devices can also be readily designed. Although the early demise of conventional fibre predicted by some, may be rather exaggerated, the introduction of microstructured optical fibre has offered extended versatility to the range of applications open to the fibre optic engineer. *Foundations of Photonic Crystal Fibres* provides a sound theoretical and mathematical background to engineers and physicists, to the novice and the experienced researcher, who want to design and predict all the linear properties of these exciting devices, described above. The authors provide a strong foundation and introduction to the field leading on to the numerical tools and methods essential to the complete modelling of the performance of current and future fibre formats, hence a vital tool in this vibrant field in photonics.

Contents

Preface

The purpose of this book

The purpose of this foreword is to explain not only what this book is but also what it is not. The last few years have seen an exponential development of research on photonic crystals, *i.e.* periodic structures, used to control the propagation of light. From the solid state physicist's point of view, such dielectric structures mimic for light what happens in a silicon crystal for electrons, while for the classical optician, this is extending to 2D or 3D structures what is done in Bragg mirrors and in gratings! Anyway, in such structures, propagation may be forbidden in any direction for intervals of frequencies called *photonic band gaps*. The category of microstructured optical fibres became quite recently a substantial subtopic with promising, amongst others, technological applications of the concept of the photonic band gap.

With the invention of lasers and the development of optical fibres able to carry light over hundreds of kilometres came the advent of a new era of worldwide high speed telecommunications, with consequences reaching from benefits in everyday life to the reshaping of the world economy. The first visible effects were cheaper transcontinental phone connections, with the latest being the "Internet revolution". Although one can bemoan the fact that the technologically unprecedented possibilities for people and populations to communicate don't yet seem to have brought better understanding and tolerance between cultures, the optimist will say that this is only a matter of time. Indeed it is easy to forget (especially for younger generations) that optical fibres "transparent" enough to make long haul data transfer possible were invented only about 30 years ago or that the first transatlantic optical fibre was laid less than 20 years ago.

For a decade, the speed (or bandwidth) at which a single optical fibre could carry data seemed pretty much unlimited. Every now and then the bandwidth of telecommunication networks became insufficient, but it appeared that the limitation was not so much due to the intrinsic limits of the fibre, but much more to the limited speed of the signal sources and receivers, so that increasing the bandwidth was for a long time a matter of improving sources and receivers. Intrinsic limits to the bandwidth of optical fibres had been predicted as early as the 1970s, but seemed as theoretical and out of reach as the speed of light for aeroplanes. Research on increasing the bandwidth of optical networks in those times was more a question of how to get the best out of fibres through injecting more information, *e.g.* through multiplexing, than a question of improving optical fibres. Of course the new data injection principles that were developed required new types of fibres, but these were merely more or less adapted versions of the original step index fibre.[1] Given the incredible possibilities offered by existing fibres, there was little eagerness to try to find something radically new, as there was little reason to believe that anything better could ever exist.

It is only in recent years, with the exponentially increasing demand for bandwidth (essentially due to the popularization of the Internet and multimedia contents) along with the progress made in high speed electronics and optoelectronics, that the intrinsic limits of optical fibres have been reached. New, higher density data injection methods are now avaible, but because of non-linear effects, polarization mode dispersion and other effects that we will describe in more detail in the following, good old optical fibres can't keep pace anymore.

Fortunately, at about the same time, the principles of photonic crystals were discovered, leading to the suggestion of radically new mechanisms of light guidance. In the early 1990s the idea of optical fibres using photonic crystal claddings emerged, and after a few years the first photonic crystal fibre was demonstrated. The very first experimental work on these fibres already showed that they could have unprecedented properties and overcome many limitations intrinsic to step index fibres. With photonic crystal fibres, almost everything seems feasible, from *guiding light in vacuum*, hence overcoming all limitations inherent to interactions between light and matter, to achieving dispersion properties unthinkable with step index

[1] What we claim here is that the fibres were based on the same principle of guidance as step index fibres, which doesn't in any way detract from the work and ingenuity needed for the task of designing such fibres.

fibres, from enhancing non-linear effects through to extreme confinement of light and minimizing the same non-linear effects through very large core single mode fibres. The discovery of these possibilities brought prospects of totally new fields of application for fibre optics, such as optical fibres for high power applications, optical fibres for non-conventional wavelength ranges (*e.g.* far infrared, ultra violet), revolutionary optical fibre sensors, particle guidance through hollow core optical fibres, extremely versatile dispersion management, compact high precision metrology, and low-threshold non-linear optics.

Unsurprisingly, the field of photonic crystal fibres became extremely popular, and soon numerous research groups around the world started drawing all kinds of photonic crystal fibres, with hollow or solid cores, with regular or irregular structures, using silica or polymers. Inevitably the pioneers of the field all gave those fibres different names – photonic crystal fibres (PCF), microstructured optical fibres (MOF), crystal fibres (CF), holey fibres (HF) – each having different connotations; we will discuss the meaning of each of these in the next Chapter.

We felt that it is important to produce a document that can be used as a short introduction to this research topic for new or not so new researchers in the field. This book is not a complete catalog of all the existing computation methods. The main reason is that we want to keep it short and not loose the reader in an endless enumeration of all the variants. Once you have understood a few methods, the new ones can be more easily learned. The methods presented here are clearly a deliberate choice of the authors and may be therefore questionable but this choice is committed since they are precisely the methods that the authors use in their own research. But before presenting the solution methods, a preliminary task is to clearly explain the problem. Roughly speaking this can be stated as solving the Maxwell equations in dielectric structures with an axis of invariance. In fact, we seek propagating solutions *i.e.* such that t (time) and z (coordinate along the invariant axis) dependencies are of the form $e^{i(\omega t-\beta z)}$ where ω and β are constants (respectively the pulsation and the propagating constant). At this moment the fundamental problem appears: not all the (ω, β) pairs are feasible! The set of pairs for which a solution of the propagation problem exists are called *dispersion relations* and can be drawn as dispersion curves in the (ω, β) plane. The electromagnetic field associated with a feasible pair is called a *propagating mode*.

The usual way to find a pair is to choose one of the numbers, either ω or β and search for the other one. This kind of problem where the existence of a solution is conditional on the value of a parameter often leads as in the case of this book, to what is called a *spectral problem*. A subtle question appears here: what do we call *a solution* to a problem *e.g.* a partial differential equation with specific conditions (boundary conditions, imposed behaviour at infinity, etc.)? For physicists, any function satisfying the equation and the boundary condition is usually sufficient and it may be given in various forms: explicit analytic closed form, infinite series and in fact any numerical or algorithmic form that gives you the values. Further mathematical sophistication is very often considered as pedantry. In our problem and especially when a dielectric waveguide in open space is considered, a propagating mode must be carefully defined as it must be associated with an eigenvalue *i.e.* a value in the discrete spectrum. Here the use of the framework of functional analysis where functions are elements of infinite dimensional vector spaces is compulsory and the question of whether a number is in the discrete, continuous or residual spectrum is physically meaningful. In the case of holey fibres, the treatment of the propagation of a signal in a low index part of the fiber raises the problem of the *leaky modes*. These are associated with complex propagation constant and here again the very definition of a propagating mode must be generalized and made mathematically accurate. And yet another problem appears: mode propagation is associated with translational invariance of the fiber, so what is a mode propagating in a curved fiber or in a twisted fiber where the translational invariance is lost? Once again, the definition of a propagation mode must be reexamined rather than just given arbitrarily, but this in turn is related to the possibility of performing a Fourier transform of a function that is constant along a co-ordinate... Therefore, the first part of the book is devoted to defining the problem as simply, precisely and as clearly as possible. In order to avoid an intricate mathematical formalism, as many features as possible will be discussed as one-dimensional cases. To complete the discussion, it must be said that if photonic crystals are periodic structures, interesting things happen in defects *i.e.* where the periodicity of the structure is broken, just as doping makes semiconductors more interesting. Moreover, real structures are of finite extension. Nevertheless, the study of ideal crystals is of the highest interest as it allows the determination of forbidden gaps and may be performed by making use of the Floquet-Bloch theory that will be added to our theoretical tools: solutions to wave propagation problems in periodic media are in the form of quasi-periodic functions obtained by multiplying a function having the same periodicity as the media by the wave-function of a plane wave. So much theory may seem superfluous but correctly defining a problem is half of the solution and it is worth the effort. Throughout the book, we try to be as accurate as possible but avoid drowning the reader in mathematical pedantry.

Once the problem is correctly set up, the next step is to solve it *i.e.* finding the (ω, β) pairs together with the corresponding eigenmodes and drawing the dispersion curves, which contain much information about the physics of the system. As stated above, only two methods are presented in this book: the *Multipole Method (MM)* where the fields are expanded in Fourier-Bessel series and a generalized scattering problem is solved, and the *Finite Element Method (FEM)* based on a weak formulation of the problem together with an expansion on a basis of elementary functions (typically low order polynomials) with bounded supports (and the sizes of these supports are small with respect to the size of the problem). How can we justify the choice of these two methods? The advantages of having two completely different methods to solve the same problem is that comparing the results gives some confidence in their accuracy if they match and also that the methods may have complementary advantages so that they can rescue each other if one of them encounters a difficulty in the solution of a particular problem. Therefore, it seems natural to choose two methods that are as different as possible. This is quite the case between the multipole and the finite element methods. As a common feature, the two methods share the fact that the unknown fields or functions are approximated using a linear combination of given basis functions. Nevertheless, one can say that the approach of the multipole method is global while the finite element method is local. On the one hand the basis functions in the Multipole Method are solutions of the Helmholtz equations to be solved (except at the source point, but that is not included in the domain where the solution is considered) as they are the field produced by a multipolar radiating source. Their linear combinations are themselves solutions and the problem is to pick the good one by choosing the right coefficients (*i.e.* the values of the equivalent multipolar charges). This is performed via the numerical expression of local conditions: continuity conditions of the tangential fields at the interfaces. On the other hand the basis functions in the finite element methods are simple (*e.g.* piecewise linear) functions with bounded support (of small size). In the case of the Whitney family of finite elements, if basis functions do not match the Helmholtz equation, they locally respect the continuity conditions at the interfaces *e.g.* tangential continuity for electric and magnetic fields represented by edge elements (that is the geometrical

nature of the fields that determines the kind of element to be used) and so do their linear combinations. In this case, the Helmholtz equation itself (in weak variational form) is used to compute the numerical coefficients of the approximation.
Let us compare the respective advantages of the methods that show their complementarity. On the one hand, the Multipole Method is specific to structures including matrix with homogeneous linear characteristics and inclusions with linear characterictics, furthermore the method is extremely efficient in case of circular boundaries for which analytical formula are available. It allows fast and accurate computations of the dispersion curves necessary for parametric study and optimization of the fibres. On the other hand, the finite element method is extremely flexible from both the geometric point of view, since it allows an easy treatment of inclusions of any shape (and permits also a 2D treatment of twisted fibres in helicoidal coordinates), and from the material point of view as it allows anisotropic, inhomogeneous (and possibly non-linear) characteristics to be incorporated.

There are of course many other methods that we discuss briefly here to explain why they are less adapted to our problem. First of all, we must agree on what we call a "method": for us, it is a complete algorithmic procedure that allows you to compute the values of the electric and/or magnetic fields at any point (at least in the vicinity of the structure).

- With respect to this definition, the *Effective Index Method* is not a method but rather an approximation. It tells you to replace an inhomogeneous structure (*e.g.* a holey part of the fiber) by an equivalent homogeneous index. Although, this kind of model may be extremely useful in some cases, it is not adapted to the study of microstructured fibre. Moreover, it is not a method but an ancillary trick as it just gives a new simplified structure to be computed by a numerical method of your choice.
- To some extent, the problem is the same with the very popular *Supercell Method.* Imagine that you have to find the modes propagating in a defect of a periodic structure. For the periodic structure itself, the Floquet-Bloch theory gives you the general form of the solution as a quasi-periodic Bloch function. The solution may be computed on the cell using special Bloch boundary conditions. As a consequence of the simple geometry of the cell, the solutions themselves may me computed as Fourier series. In the case of a defect (*e.g.* a missing element in the infinite repetition of a pattern), the

structure is no longer periodic and the Floquet-Bloch theory can not be applied. Nevertheless, one arbitrarily decides that this defect may be correctly represented by an empty cell surrounded by a finite set of layers of regular cells. Moreover, in order to get a convenient problem, one decides again arbitrarily that this defect and its neighbourhood may be repeated infinitely in order to recover a periodic structure where the Floquet-Bloch theory applies but now with a much larger cell that contains the defect and its neighborhood: a supercell! This procedure is quite easy to implement and provides interesting numerical results but we do not use it as it is based on several unjustified hypotheses and it is therefore impossible to prove convergence to the requested solution (since, if it is stable and converges to the solution of the artificial periodic problem, there is no guarantee that this result is a solution of the original problem!). Besides, we cannot hope to estimate the loss of the leaky modes: an electromagnetic field in a supercell cannot run away! Here again, the supercell method is merely a trick, replacing a non-periodic structure by an artificial periodic one, that is usually associated with a Fourier method but could be used with other numerical procedures able to find the Bloch solutions (*e.g.* the FEM).

- The basic principle of the *Plane Wave Method* is to express the unknown fields as a sum of plane waves with various propagation vectors. The only difference between this method and the multipole method is the choice of the basis functions. Plane waves are also solutions of the Helmholtz equations. As they are complex exponential of linear forms of the coordinates, this method is equivalent to a bare Fourier method. Therefore, why choose the MM which involves more tedious computations with Bessel functions? Simply because the plane waves are completely non localized and they slowly converge to localized wave packets (such as the propagating modes). On the contrary, multipole series require a small number of terms to get a reasonable accuracy. Note that a similar situation occurs in the computation of electronic band structures in solid state physics. For conduction bands with nearly free electrons, a plane wave method is used since plane waves correspond to free electrons while a Linear Combination of Atomic Orbitals (LCAO) method is used for valence bands.
- The *Boundary Element Method* (BEM) may be considered as an

intermediate between the finite element method and the multipole method. On the one hand, the geometry is divided into small elements of given shapes where some unknown functions are supposed to have a given behavior (*e.g.* low order polynomial) just like in the finite element method. On the other hand, the problem is set in the form of an integral equation involving the fundamental solution (Green function) of the differential problem. The main difference with the finite element method is that instead of the geometrical domains of the problem, only the boundaries have to be discretized. This leads to a dramatic decrease of the number of unknowns in the problem. As the physical interpretation of the Green function is the field produced by a point charge (Dirac distribution), the BEM can be considered as a member of the equivalent charge method family just as the multipole method (usually one speaks of equivalent or fictitious charge methods when the charges, monopoles or multipoles, are not on the surfaces of discontinuity in order to avoid singular integrals). The BEM is as flexible as the finite element method as far as the geometry is concerned and it allows complex geometries to be dealt with more easily since only the boundaries have to be meshed. Unfortunately the method is well adapted only to homogeneous (and therefore linear) media. From a practical point of view, one of the most difficult steps in the method is the numerical computation of a singular (multiple) integral involving the Green function and/or its derivatives, and some of those integrals are even defined as Cauchy principal values or Hadamard finite parts. The real bottleneck of the boundary element method is the fact that it produces large full matrices. Even if the number of unknowns is much lower than in FEM, the sparsity of matrices in FEM makes it often much more tractable. Nevertheless computational linear algebra is an extremely active field of research and the ever increasing power of computers prevents us from drawing any definitive conclusion on this subject. Anyway, in the present state of the art, the FEM is more flexible than the BEM (since FEM allows almost any kind of material properties and coordinate systems to be dealt with) and the drawback of having to mesh domains is not a heavy one since pretty good free tools are now available that do this work so that it appears as an automatic step after the boundary meshing.
There are two main settings of BEM: collocation methods where the

algebraic equations are obtained by writing the integral equation for particular points of the surface and the Galerkin method where the integral equation is used in a weighted residual process similar to that directly applied to the differential equation to obtain the FEM.
In the context of high frequency electromagnetism, the BEM and other related equivalent charge methods are often called *Method of Moments*. An extremely active field of research is the *Fast Multipole Method (FMM)* where the algorithmic efficiency is optimized using multipole decomposition of the remote action of sets of charges and the idea that the higher the moment the faster it decreases with the distance. The adaptation of such methods to "mode chasing" may be a promising path for future research.

- The *Finite Difference Time Domain Method (FDTD)* is one of the most powerful methods designed to tackle electromagnetic wave propagation. The principle of the finite difference method is to replace the derivatives of functions in differential operators by finite differences *i.e.* by differences of the values of the function computed on two ends of a small interval (of time or of space). A fundamental question in using the method is of course "How small?". Doing this, algebraic equations, instead of differential equations, can be obtained directly, and these can be solved via some computational algebra. The FDTD algorithm proposed by Yee in 1966 is in fact much more than that since it relies on the deep structure of Maxwell equations. Instead of using a single grid to produce the finite differences, two dual grids are used. First take a 3D rectangular grid, the primal grid, and then build the dual grid so that its nodes are the centers of the parallelepipeds of the primal grid. Note that there is a cross-correspondence not only between nodes and parallelepipeds of the two grids but also between edges and faces. The unknowns of the problem are the components of the electric fields along the edges of the primal mesh and of the magnetic field along the edges of the dual mesh. A development of the Maxwell equations to the first order provides the algebraic system. As for the time variation, duality is also used. One associates the electric field with equally spaced instants (integer index time steps) and the magnetic field with the middle of the time intervals between those instants (half integer index time steps). Here also a first order development provides the finite difference algebraic equations.

The time integration scheme thus obtained is called "leapfrog" as one "jumps" from the values of electric field for an integer time step to the values of magnetic field for the following half integer time step and so on... The algorithm is extremely powerful as it is purely explicit: no matrix storage and computation is necessary! This method is probably the most used currently to solve very large electromagnetic propagation problems. The main drawback is the use of a structured mesh that makes the accurate representation of complex geometries difficult. The modern analysis of the FDTD method and of edge FEM shows that the two methods are in fact very close in spirit: the unknowns are associated with the edges of a mesh covering the whole domain and they rely on duality. Current research is still looking for FDTD on an unstructured mesh *e.g.* by lumping finite element matrices on the diagonal to avoid heavy matrix computations in time domain FEM. Nevertheless, in our case, the FDTD suffers a major flaw: it is intrinsically a time domain method, but we need a frequency domain method to directly obtain the modes (even though the Fast Fourier Transform is a magic algorithm to convert time domain results to frequency domain results).

- There are many other variants of the numerical methods. For instance, giving up the inter-element continuity in "finite-element-like" methods leads to the *Discontinuous Galerkin Method* that is receiving an increasing amount of interest for its efficacy in the solution of hyperbolic problems. FE-FD methods lead to similar schemes starting from different points of view. Another approach is to emphasize conservation laws for fluxes, leading to the *Finite Volume Method.*

  There are also the meshless methods or fuzzy element methods where the approximation is based on a linear combination of functions with overlapping supports.

  Another category of methods comprises those close to geometrical optics and ray tracing such as the *Geometrical Theory of Diffraction* (GTD). They are applicable when the wavelength is small with respect to the geometry but this is not the case in our problem.

This brief overview of numerical methods in electromagnetism shows that it is hopeless to give any complete account. The two main methods presented in this book are naturally suited to the problem of mode determination in microstructured fibres are, quite complementary, and cover a large part of the basic techniques necessary to understand numerical modelling in this context.

The adaptation of those methods to periodic problems in the context of Floquet-Bloch theory is also discussed in detail.

In the final chapter, some numerical examples demonstrate how the methods can be used to study microstructured fibres and they also emphasize some physical phenomena at the edge of the current understanding of microstructured fibres.

We hope we have convinced the reader that it is not a waste of time for her/him to read the rest of this book and wish her/him to enjoy it.

Vade mecum

From the start, our intention was to write a book for an audience as wide as possible, say from graduate students to researchers. We hope that we have achieved our aims! Be that as it may and despite all the efforts of the authors especially on the homogeneity of the notations some chapters are merely aimed at graduate students and some others are intended for experienced researchers... Additionally, our aim was to gather not only practical features of a certain class of new fibres but also theoretical aspects associated with such fibres. This short section should allow the reader to find efficiently what he is looking for.

Chapter 1 Everybody should start with this chapter. After a short review of conventional optical fibre properties (section 1.1). It briefly defines what is a photonic crystal (section 1.2), it also sets the terminology used (microstructured optical fibres and photonic crystal fibres). The skilled reader may directly start with section 1.3. This section explains how it is possible to guide light in a fibre using photonic crystal. Section 1.4 is devoted to solid core microstructured optical fibres dealing with their guidance mechanism and some of their properties. This chapter finishes with section 1.5, this section details what is a leaky mode through several approaches. Some parts of this section may require careful thought for the beginner but it is worth it.

Chapter 2 This chapter is aimed at all readers. Section 2.1 is a brief survey of Maxwell equations both *in vacuo* and in idealized matter. Section 2.2 is devoted to one-dimensional structures; these simple structures permit the adoption of a pedestrian approach to the concepts encountered in the rest of this book in a more general context. Amongst other things, both structures of finite and infinite extent are considered. The former are studied in two different ways: through the poles of the scattering matrix and trough the spectrum of an operator. The latter are studied by making use of the so-called Bloch wave theory. Section 2.3 is more essential ; relations between transverse and axial components are derived in quite a general context. Section 2.5 probably will interest readers to which the theoretical aspects of Physics appeal. Finally, we found it necessary to address Bloch wave theory; in the section 2.6 attention has been drawn especially to the Bloch wave decomposition and the so-called Wannier transform.

Chapter 3 This chapter is a general presentation of the finite element method in electromagnetism together with a detailed account of its applications to inhomogeneous waveguides. Section 3.1 is a very general presentation of the finite element method. It starts with a one dimensional example for absolute beginners. It also includes more advanced topics such as mixed formulations and eigenvalue problems that are more mathematically demanding, as this section relies strongly on functional analysis. Some abstract material is relegated to appendix A to lighten the presentation. Section 3.2 is a kind of fresh look at electromagnetism and discrete methods that can be read almost independently from the rest of the book. It presents the geometrical point of view on how FDTD and FEM fit together in this framework. Sections 3.1 and 3.2 are quite parallel and lead to the same conclusion: the necessity to use edge elements to interpolate the electric and magnetic fields. Section 3.3 contains a discussion of some technical problems at the heart of finite element modelling. Each subsection can be read more or less independently (even though there is some logical progression in their ordering) when the reader feels the need to have a deeper look at the considered problem. Section 3.4 is in fact the first section specifically devoted to waveguides and optical fibres. The reader acquainted with finite elements should start reading here. It is a detailed description of the full wave model using an electric formulation and edge elements followed by a short presentation of the variants found in the abundant literature on the subject. Section 3.5 explains how the finite element method can be combined with the Floquet-Bloch theory to study wave propagation in periodic structures. Section 3.6 on twisted fibres is an illustration of the versatility of the finite element method together with a discussion on how the concept of a propagation mode can be generalized when the translational invariance is lost.

Chapter 4 This chapter is intended for all the readers and describes in some detail one of the two main numerical methods used in this book to study MOFs: the Multipole Method (MM). We define the foundations of MM in the first parts of section 4.2. Then we present a simplified approach of the method in order to clearly describe its key ideas and to define the way the electromagnetic fields are expressed. Then, we give the full method. The related sections are more difficult but we try to make it accessible to the majority of interested readers, nearly all the delicate or tedious calculus being done in Appendix B. In section 4.3, we sum up the theoretical work of McIsaac relating to waveguide symmetries and the properties of modal fields, these results being also very useful to classify MOF modes. We give several examples which cover most idealized MOF structures. In the next two sections 4.4 and 4.5, we show how the Multipole Method can be implemented and how it can be validated. In section 4.6 of this chapter, we give complete and accurate results obtained with the Multipole Method for the electromagnetic fields and Poynting vector for three few-hole MOF: a hexagonal MOF with six identical holes, a six hole birefringent MOF and a square MOF with eight holes.

Chapter 5 This chapter presents the analysis of electromagnetic waves propagating through a doubly periodic array of cylindrical channels in oblique incidence. The change of the Multipole Method described in the previous chapter to periodic structure give us the necessary tools. This change amounts to reducing a spectral problem for partial differential equations set on a basic cell with Floquet-Bloch boundary conditions, to a certain algebraic problem of the Rayleigh type. We obtain a formulation in terms of an eigenvalue problem that enables us to construct *dispersion curves* and thereby to look at *photonic band gap* structures in oblique incidence. The mathematics behind the Rayleigh algorithm depend upon the resolution of an integral equation involving a Green's function which has sources at all the lattice points within the array. The idea is then to expand both the electromagnetic field and the Green's function in terms of a basis of functions which are appropriate to the geometry of the problem. Since the cylinders which we consider have circular cross-sections, we expand the field in terms of Bessel functions rather than in a Laurent series contrarily to Rayleigh. We show that these so-called multipole-expansions are absolutely convergent within a prescribed annulus. Projecting these expansions back onto the surface of each cylinder and using the boundary conditions (tangential continuity of the fields) we are led to a *scattering matrix*. It then remains to take into account the contribution of the superposed effects of all singular sources arising from all the other cylinders in the array.

This is done by introducing the so-called *lattice sums* which require special treatment in order to avoid a conditional convergence. Once this is done, we end up with an infinite linear system which, after suitable normalisation, can be truncated to a certain multipole order with a fairly high accuracy.

Chapter 6 This chapter tackles a natural question when dealing with all methods based upon scattering matrices like the Multipole Method. As a result, this part is merely intended for all readers interested in the numerical implementation of the method. The question posed is the following: are we doomed to find the complex poles of the scattering matrix in a point-by-point fashion or else, following the example of the *Brave Little Tailor*, can we obtain seven (or more) poles at one stroke? The answer is affirmative and this fairy tale is told in this Chapter.

Chapter 7 This chapter is intended for all the readers and is the key chapter of the book as far as MOF properties are concerned. We start with a brief review of the basic properties of MOF losses (section 7.1). In section 7.2, we expound the definitive answer concerning the single-modedness of plain core MOF through the study of the description of a cut-off of the second mode. In the following section, 7.3, we show that the fundamental mode exhibits another kind of cut-off. We then explain how using the results of the previous sections we can construct a diagram of the different operational regimes of plain core MOFs. We also give simple physical models of conventional optical fibres which are valid approximations of MOF outside a transition region where MOF-peculiar properties arise. In section 7.4, we describe plain core MOF chromatic dispersion properties and we explain how the chromatic dispersion can be managed. In the last section, we illustrate one of the most striking properties of MOF, the possibility to guide a mode in a hollow core in a way that allows us to systematically reduce the losses. This last section allows us to couple the finite element method for band diagram computations and the multipole one for the study of finite structure.

Appendix A This appendix gives the general abstract framework for mixed finite element methods and discusses the fundamental inf-sup criterion. It can be read as a fruitful complement to Section 3.1.3.

Appendix B This appendix contains some details concerning the formulation of the Multipole Method (MM). Section B.1 deals with the Wijngaard identity. This identity is the crux of the method. In the full derivation of the MM several changes of basis for the Bessel and Hankel functions are required. Section B.2 is devoted to this topic. In the presented implementation of the MM, only circular inclusions are used since in this case analytical results are available for the inclusion scattering matrices even when conical mounting is considered. These results are described in section B.3.

Appendix C This appendix is a short survey of the mathematical tools used in this book. Main notations and definitions of mathematical concepts are given there together with some important theorems and a few fundamental examples. This is not a course but rather an informal lexicon. Vector spaces appear as a powerful unifying concept and it is worth the effort for the physicist to grasp the main ideas of this abstract theory. A first application is the functional analysis that appears as the theory of infinite dimensional vector spaces and is a natural framework to set up distribution theory, spectral analysis, and discrete (*i.e.* finite dimensional) approximations. A second application is differential geometry. Although classical theoretical electromagnetism is most frequently presented using vector analysis, differential geometry gives a sharper view (see Section 3.2) even if vector analysis in Cartesian coordinates remains a valuable tool to perform some explicit computations. Even if this is not really necessary here, we could not resist the pleasure of adding a few words on how the distribution theory merges with differential geometry in the theory of de Rham currents and recalling the basics of complex analysis as an application of differential geometry.

Acknowledgements

We would like to thank:

Ross McPhedran for honouring us by writing the preface of this book and also for all the friendly and inspiring conversations on the subject and on many other topics including Shakespeare. This book owes a lot to the Marseille-Sydney collaboration.

Our colleagues and friends from the ARC Center for Ultrahigh Bandwidth Devices for Optical Systems (CUDOS) at the University of Sydney and University of Technology, Sydney, who developed the Multipole Method for MOFs with us. We are particularly indebted to Tom White who was audacious enough to dive into the -in these early days still quite hazardous-task of the very first simulations of MOFs using the Multipole Method; to the wizard of matrices (and more) Lindsay Botten; and to C. Martijn de Sterke – *de facto* and much appreciated co-supervisor of one of the authors. In this context we would also like to acknowledge the travel support received from the French and Australian governments and from the French embassy in Canberra and the University of Sydney under the PICS, IREX and *cotutelle* schemes, without which our very fruitful collaboration with our Australian colleagues would have remained wishful thinking.

Daniel Maystre, who got two of us started on the Multipole Method and whose wisdom and knowledge reach unfathomable depths.

Part of the writing of this book was undertaken when one of us was working with Sasha Movchan in the Department of Mathematical Sciences at Liverpool University. The book came to fruition whilst this author was working with John B. Pendry, FRS, in the Physics Department at Imperial College London. Henceforth, we wish to thank them both warmly for their invaluable human and scientific support.

John Pottage who demonstrated that it is possible to read this book within a week.

Niels Asger Mortensen for interesting discussions on modal cutoffs.

C. Geuzaine for his invaluable help as an infallible GetDP guru and, with him, all the present and former colleagues of the Department of Electrical Engineering of the University of Liège led by W. Legros: J.-F. Remacle, F. Henrotte, P. Dular, D. Colignon, H. Hédia, F. Delincé, A. Genon, B. Meys, R. Sabariego, Y. Gyselinck, J.-P. Adriaens, M. Umé, P. Scarpa, the late J.-Y. Hody, T. Ledinh, J. Mauhin, V. Beauvois, L. Brokamp, N. Bamps, M. Paganini, W. Legros, and many others...

A. Bossavit, E. Tonti, and P. Kotiuga for fascinating conversations on numerical electromagnetism and particularly on the applications of differential geometry and algebraic topology.

Chapter 1

Introduction

This Chapter is a short technical introduction to microstructured optical fibres. Our aim is to introduce the notation and the fundamental concepts concerning optical fibres that will be encountered throughout the book. First, we review the principles and main properties of conventional optical fibres. Then we give a short overview of photonic crystals and light guidance in the defects that can be incorporated within a host lattice. This leads us to introduce the principles of guidance in photonic crystal fibres and more generally in microstructured optical fibres (MOFs). We also mention some of their advantages and applications. To conclude, we come back to a crucial issue: the nature of mode propagation in MOFs. The finite extent of the cladding structure of MOFs implies that all of the so-called guided modes are in fact leaky.

1.1 Conventional Optical Fibres

1.1.1 *Guidance mechanism*

Conventional optical fibres [SL83; Agr01] rely on total internal reflection to guide light. The simplest optical fibre – the step index fibre – consists of a dielectric core with refractive index n_{CO} (the refractive index is the square root of the relative permittivity) surrounded by another dielectric (called *cladding*) with refractive index n_{CL}. Using a ray approach and the Snell-Descartes law, it is easy to see that if $n_{\mathrm{CO}} > n_{\mathrm{CL}}$, light propagating in the core reaching the core/cladding interface is totally reflected back into the core as soon as the angle between the direction of propagation and the core/cladding interface is small enough.

More rigorously, we consider the infinitely long step index fibre with circular cross-section (radius ρ) depicted in Fig. 1.1. We suppose that the light propagating along the axis of the fibre (z-axis) has free-space wavelength λ and wavenumber k_0 ($k_0 = 2\pi/\lambda$). The studied system is invariant under any translation in the z-direction, which implies that the electromagnetic fields associated with an individual mode can have a dependence of $\exp(i\beta z)$ along z. β is called the *propagation constant.*[1] Therefore β is the common z-component of the wave vectors in the core and in the cladding. Since the norms of the wave vectors in the core and in the cladding are $n_{\rm CO}k_0$ and $n_{\rm CL}k_0$ respectively, β must be less than or equal to $n_{\rm CO}k_0$ to propagate in the core and less or equal to $n_{\rm CL}k_0$ to propagate in the cladding. If $n_{\rm CL} < \beta/k_0 < n_{\rm CO}$, light can propagate in the core, but not in the cladding: the light is trapped in the core. We will sometimes also use the *perpendicular wavenumber* $k_\perp$, defined by

$$k_\perp^2 + \beta^2 = n^2 k_0^2, \tag{1.1}$$

where n is the local refractive index (equal to $n_{\rm CO}$ or to $n_{\rm CL}$). Since β is real in the case of conventional optical fibres with $n_{\rm CO} > n_{\rm CL}$, in the region where $\beta < nk_0$ we simply have $k_\perp = \sqrt{n^2k_0^2 - \beta^2}$ and $k_\perp = i\sqrt{\beta^2 - n^2k_0^2}$ elsewhere.

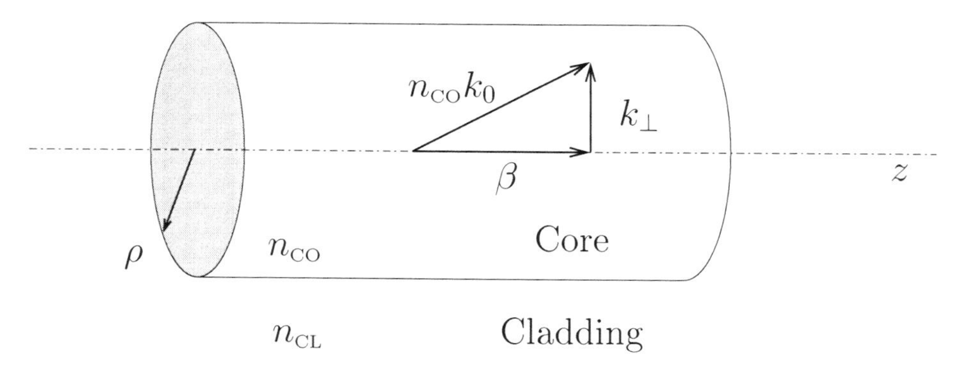

Fig. 1.1 Conventional step index fibre.

[1] Actually, a more general z dependency can be assumed but this is beyond the scope of this introduction.

1.1.2 *Fibre modes*

Considerations relating to total internal reflection or, equivalently the possible propagation constants in the core and the cladding give a necessary but not sufficient condition on β for light to be guided. Indeed, we have seen that the variation that the fields can have along z is given by the phase factor $\exp(i\beta z)$. Between two arbitrary values z_1 and z_2 of z, the fields propagate and undergo reflection at the core/cladding interface, but the transverse field distribution at $z = z_1$ and $z = z_2$ must only differ by the phase factor $\exp[i\beta(z_2 - z_1)]$. This defines a resonance condition, and only a discrete set of transverse field distributions and associated β values fulfill this condition. Mathematically speaking, the values β and their associated transverse field distributions are eigenvalues and associated eigenfunctions of the propagation equation [SL83]. The value of β together with its corresponding field distribution constitute a *mode* of the fibre. There are a countable number of modes with $n_{\mathrm{CL}} k_0 < \beta < n_{\mathrm{CO}} k_0$, and these are called guided modes.

Note that the propagation equation also has solutions with $0 < \beta < n_{\mathrm{CL}} k_0$, *i.e.* outside the range where total internal reflection occurs. These modes can propagate in the cladding, and are called *radiative modes* [Mar91]. For these modes, the set of β values is infinite and continuous.

Each mode has a specific field distribution, with its own specific symmetry properties. From symmetry considerations it can be shown that for circularly symmetric fibres, modes can be either non degenerate or twofold degenerate [Isa75a; Isa75b]. In the latter case two field distributions with complementary symmetries are associated with the same propagation constant.

1.1.3 *Main properties*

1.1.3.1 *Number of modes*

The number of guided modes of the step index fibre depends on the dimensionless fibre V-parameter [SL83].

$$V = k_0 \rho (n_{\mathrm{CO}}^2 - n_{\mathrm{CL}}^2)^{1/2}. \tag{1.2}$$

The smaller this parameter, the smaller the number of guided modes a fibre can carry. If at a given wavelength $V < 2.405$, a single degenerate pair of modes[2] is guided by the fibre. When there is only one possible value of β, the fibre is said to be *single-mode*. Either one of the pair of guided modes, associated with the same value of β, is referred to as the *fundamental mode*. Note that for a fibre to be single-mode, it needs to be designed with a combination of small core size to wavelength ratio and small difference in refractive indices between core and cladding. Conversely, a given fibre is always multi-mode for sufficiently small wavelengths. It is also worth noting that however small the parameter V is, there is always a fundamental guided mode of the fibre.

1.1.3.2 *Losses*

In the ideal step index fibre, the attenuation of a guided mode while propagating is solely due to material absorption so if the material is lossless then the guided mode is not attenuated at all. The intrinsic material absorption of pure silica for wavelengths between approximately 0.8 μm and 1.8 μm is very small (however in fabricated fibres, the absorption by the hydroxyl group OH is not negligible [Oko82] especially around its main peak centred at 1.38 μm), and in theory light in this wavelength range could be carried over hundreds of kilometers without noticeable loss. Nevertheless, until the early 1970s material absorption in fibres was considerable, due to contamination by water or metallic ions of the silica used to draw fibres. Since the work of Keck and his colleagues [KMS73], great improvements have been achieved in avoiding contamination and nowadays fibres can be drawn in which the attenuation of modes is no longer limited by absorption, but by Rayleigh scattering from nanoscopic fluctuations of the refractive indices [TSGS00]. This kind of fibre can have loss coefficients as low as 0.18 dB/km at $\lambda = 1.55$ μm, allowing the transmission of information over hundreds of kilometers without amplification. An historical survey of optical communication can be found in the first chapter of reference [Oko82].

Outside the low absorption wavelength range mentioned above, silica fibres are quite lossy. This is of no concern for telecommunications, where the minimum loss band has determined the wavelength used, but other applications in which guided optics beyond the minimum loss band would prove very useful, suffer from these limitations. Optical fibres operating with acceptable losses at the carbon dioxide laser wavelength (10.6 μm)

[2]This means that for a fixed wavelength one gets two different field maps associated with a single value of β. Waveguide symmetry and mode degeneracies will be studied in more details in Chapter 3.7.

would for example revolutionize industrial and surgical laser applications. Note that for high power light guidance, the limiting factor is not so much the power lost between the source and the target, but rather the temperature elevation in the fibre if losses are due to absorption, a high value of which can result in destruction of the waveguide.

1.1.3.3 *Dispersion*

In telecommunication networks, information is transmitted as binary data, taking the form of light pulses in optical fibres. In the field of optical waveguides, *dispersion* is a generic term referring to all phenomena causing these pulses to spread while propagating. There are essentially four causes of dispersion [Oko82; SL83]:

- *Inter-modal dispersion*: In a multi-mode fibre, for a given wavelength, different modes are associated with different values of β and hence different propagation velocities. This results, for a signal exciting more than one mode, in pulse spreading or echoing, depending on the propagation length. The obvious solution to avoid inter-modal dispersion is to use single-mode fibres.
- *Material dispersion*: All materials are intrinsically dispersive, *i.e.* the refractive index is wavelength dependent. Spectrally, a pulse of light is associated with a superposition of a range of frequencies, centered on the frequency of the modulated light source. Due to material dispersion, each spectral component of the pulse will propagate at different speeds, resulting in pulse spreading and deformation.
- *Waveguide dispersion*: Even without material dispersion, the solutions of the propagation equation are wavelength dependent: the propagation constant of a given mode is wavelength dependent. This leads to pulse spreading and deformation for the same reasons as above.
- *Polarization mode dispersion* [RU78; PW86; SP01]: It is in fact the same phenomenon as inter-modal dispersion, but the relevant modes are here originally degenerate. We have seen that a single-mode fibre in fact carries two degenerate modes. Because of anisotropic perturbations (stress, bends, torsion...) the degeneracy between these modes is *de facto* lifted, and inter-modal dispersion occurs. Until recently, the effects of polarization mode dispersion were negligible, but with the bit-rates and propagation lengths

aimed at today, and since other sources of dispersion can relatively easily be compensated for, polarization mode dispersion becomes a significant problem.

Chromatic dispersion (the so-called dispersion parameter D),[3] is the dispersion resulting from the combined effects of material and waveguide dispersion. When the chromatic dispersion D is positive (the dispersion regime is said to be anomalous), shorter wavelengths propagate faster than longer wavelengths. In the opposite case of negative D, the dispersion regime is said to be normal.

In optical telecommunications, to maximize bandwidth it is essential that light pulses keep their initial widths. Indeed, if light pulses spread, they eventually overlap and cannot be distinguished by the receiver. There are several ways of constraining the pulses to keep their initial widths. The first and most obvious is to design dispersionless fibres, through compensating material dispersion with waveguide dispersion. This is generally possible at only one wavelength, so that all information must be carried within a very narrow range of wavelengths. Shifting the zero-dispersion wavelength in silica fibres while keeping single-mode behaviour is generally achieved through sharp triangular index profiles of the core, with additional layers of different refractive indices between the core and the cladding ([AD86] and chapter 7 of reference [Oko82]). However, through such designs it is only possible to shift the zero-dispersion wavelength of 1.3 μm towards longer wavelengths.

A second way of keeping a constant pulse width during propagation is to use solitons, high power pulses for which the nonlinearity (more precisely the self phase modulation) of the fibre exactly compensates the dispersion. For solitons to exist, the fibre must have a relatively small anomalous dispersion.

A third method of keeping the pulse width constant is to retain a small, well known normal dispersion in the fibres, and to add dispersion compensating devices at each repeater. The latter can take the form of optical fibres with strong anomalous dispersion, which exactly compensate for the normal dispersion resulting from fibres between the repeaters. The reason for preferring the latter solution to optical fibres with strict zero dispersion

[3] Expanding the propagation constant β in a Taylor series around the pulsation ω_0, one obtains $\beta(\omega) = \beta_0 + \beta_1(\omega - \omega_0) + 0.5\,\beta_2(\omega - \omega_0)^2 + \ldots$ with $\beta_n = (\partial^n \beta / \partial \omega^n)_{\omega_0}$. β_0 is the inverse of the phase velocity, β_1 is the inverse of the group velocity, and β_2 is the group velocity dispersion parameter. The chromatic dispersion D is related to β_2 through the relation: $D = -(2\pi c \beta_2)/\lambda^2$.

is twofold. First, through careful design it is possible to obtain an effective zero-dispersion wavelength range (*i.e.* the range in which the combined effects of dispersion and dispersion compensation result in a negligible overall dispersion) that is much wider than the effective zero-dispersion wavelength range of zero-dispersion fibres. The wider available wavelength range enables wavelength multiplexing of information, *i.e.* encoding the signal using carrier waves having different wavelengths (*channels*) in a single fibre, which multiplies the bandwidth. Secondly, pulses carried in zero-dispersion optical fibres are subject to non-linear interactions, even if their power density is not large. Indeed since dispersion is negligible at the carrier wavelength, neighboring wavelengths will have a very long coupling length, thus increasing the effect of non-linear interactions. When using multiple wavelength channels, these undergo non-linear interactions (especially four-wave mixing) leading to channel crosstalk and information loss. State-of-the-art long haul telecommunication networks use fibres having small but non-zero, normal dispersion around $\lambda = 1.55$ μm,[4] and dispersion compensators at each repeater [Val01]. The operating wavelength $\lambda = 1.55$ μm was chosen for two reasons: it is the wavelength at which losses in silica are smallest, and it corresponds to the wavelength range in which erbium doped fibre light amplifiers are most effective. These light amplifiers [MRJP87; Des02] were invented relatively recently; they now replace electronic repeaters, enabling all-optical signal processing from source to receiver.

Regarding polarization mode dispersion, the usual way to avoid it is to use single-mode polarization maintaining fibres. Polarization maintaining fibres are optical fibres in which a birefringence has been introduced, generally through applying stress to the fibre during the drawing process or through using elliptical cores. The degeneracy between the two fundamental modes is then lifted, and since both modes have different polarization properties, they can be separated with polarizers by the receiver, eliminating polarization mode dispersion. In practice the whole process is not as straightforward as it sounds, and polarization maintaining fibres have numerous drawbacks. In order to keep a good coupling efficiency between two fibres or between fibres and devices, care must be taken to keep a precise alignment of the polarization axes, which is tricky given the approximate circular symmetry of the fibres. Further, even with the birefringence, with long propagation distances crosstalk between the two fundamental modes appears because of residual imperfections, so that polarization mode dis-

[4]These fibres are called Non-Zero Dispersion Shifted Fibres, or NZ-DSF.

persion is not completely eliminated. Single-mode single polarization fibres are fibres which carry only one polarization. They are produced in much the same way as polarization maintaining fibres, but they need a larger birefringence. Their main drawback is that the same physical effect that eliminates one of the two polarizations of the fundamental mode gives rise to leakage for the other polarization, so that low loss single polarization operation is not achievable with conventional optical fibres.

In the field of telecommunications, current work to further improve dispersion management – and hence bandwidth – is concentrated on obtaining fibres with flat near-zero dispersion, improvement of dispersion compensating devices and management of polarization mode dispersion.

For other fields of application other dispersion properties are sought. Anomalous and zero-dispersion at wavelengths below $\lambda = 1.3\ \mu$m in single-mode fibres can be useful for super-continuum generation [WKOB+00; RWS00; CCL+01; DPG+02; HH01], ultrashort pulse compression, soliton generation and propagation.

1.1.3.4 *Non-linearity*

Non-linear optical effects always appear when the power density of light is large enough, regardless of the material. Since in optical fibres light is well confined in a narrow core, non-linear effects can appear even in ordinary silica at relatively modest injection powers [Agr01]. These effects are generally a nuisance in long haul telecommunication networks, but can also be useful for some applications [Val01].

As far as the negative aspects are concerned [Chr90], there are three major non-linear effects in optical fibres arising from the Kerr effect (self phase modulation, cross-phase modulation, and four wave mixing), and two due to stimulated inelastic scattering (stimulated Brillouin and Raman scattering). We have already mentioned that four wave mixing leads to channel crosstalk in wavelength multiplexed systems. Self phase modulation and cross phase modulation result in chirping of the pulse frequency, which, combined with dispersion, gives rise to signal deformation and spreading, and hence crosstalk. Four wave mixing can be reduced through increasing the magnitude of dispersion in the fibre, whereas the penalty due to phase modulation effects tends to be less important if the dispersion is as small as possible.[5] Both stimulated scattering effects result in pulse power being

[5]Note that with large chromatic dispersion, interaction between channels is diminished and hence crosstalk modulation is reduced as well; there is no easy general rule

scattered into frequency-shifted waves. In the case of stimulated Brillouin scattering, the scattered wave propagates in the opposite direction to the incoming signal, and above a certain power threshold more power is scattered back than propagated forward. The stimulated Brillouin scattering threshold sets the upper limit of power which can be propagated efficiently through an optical fibre. In the case of stimulated Raman scattering, the scattered, frequency-shifted wave propagates in the same direction as the incoming signal. In wavelength multiplexed systems, the Raman scattered signal of one channel can hence overlap with another channel. Note that both scattering phenomena generally increase with fibre doping. All these effects are of course power dependent, and smaller power densities at the same injection powers can only be of benefit.

Most of the previous effects can also be exploited to create devices. When controlled, four wave mixing and cross phase modulation can be used for all-optical switching, frequency conversion, pulse reshaping and other forms of optical signal processing. Self phase modulation is essential for soliton propagation. Stimulated Raman scattering can be used for signal amplification [SI73]. The main restriction of applications using non-linearities of conventional step index optical fibres is that the material interacting with the light must be able to be drawn into a usable fibre. In practice this severely limits candidate materials, and most non-linear fibre applications either use the nonlinearities inherent in silica or nonlinear material that can be introduced into silica through doping.

1.2 Photonic Crystals

The idea of photonic crystals originated in 1987 from work in the field of strong localization of light [Joh87] and in the inhibition of spontaneous emission [Yab87]. It was subsequently shown that in periodic arrangements of –ideally lossless – dielectrics, the propagation of light can be totally suppressed at certain wavelengths, regardless of propagation direction and polarization. This inhibition does not result from absorption but rather from the periodicity of the arrangement, and is quite fundamental: in the frequency range where no propagation is possible (the so-called *photonic band gap*), the density of possible states for the light vanishes, so that

regarding the ideal dispersion properties to minimize cross phase modulation: Predicting and compensating cross phase modulation is far from easy and is a topic of current research.

even spontaneous emission becomes impossible. Such periodic arrangements of dielectrics have been called *photonic crystals*, or photonic band gap materials.[6] More background information can be found in the following book [JMW95].

1.2.1 *One dimension: Bragg mirrors*

The simplest device using the principles of photonic crystals is the one-dimensional photonic crystal, well known under the name of the Bragg mirror or the multilayer reflector. It consists of a periodic stack of two alternating dielectric layers. Light propagating in a direction normal to the layers undergoes successive reflection and transmission at each interface between adjacent layers. With an appropriate choice of layer thickness and refractive indices, waves reflected from each interface are in phase, whereas waves transmitted are out of phase. In that case, the transmitted wave components cancel each other out, and only the interference of the reflected components is contructive: the light is totally reflected. This works for a range of wavelengths. Bragg mirrors have been in use for decades, but it is only recently that they have come to be regarded as a special case of photonic crystals. The classical way of analysing Bragg mirrors with a finite number of layers, uses reflection and transmission matrices for each layer, and it is then quite straightforward to prove through recurrence relationships that reflection can be perfect with an infinite number of layers. There is nevertheless another approach to deal with a stack having an infinite number of layers, originating from solid state physics. If the stack is infinite, it has a discrete translational symmetry. The Bloch Theorem then applies, and solutions to the propagation equation in the stack are Bloch waves. Hence two wave vectors differing by a vector of the reciprocal lattice associated with the periodic stacking are physically the same: the dispersion diagram "folds back" along the limits of the Brillouin zone. At the edge of the Brillouin zone, two solutions exist having the same wave vector but different frequencies, and in between those two frequencies no solutions exist at all. The gap of frequencies for which no solutions exist is called a photonic band gap. Note that, until recently, reflection from Bragg mirrors was thought to be possible only within a relatively narrow range of angles of incidence. Recent work by Fink *et al.* has demonstrated the feasibility of omnidirectional reflection with Bragg mirrors [FWF+98].

[6] Note that photonic crystals can also result from periodic arrangements of conductors.

1.2.2 *Photonic crystals in two and three dimensions*

Photonic crystals with two or three-dimensional periodicity can be seen as a generalization of Bragg mirrors. The simple approach with reflection and transmission matrices cannot be applied analytically here, and this is probably why their properties were discovered relatively recently, although, for example, important work on stacked grids for filtering in the far infrared was carried out by R. Ulrich in the 1960s [Ulr67; Ulr68]. The Bloch approach can be used similarly, and shows that band gaps can still open up. The point of using periodicities along two or three-dimensions is to open up an omnidirectional band gap: for the Bragg mirror, band gaps usually only exist for a narrow range of angles of incidence, and propagation parallel to the Bragg layers can never be inhibited. With photonic crystals having a two-dimensionally periodic arrangement of parallel rods, band gaps can exist for all directions of propagation in the plane of periodicity, and for photonic crystals with three-dimensional periodicity, propagation of light in all directions can be prohibited. When a band gap exists regardless of direction of propagation and polarization, one speaks of a total photonic band gap.

Photonic crystals with two-dimensional periodic arrangements are usually either made of parallel dielectric (or metallic) rods in air, or through drilling or etching holes in a dielectric material. In the field of integrated optics, holes of a fraction of a micrometer etched in slab waveguides are very promising for integrated photonic circuits, and have been succesfully demonstrated experimentally. Photonic crystals with three dimensional periodicity are a bit more tricky to achieve.[7] Yablonovitch suggested drilling an array of holes at three different angles into a dielectric material [YGL91]. The so-called wood-pile structure has attracted much attention [SD94; FL99; GdDT+03], and recent progress with artificial inverse opals is promising [BCG+00; VBSN01].

Note that the term photonic crystal was originally introduced to refer to materials having a photonic bandgap. It seems that it is now progressively more often used to refer to any kind of perodic arrangement of dielectrics or metals, with or without photonic band gaps. The latter generalization of the term makes sense considering that in solid state physics, a crystal is named so on account of the periodicity of its lattice, with band gaps ap-

[7] Nature, as so often, has demonstrated its supremacy in achieveing three dimensional photonic crystals billions of years before man. They can be found in opals. Note that one, two and three dimensional photonic crystals are also found elsewhere in nature: they give bright colours to beetles, butterflies, sea-mice and birds.

pearing in certain cases. Usual practice is then to reserve the term *photonic band gap material* for a photonic crystal having a photonic band gap. In the remaining chapters of the book we will avoid any confusion and speak of microstructures, since sometimes the dielectric structures may not even be periodic.

1.3 Guiding Light in a Fibre with Photonic Crystals

For frequencies within a total photonic bandgap, no propagation is allowed in an infinite photonic crystal. If a defect is introduced in the infinite lattice, localized defect states for isolated frequencies within the band gap can emerge, similar to bound states associated with defects in semiconductors. For three dimensional photonic crystal lattices, this can be a single point defect: in that case light emitted from within the defect will remain confined in the vicinity of the defect. It could also be a linear defect, in which case light remains in the locality of the defect but can propagate along it. Another way of looking at defect states is to consider the photonic crystal to be a perfect mirror in a certain frequency range. If one drills a hole all the way through the photonic crystal, light injected in the hole will be reflected at the borders of the hole and will propagate within it, in a similar way to that of light propagation in an optical fibre.

Given that photonic crystals can have a high reflection coefficient even with a relatively small number of periods, the width of the photonic crystal around the defect can be reduced to a few layers: we can hence imagine optical fibres consisting of a micrometric core surrounded by a photonic crystal cladding only a few times wider than the core. The resulting optical fibre is called a *photonic crystal fibre* and has an important difference from conventional optical fibres: in the latter, the core, in which light is guided, has to be of higher refractive index than the cladding. Using a photonic bandgap material for the cladding, reflection at some frequency is guaranteed regardless of the refractive index of the material inside the defect. A defect in a photonic crystal can hence confine and guide light in low refractive index media, such as a gas (air for example) or vacuum. This opens up possibilities never dreamt of before. An optical fibre guiding light in a vacuum would have absorption losses and non-linear effects reduced by orders of magnitude[8] compared with solid core fibres, paving the way for

[8] Material absorption, non-linear effects and material dispersion wouldn't *completely* vanish because of the evanescent parts of the fields remaining in the photonic crystal.

high power light guidance applications; material dispersion would become negligible, giving rise to completely new forms of dispersion management; guiding highly confined light in gas or liquids would enable the production of new types of non-linear fibres as well as a whole new family of fibre sensors; even guidance of atoms, molecules or cells through hollow core optical fibres would become possible [BKR02].

If one seeks to guide light along a linear defect, it is not necessary to use three dimensional periodicity or a total photonic band gap. Considering Fig. 1.1 but with the cladding now being a perfect mirror, $k_\perp$ will be set by the size of the defect, and β will follow from Eq. (1.1). If for the range of wave vectors given by these considerations no propagation is possible in the photonic crystal, a guided mode will exist. This can be achieved with two dimensional and even one dimensional radial (concentric) periodicities.

1.3.1 *Bragg fibres*

The idea behind the Bragg fibre is to use a Bragg mirror as a cladding of an optical fibre [YYM78]. Accordingly, a Bragg fibre is made out of a concentric arrangement of dielectrics, wound around a core which may or may not be hollow. Theoretical studies have shown only recently that hollow core Bragg fibres could be feasible, given the possibility of attaining large refractive index contrasts (*e.g.* 3 to 1.1 or 2.4 to 1.6, see Ref. [FWF+98]) between successive layers. Indeed, the first working experimental hollow core Bragg fibre has already been demonstrated, and had in some wavelength ranges losses that are orders of magnitude smaller than any conventional optical fibre [THB+02]. Furthermore, Bragg fibres can be single-mode, single-polarization even with full circular symmetry and without birefringence [Arg02; BA02].

1.3.2 *Photonic crystal fibres and hollow core microstructured optical fibres*

Strictly speaking, Bragg fibres are a special case of photonic crystal fibres. The term photonic crystal fibre is however mostly used to refer to fibres with a cladding consisting of a two dimensionally periodic array of inclusions. Their most immediate advantage over Bragg fibres is that lower refractive index contrasts are needed to achieve photonic bandgaps: a lattice of air (or vacuum) holes in silica or polymer is sufficient.

Fibres with a lattice of microscopic holes running along the fibre axis can be manufactured by drawing (*i.e.* heating and axially pulling) a preform containing macroscopic holes. The holey (*full of holes*) preform can readily be obtained either by stacking capillaries together, or through drilling (mainly for polymers) or by extrusion [vELA+01; Rus03]. In the drawing process, the overall shape of the preform is generally maintained, but the diameter of the cross-section is scaled down from centimetric to micrometric dimensions. Note that this process can be used to produce fibres with regular arrays of holes as well as fibres with non-periodic arrangements of holes, either with a solid or a hollow core (see Fig. 1.2). The general terms *holey fibre* and microstructured optical fibre (MOF) refer to any kind of fibre with a set of inclusions running along the fibre axis, whereas the term photonic crystal fibre is generally used to refer to MOFs in which guidance results from a photonic band gap effect. Note, however, that some authors also use the term *photonic crystal fibre* (PCF) for referring to MOFs in which the inclusions form a subset of a periodic array, but in which guidance may or may not result from photonic band gap effects[BBB03].

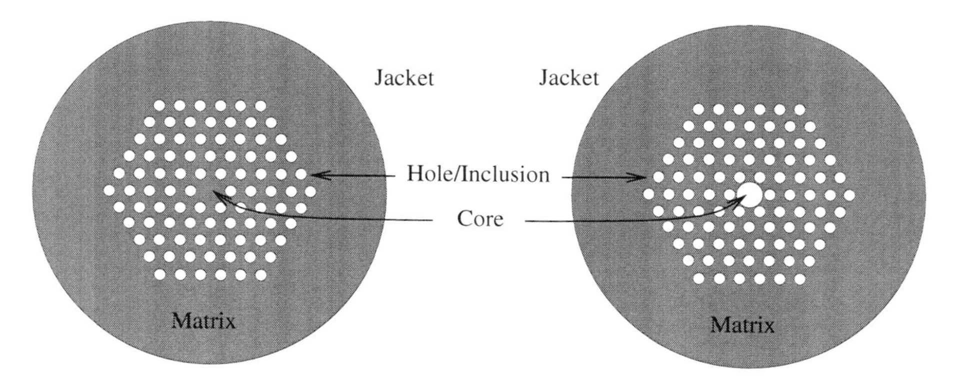

Fig. 1.2 Schematic representation of the cross-section of a typical solid core MOF (left) and hollow core MOF (right) with holes on a triangular lattice and a single central core. We will often refer to the microstructured part as the cladding. The jacket represents the physical boundaries of the MOF, and can be a solid jacket *e.g.* for mechanical protection or simply air. Note that the triangular lattice is also referred to as an hexagonal lattice.

1.3.2.1 *Hollow core MOFs*

Light guidance in hollow core MOFs can only be achieved by using the photonic band gap effect. Hollow core MOFs are hence necessarily photonic crystal fibres. We must consider a 2D photonic crystal with a 1D defect (due to the assumed translational invariance along the fibre axis z). The results concerning the band gaps are then quite different from those obtained in the usual case in which the longitudinal component of the wavevector, β, is equal to zero [MM94] as in thin-film nanophotonics in which the guidance occurs in a plane perpendicular to the axes of the inclusions. For hollow core MOFs, the most common configurations correspond to $|\beta|$ larger than both $|k_x|$ and $|k_y|$. It is then no longer necessary for the spatial period of the inclusion lattice to be of the order of half the wavelength: it can be much bigger, thus enlarging the possible parameter ranges yielding useful band gaps. For these photonic crystal fibres, the band gap is no longer a frequency range in which no propagation is possible irrespective of k_x and k_y but is instead a band gap in which for a given value of β no tranverse propagation is allowed.[9] For a silica matrix with a triangular lattice of circular air holes, there is no complete photonic band gap in the transverse propagation plane due to the low contrast in index between the two regions. Nevertheless, for $\beta \neq 0$ band gaps appear allowing the use of silica hollow core MOFs. Inserting a defect in the middle of the photonic crystal structure (which is infinite in most theoretical studies) will make possible the existence of a propagating mode in the perturbed crystal. If the propagation constant of this mode coincides with a band gap in the transverse plane then the mode will be confined in the locality of the defect, which constitutes the hollow core of the MOF.

These true photonic crystal fibres offer the whole range of benefits of hollow core fibres that we have described. Light guidance being possible solely within a photonic band gap, the wavelength range in which these fibres guide light is very narrow, only a few tens of nanometres for guidance in the infrared or the visible spectrum. Furthermore the accuracy of the periodicity of the lattice required to obtain a clear band gap effect makes the fabrication of these fibres challenging. Experimental hollow core MOFs have nonetheless been demonstrated using a cladding consisting of a triangular array of large air holes in silica [CMK+99], and the first the applications [BKR02] also make use of this basic structure.

[9] This explains why in hollow core MOF studies the band diagrams are given as a function of the (normalized) wavevector longitudinal component β.

1.4 Solid Core MOFs

1.4.1 *Guidance mechanism*

As we have already stated, photonic crystal fibre is only a particular case of MOF. The most common type of MOF is solid core MOF due to their (relative) ease of fabrication. For solid core MOFs it is often argued that guidance is due to *modified total internal reflection*: in a solid core MOF, the "average" refractive index of the cladding is lower than that of the core refractive index, leading to an equivalent geometry similar to those of conventional step index fibres. Following this argument, there is no need to invoke the concept of the photonic band gap, and any arrangement of holes – periodic or random – around a silica core results in a wave-guiding structure. It is easy to see the limits of such an interpretation of guidance, raising the question of what "average" means in this context: in the extreme case of a random hole distribution [MBBR00] with holes concentrating around a few isolated spots, leaving wide straight pathways for light to escape, it would be very surprising to find any kind of guidance. In fact, the "average" refractive index referred to when explaining guidance with modified total internal reflection is not a geometric average. It is actually not an average at all, but is a value extracted from the band structure of the surrounding arrangement of holes.[10] It corresponds to an "effective index" associated with the largest possible value of the propagation constant β for a given frequency in the microstructure: at a given frequency, light with a component β of the wave vector along the axis of the holes larger than a specific value β_{MAX} cannot propagate in the microstructured part of the fibre. This is analogous to total internal reflection in the case of step index fibres, in wich light with $\beta > n_{\mathrm{CL}} k_0$ cannot propagate in the cladding. The average, or effective, index of the cladding of a MOF is then given by β_{MAX}/k_0. But since β_{MAX} is a band property, it is slightly contrived to distinguish the band gap between β_{MAX} and $\beta = \infty$ from any other band gap bounded by finite values of β. Modified total internal reflection can therefore also be seen as a specific case of band gap guidance. The only true difference with band gap guidance using other band gaps is that the band gap between β_{MAX} and $\beta = \infty$ always exists, regardless of frequency or the exact structure of the cladding, so that guidance relying on this band gap is much easier to achieve. Furthermore, this clarifies the idea that the effec-

[10]This kind of homogenization procedure implies to the artificial periodicising of the hole region.

tive index is in fact a property of the infinite periodic lattice surrounding the core, and thus the concept of modified total internal reflection must be used with circumspection when the holes around the core are not arranged periodically. The argument that modified total internal reflection guidance is not fundamentally different from band gap guidance is further discussed in Ref. [FSM+00a].

1.4.2 *Main properties and applications*

Guidance due to modified total internal reflection in solid core MOFs is much easier to achieve than band gap guidance, and indeed the first MOF in which guidance was demonstrated had a solid core [KA74; KBRA96]. All the new possibilities offered by photonic crystal fibres hitherto mentioned were based on the fact that guidance could be achieved in a hollow core, and guidance using photonic crystals in solid cores might seem uninteresting at a first sight. Nevertheless, the study of the first experimental solid core MOFs showed that these possess unique properties of their own, unachievable by conventional optical fibres. The most striking among these is certainly their ability to be single mode over an infinite range of wavelengths [KBRA96]. In other words, for some solid core MOFs, however small the wavelength is compared with the core size, only a single-mode is guided. This is fundamentally different from conventional fibres where, at small wavelength to core size ratios, multi-modedness is unavoidable. The importance of this property is not so much linked to the possibility of having single-mode guidance over a large range of wavelengths in the same fibre – most of the time the wavelength range at which a fibre will be used is quite narrow – but rather remains in the converse property: for a given range of wavelengths a solid core MOF with arbitrarily large core can be single-mode. Possibilities offered by the resulting large core single-mode fibres are unprecedented: in the field of telecommunications for example, where single-mode guidance is essential, if the core is larger then light can be injected with higher power without the power density reaching levels at which non-linear effects become problematic, so that the distance between repeaters can be greatly increased.

On the opposite side of the core size scale, it appears that because of the large index contrast modes are very well confined in the core, even when the wavelength to core size ratio is not small. This again differs from conventional fibres, in which the fraction of the field in the cladding at large wavelength to core size ratios is far from being negligible because of the very small difference in refractive index between the core and the cladding. Good confinement in small cores enables higher power densities and hence accentuated non-linear properties.

The large available parameter space of solid core MOFs (positions, sizes, and shapes of the holes, or refractive indices of the inclusions if they are not holes) makes the waveguide dispersion, which can have strong effects due to the high index contrast, highly configurable. Almost any dispersion curve seems accessible to MOFs with the correct design. The combination of endlessly single-mode guidance and adjustable dispersion has led to solid-core, single-mode MOFs with anomalous dispersion, to single-mode fibres with a zero-dispersion wavelength shifted down to the visible, as well as to single-mode fibres with ultra-flat normal or anomalous dispersion over a large wavelength range [FSMA00; KAB+00; RKR02; KRM03; RKM03]. With the additional possibility of good mode confinement, the configurable dispersion also gave rise to promising non-linear applications, either impossible to achieve with conventional fibres or having much lower power thresholds than in conventional fibres. These include temporal soliton formation and propagation [HH01; PBM+02; HGZ+02; WRW02], super-continuum generation [DPG+02; HH01; CCL+01; HGZ+02] and the formation of new types of stable spatial solitons [FZdC+03].

Finally, given the ease with which a defect may be introduced in the MOF lattice at the preform fabrication stage, MOFs with multiple cores or MOFs with large birefringence [OBKW+00] are straightforward to produce. Most MOFs consist of an array of holes in which one hole has been left out, playing the role of the core. With the symmetries of the lattices of holes that are generally used in practice (mostly six- or four-fold symmetries), this results in the fundamental mode being doubly degenerate, as in conventional optical fibres [SWdS+01]. If the core is now extended to two adjacent missing holes, or if the symmetry around the core is reduced to two-fold symmetry (*e.g.* through changing the sizes of two diametrically opposed holes [SKK+01], or with elliptical holes [SO01]), the degeneracy is lifted and the fibre becomes birefringent: the MOF becomes polarization maintaining. The resulting birefringence can be orders of magnitude larger than stress-induced birefringence in conventional polarization maintaining fibres, so that coupling between the two modes is greatly reduced. Single-mode single-polarization fibres can be achieved in the same way. We have already mentioned how important this could be in the elimination of polarization mode dispersion.

The main advantage of MOFs with multiple cores compared to conventional multiple core fibres is the ease with which they can be produced. If the cores are separated by a large number of lattice periods, the cores become independent, with negligible crosstalk, enabling, for example a spatial multiplexing of signals.[11] If on the contrary the cores are separated by only a few periods, modes guided in different cores are coupled. This can be useful *e.g.* for sensors, and bend sensors relying on multiple core MOFs have already been demonstrated [BBE+00; MGM+01]. Furthermore, through filling one or more holes with polymers, liquids, or gases, or through writing gratings into the MOF core, fibres with *in situ* adjustable properties as well as a great range of sensors and other devices can be obtained [EKW+01].

To sum up, although solid core MOFs do not seem as radically different from conventional fibres as hollow core MOFs at first sight, the range of new prospects they offer and the numerous fields in which they could exceed the performance of conventional fibres is at least as exciting as hollow core guidance. Furthermore, since guidance in solid core MOFs relies on modified total internal reflection[12] and not on the use of a very narrow band gap, solid core MOFs are also much easier to realize than hollow core MOFs. Solid core and hollow core MOFs are both extremely promising new types of fibres, with completely different properties and possible applications; given their huge differences there is not much point in comparing them directly.

1.5 Leaky Modes

1.5.1 *Confinement losses*

In the solid core MOFs that we will study, light guidance is due to modified total internal reflection between the core and a microstructured cladding consisting of inclusions in a matrix. The core and matrix material are generally the same, and hence have the same refractive index. In practice, the cladding has a finite width, as it consists of several rings of inclusions. Beyond the microstructured part of the fibre, the matrix extends without any inclusions until the jacket is reached. If we consider the jacket to be far from the cladding and core, and hence neglect its influence, guidance

[11] Note that the theoretical and practical feasibility of such multiplexing remains to be demonstrated.

[12] ...*i.e.* on a wide band gap existing for all materials.

in the core is solely due to a finite number of layers of holes in bulk silica extending to infinity. *A priori*, the cladding does not "insulate" the core from the surrounding matrix material since the holes do not merge with their neighbours and consequently the matrix is connected between the core and the exterior. Physically, we can imagine the light leaking from the core to the exterior matrix material through the bridges between holes, and thus expect losses. In the modified total internal reflection model of guidance, in which the microstructured part of the fibre is replaced by homogeneous material with an effective refractive index lower than the core, the core is completely surrounded by the cladding. The exterior matrix material and the core are then no longer directly connected. Nevertheless the width of the "effective cladding" is finite, and hence tunneling losses are unavoidable. Regardless of the approach one uses to explain guidance in MOFs, as long as guidance is due to a finite number of layers of holes, leakage from the core to the outer matrix material is unavoidable. We will call the losses due to the finite extent of the cladding *confinement losses*, or *geometric losses*.

1.5.2 *Modes of a leaky structure*

Confinement losses being unavoidable, modes of a MOF decay while propagating. They are no longer called guided modes, but *leaky modes* [Mar91; SL83]. The equations satisfied by the fields being linear, losses are proportional to the field intensity. Simple mathematics show that in that case the decay of the fields must be exponential along the direction of propagation. This is reflected by the propagation constant β taking an imaginary part. We have seen that modes are characterized by a transverse field distribution invariant with z, which is modulated by a phase factor taking the form $\exp(i\beta z)$. For leaky modes, β is complex, and the phase factor becomes $\exp(i\Re e(\beta)z)\exp(-\Im m(\beta)z)$ ($\Re e$ and $\Im m$ denoting the real and imaginary parts respectively). For the mode to decay in the direction of propagation (which is the case with leakage), the real and imaginary parts of β must be of the same sign.

This simple and elegant way of accounting for the decay of modes through a complex propagation constant is nevertheless only the tip of the iceberg of dealing with leaky modes. Indeed, using a complex propagation constant leads to more difficulties than simplifications. First, β being complex, the value of the perpendicular wavenumber $k_\perp$ (Eq. (1.1)) becomes complex as well. This has two consequences: it complicates the choice of the square root in Eq. (1.1), and the fields of the modes become divergent at large distances from the core. This in turn renders the modal field distributions non square-integrable, so that dealing with them becomes mathematically very delicate. As a consequence, leaky modes are not orthogonal in the usual sense and their completeness is not self-evident [SL83]. The very notion of modes becomes unclear, since the lack of orthogonality implies that two modes could interact: if for example the fundamental mode and the second mode have an intrinsic crosstalk, their distinction would become totally arbitrary. Finally we have seen that confined guided modes in the case of conventional step index fibres are found for values of β satisfying $n_{CL}k_0 < \beta < n_{CO}k_0$, and that for $0 < \beta < n_{CL}k_0$ there is a continuum of radiative modes. Since there is no natural ordering in the complex plane, we can no longer use such considerations to locate leaky modes. Furthermore, the continuum of radiative modes still exists and the range of β associated with radiative modes is not *a priori* disjoint from the range of β of leaky modes. This seriously complicates the task of finding leaky modes, since a leaky mode can be "lost" amongst a continuum of radiative solutions to the wave equations. Some of the difficulties pointed out here have not yet been overcome, and are the topic of current research (*cf.* paragraph 1.5.4). Since we refer to leaky modes in several parts of this book, we will nevertheless explain in more detail what leaky modes represent physically and mathematically, how some of the difficulties linked to their use can be circumvented, and more generally how we can justify our approach.

1.5.3 *Heuristic approach to physical properties of leaky modes*

To fully understand the slightly disconcerting properties of leaky modes, we have to keep in mind that modes are defined for a fibre that is *infinitely long*. The exponential decay of the mode with increasing z being equivalent to an exponential growth with decreasing z, the modal fields diverge when z approaches $-\infty$. We show here how this implies that the fields have to diverge radially along a cross-section of the fibre.

For the sake of simplicity, we consider a step index fibre consisting of a core with refractive index n_{CO} and a cladding of finite size with refractive n_{CL} surrounded by the same material as the core: beyond the cladding, the refractive index keeps a constant value n_{CO} everywhere (Fig. 1.3). In such a fibre, all modes are leaky because of tunneling losses [MY77; FV83]. We consider the fundamental (leaky) mode of that structure, prop-

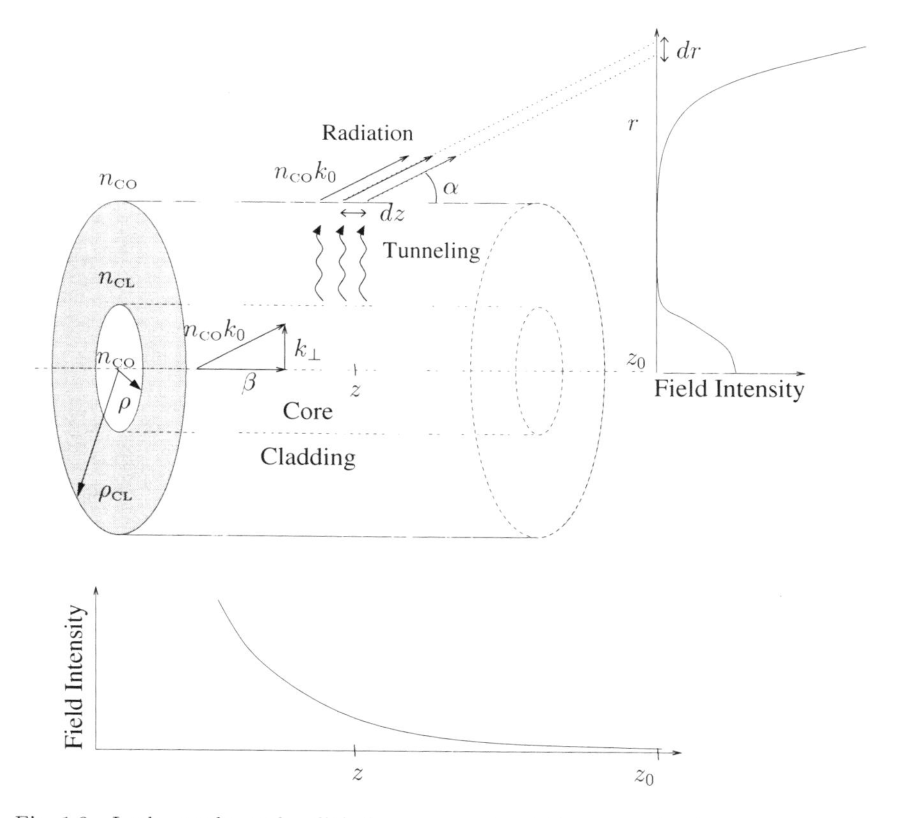

Fig. 1.3 Leaky modes and radial divergence, a heuristic approach. See text for details.

agating in the direction of increasing z. In the core, its power density distribution is similar to that of the fundamental mode of a lossless step index fibre. It is centrosymmetric with maximum value at the centre of the core. In the cladding the fields are evanescent, the power density decays exponentially with increasing distance from the core, until the exterior boundary of the cladding is reached. The propagation constant and the norm of the wave-vector being the same in the core and in the matrix surrounding the cladding, the amount of power which has reached the exterior cladding boundary can radiate away, and it does so at an angle of

$$\alpha = \cos^{-1}\left(\frac{\Re e(\beta)}{n_{CO}k_0}\right) . \tag{1.3}$$

In terms of rays, a radiated ray originating from the cladding at z will arrive at a reference position $z_0 > z$ at a radial distance

$$r(z) = \rho_{\text{CL}} + (z_0 - z) \tan \alpha \tag{1.4}$$

from the core centre. The whole power emitted from the cladding boundary in an infinitesimally long cylinder of length dz at z is found in the cross-section located at z_0 on an annulus with radius $r(z)$ and width $dr = dz \tan \alpha$. The power density in that annulus is hence the total power radiated from dz at z divided by the area of the annulus $2\pi r dr$. The total power radiated from the infinitesimal cylinder being proportional to the total power density at z and hence to $\exp(-2\Im m(\beta) z) dz$,[13] outside the cladding the power density $S(r)$ is proportional to

$$S(r) \propto \frac{1}{r \tan \alpha} \exp \left[-2\Im m(\beta) \left(z_0 - \frac{r - \rho_{\text{CL}}}{\tan \alpha} \right) \right]. \tag{1.5}$$

The power density at the center of the core $r = 0$ at z_0 being proportional to $\exp(-2\Im m(\beta) z_0)$, the normalized power density outside the cladding at z_0 is given by

$$\frac{S(r)}{S(0)} \propto \frac{1}{r \tan \alpha} \exp \left[2\Im m(\beta) \left(\frac{r - \rho_{\text{CL}}}{\tan \alpha} \right) \right]. \tag{1.6}$$

$\Im m(\beta)$ being positive, we see that the power diverges exponentially with increasing radial distance.

Counter-intuitive though it may seem, the radial exponential growth of field distributions is a fundamental property of leaky modes. Of course such a field distribution is impossible to obtain in practice. Real fibres are always of finite length, and hence by a reasoning similar to that given above will have to be limited by the value z_s of z at which the fibre starts. Upon incorporating this modification, we would find an exponential growth within the cross-section of the fibre, but the exponential growth would stop at $r(z_\text{s})$, and for larger values of r the field would be strictly zero. The total power flowing through a cross-section at any given z would be equal to the power flowing through the cross-section at z_s. The exponential growth in the cross-section is not in contradiction with energy conservation: it is a direct consequence of it. This remains true in the case of infinitely long fibres, but the power flow through any cross-section is then infinite.

[13]The factor 2 in the exponential comes from the power being a square function of fields.

1.5.4 *Mathematical considerations*

In the example that we have considered above, we were led by intuition to the conclusion that the losses due to tunelling were small, and that their influence on the mode should thus remain small. Implicitly, we adapted the truly guided modes of a fibre with infinite cladding to kae amount of their lossy nature through letting β take a small imaginary part. Mathematically, this would amount to letting the lossless boundary conditions become slightly lossy without reformulating the whole problem. However, lossy boundary conditions, usually refered to as *open boundary conditions* are not mathematically straightforward to deal with.

Until recently, the only rigorous way of treating open boundary conditions was to avoid them: instead of considering the system as consisting solely of the core and the cladding, with lossy boundary conditions at the outer cladding boundary, we would consider the system consisting of the core, the cladding, and the rest of the universe, so that energy conservation is satisfied. The drawback of this approach is that the only modes which are solutions to the problem are associated with a continuum of real β values. The leaky modes, which allow us to analyse the physics of fibres with finite and infinite claddings along parallel lines, are not natural solutions to the physical problem encompassing the fibre and its exterior.

In contrast, the solutions to the true open boundary problem are the leaky modes. However, with open boundary conditions the system (the fibre alone) does not satisfy energy conservation. The mathematical operators used are then no longer hermitian, and the mathematics of non-hermitian operators is unpleasant. Firstly, eigenfunctions of non-hermitian operators do not form a complete orthogonal basis, but a set of non-orthogonal functions which may or may not be complete. Decomposing a field on this set is hence not straightforward, and the usual tools involving modal decomposition (*i.e.* almost all techniques in the theory of guided optics) cannot be used. Secondly, deriving rigorously the solutions to the physical problem raises difficulties. The work presented in this book is no exception in this regard, and the derivation we give of the multipole method is in fact not mathematically rigorous: for some key steps of the derivation, we implicitly assume the fields to be square integrable, yet leaky modes are not. More specifically, in the derivation of the Wijngaard identity, Appendix B.1, we use the Green's function

$$G_e = -\frac{i}{4} H_0^{(1)}(k_\perp^{\mathrm{M}} r). \tag{1.7}$$

The value of $k_{\perp}^{\mathrm{M}}$ is not determined at that point of the derivation, but for leaky modes $k_{\perp}^{\mathrm{M}}$ will be complex with a positive imaginary part, so that the Green's function will not be square integrable. The convolution used in the remainder of the derivation then becomes dubious. The derivation of the multipole method is rigorous for guided modes, but we cannot justify its extension to leaky modes with rigorous mathematics. However, the agreement between results obtained from the multipole method and results from other numerical methods or experiments, even when involving leaky modes, somewhat legitimates the confidence we have in the method.

Open boundary problems are a topic of current research. Recent work initiated by P. T. Leung and K. M. Pang [LP96; LTY97a; LTY97b; LSSY98; HLvdBY99; LLP99a; LLP99b; NLL02] suggests that the set of leaky modes of a class of open boundary problems, similar to the one considered here, forms a complete orthonormal basis if the space of functions and its inner product are adequately defined. From the point of view adopted in their work, the solution to the open boundary problem does not in fact include a continuum of eigenvalues, but solely the discrete, complete set of leaky modes. It is not yet clear if guidance in MOFs is strictly speaking a specific case of the class of problems studied by Leung *et al.*, but it might well be that the way to a rigorous derivation of multipole methods for leaky modes has already been paved.

1.5.5 *Spectral considerations*

There is a countable infinity of leaky modes [SL83; HLvdBY99]. For a step index fibre with infinite cladding, propagation constants satisfying $n_{\mathrm{CL}} < \beta/k_0 < n_{\mathrm{CO}}$ give strictly guided modes. The propagation constant of leaky modes being complex, we cannot use this argument any more. Nevertheless, we can assume that the "most confined" leaky modes of a fibre with finite cladding are similar to the guided modes of the fibre with the same parameters but with an infinite cladding, the main difference being that the propagation constant takes a small imaginary part. These modes would satisfy $n_{\mathrm{CL}} < \Re e(\beta)/k_0 < n_{\mathrm{CO}}$. For a solid core MOF, we could replace n_{CL} by the effective index of the cladding, as long as one can define such an effective index. Otherwise, another lower bound can be used, namely the refractive index of the inclusions n_{i}. Indeed, if $\Re e(\beta)/k_0 < n_{\mathrm{i}}$, light is "likely to propagate" in the inclusions; there would be no barrier between the core and the exterior, and hence losses should be extremely high. For hollow core MOFs, light has to propagate in the hollow core and hence we must have $\Re e(\beta)/k_0 < 1$.

Note that when the imaginary part of β becomes very large, say one order of magnitude less than $\Re e(\beta)$, considerations relating to ordering become dangerous, and very leaky modes exist having values of $\Re e(\beta)$ well outside the mentioned boundaries. Finally, it is worth noting that having $n_{\text{CL}} < \Re e(\beta)/k_0 < n_{\text{CO}}$ does not imply that the imaginary part of β is small.

The fundamental mode of a step index fibre with infinite cladding is the mode with largest β. Consequently, it is also the mode with the fastest decaying evanescent tail in the cladding. For a microstructured fibre with a cladding of finite extent, we can thus expect the losses of the fundamental mode to be the smallest, so that its propagation constant would have both, the smallest $\Im m(\beta)$ and the largest $\Re e(\beta)$. When sufficient similarity between solid core MOFs and step index fibres with finite cladding exists, we can expect the same behaviour of the fundamental mode for MOFs. To locate the fundamental mode of a MOF it is therefore a good idea to start to look for values of β with small imaginary part, and with real part in the vicinity of $k_0 n_{\text{CO}}$.

Chapter 2

Electromagnetism – Prerequisites

2.1 Maxwell Equations

2.1.1 *Maxwell equations in vacuo*

Numerous mathematicians and physicists had proposed systems of equations for the effects of moving electric charges which had seemed to be satisfactory within the general Newtonian framework. In the Newtonian scheme, the forces of electricity and magnetism (the existence of both having been known from antiquity, and studied in some detail by William Gilbert in 1600 and Benjamin Franklin in 1752) act in a way similar to gravitational forces in that they also fall off as the inverse square of the distance, though sometimes repulsively rather than attractively, as shown by Coulomb's law. Here, the electric charge (and the magnetic pole stength), rather than the mass, measures the strength of the force. The first scientist to have made a serious challenge to the Newtonian picture seems to have been the English experimentalist and theorician Michel Faraday (1791-1867). Faraday's profound experimental findings (with moving coils, magnets and the like) led him to believe that electric and magnetic fields have physical reality and, moreover, that varying electric and magnetic fields might sometimes be able to push each other along through otherwise empty space to produce a kind of disembodied wave. He conjectured that light itself might consist of such waves. Such a vision would have been at variance with the prevailing Newtonian wisdom, whereby such fields were not thought as real in any sense, but merely convenient mathematical auxiliaries to the true Newtonian, point particle action, picture of actual reality. Confronted with Faraday's experimental findings, together with earlier ones by the French physicist André Marie Ampère (1775-1836) and others, and inspired by Faraday's vision, the Scottish physicist and mathematician

James Clerk Maxwell (1831-1879) puzzled about the mathematical form of the equations for the electric and magnetic fields that arose from those findings. With a remarkable stroke of insight he proposed a change in the equations, seemingly perhaps rather slight, but fundamental in its implications. This change was not at all suggested by (although it was consistent with) the known experimental facts. It was a result of Maxwell's own theoretical requirements, partly physical, partly mathematical, and partly aesthetic. One implication of Maxwell's equations was that electric and magnetic fields would indeed push each other along through empty space. An oscillating magnetic field would give rise to an oscillating electric field (this was implied by Faraday's experimental findings), and this oscillating electric field would, in turn, give rise to an oscillating magnetic field (by Maxwell's theoretical inference), and this again would give rise to an electric field and so on. Maxwell was able to calculate the speed with which this effect would propagate through space and he found that this would be the speed of light! Moreover these so-called *electromagnetic waves* would exhibit the interference and the puzzling polarization properties of light that had long been known (see Young and Fresnel). In addition to accounting for the properties of visible light, for which the waves would have a particular range of wavelengths (4 to 7.10^{-7}m), electromagnetic waves of other wavelengths were predicted to occur and to be produced by electric currents in wires. The existence of such waves was established experimentally by the German physicist Heinrich Hertz in 1888. Faraday's inspired hope had indeed found a firm basis in the equations of Maxwell. One of the main differences of Maxwell's equations compared to the Newtonian framework is that they are field equations rather than particle equations, which means that one needs an infinite number of parameters to describe the state of the system (the field vectors at every single point in space), rather than just the finite number that is needed for a particle theory (three coordinates of position and three of momenta for each particle). The so-called Newtonian field was studied in detail by Laplace in his famous electrostatic equation $\Delta V = \rho$, where V is the electrostatic potential in a dielectric body and ρ is its volumetric density of charge. More precisely, denoting respectively by the vector fields $\mathbf{E}$, $\mathbf{B}$, $\mathbf{J}$ and the scalar function ρ the electric field, the magnetic field, the electric current, and the density of electric charge,

Maxwell stated that *in vacuo*:

$$\begin{cases} \mathbf{curl}\,\boldsymbol{B} = \mu_0\varepsilon_0 \dfrac{\partial \boldsymbol{E}}{\partial t} + \mu_0 \mathbf{J} & (2.1a) \\ \mathbf{curl}\,\boldsymbol{E} = -\dfrac{\partial \boldsymbol{B}}{\partial t} & (2.1b) \\ \operatorname{div} \boldsymbol{E} = \dfrac{\rho}{\varepsilon_0} & (2.1c) \\ \operatorname{div} \boldsymbol{B} = 0 & (2.1d) \end{cases}$$

where $\mathbf{J}$ is related to the velocity of charge $\mathbf{v}$:

$$\mathbf{J} = \rho \mathbf{v} \tag{2.2}$$

and where $\mu_0\varepsilon_0$ is a constant equal to the inverse of the square of the speed of light, usually denoted by c.

2.1.2 *Maxwell equations in idealized matter*

2.1.2.1 *Mesoscopic homogenization*

We state that Maxwell's system (Eq. 2.1) still holds in an "ordinary" material. The optical wavelengths are very large compared to the atomic scale, so we are interested rather in the mean value $(<\mathbf{E}>, <\mathbf{B}>)$ where the mean value is defined via the convolution with a smooth function whose support L is large compared with the atomic scale but small compared with the wavelength:

$$\begin{cases} <\mathbf{E}> = a \star \mathbf{E} & (2.3a) \\ <\mathbf{B}> = a \star \mathbf{B} & (2.3b) \end{cases}$$

We call this first homogenization process (from the microscopic scale to the mesoscopic one), *natural homogenization*[1] [2]. It is well known that if we assume that the material consists of neutral molecules with *dipole momenta*

[1] See [Jac99]–[Sch94] for a comprehensive discussion on this topic.

[2] The term *homogenization* is somewhat confusing: the *natural homogenization* (from the microscopic to mesoscopic scale) is different from a new branch of research in electromagnetism called *homogenization of diffraction* (from mesoscopic to macroscopic scale). See [GZ00]–[ZG02], for instance.

$\mathbf{p}_i$ and *charges* q_i located at points $\mathbf{r}_i$, we can define a field $\mathbf{D}$ such that:[3]

$$\mathbf{D} = \varepsilon_0 < \mathbf{E} > +\mathbf{P} \tag{2.4}$$

and

$$\operatorname{div} \mathbf{D} = \rho_c \,, \tag{2.5}$$

where $\mathbf{P} = \sum_i \mathbf{p}_i \delta(\mathbf{r} - \mathbf{r}_i)$ is the *electric polarization vector* and ρ_c is the *density of charges of conduction.*

Similarly, we define the current of conduction:

$$\mathbf{J}_c = \sum_i < q_i \Big(\frac{\partial \mathbf{r}_i(t)}{\partial t} \Big) \delta(\mathbf{r} - \mathbf{r}_i) > \tag{2.6}$$

and the magnetic polarization vector:

$$\mathbf{M} = \sum_i < \frac{\mu_0}{2} \Big(\int_{\mathbb{R}^3} (\mathbf{r}' - \mathbf{r}_i) \times \mathbf{J}(\mathbf{r}', t)\, d\mathbf{r}' \Big) \delta(\mathbf{r} - \mathbf{r}_i) > . \tag{2.7}$$

We therefore have

$$\mathbf{curl}\, \mathbf{H} = \mathbf{J}_c + \frac{\partial \mathbf{D}}{\partial t} \tag{2.8}$$

where $\mathbf{H}$ is defined by

$$\mathbf{H} = \frac{1}{\mu_0} (< \mathbf{B} > - \mathbf{M}) \,. \tag{2.9}$$

2.1.2.2 *Dispersion relations – Kramers-Kronig relations*

We now have to give some relations between, on the one hand, the mesoscopic derived quantities $\boldsymbol{H}$ and $\boldsymbol{D}$ defined above, and the mean values of the microscopic fields $< \boldsymbol{E} >$ and $< \boldsymbol{B} >$ on the other hand: these relations are the so-called *constitutive relations* (See Fig. 2.1). We first begin with the simplest model: the model of perfect media. In that case, we assume that:

$$\mathbf{P} = \varepsilon_0 \chi_e < \mathbf{E} > , \tag{2.10}$$

[3] Note that we tackle the homogenization process within a classical framework even though atoms must be described with quantum tools. As a consequence, the electromagnetic quantities have to be derived in a phenomenological way. By this, we mean that the mesocopic electromagnetic quantities are not derived from statistical quantum physics but are derived from *ad hoc* quantities obtained from the experiments and other general concepts such as causality and some reasonable assumptions...

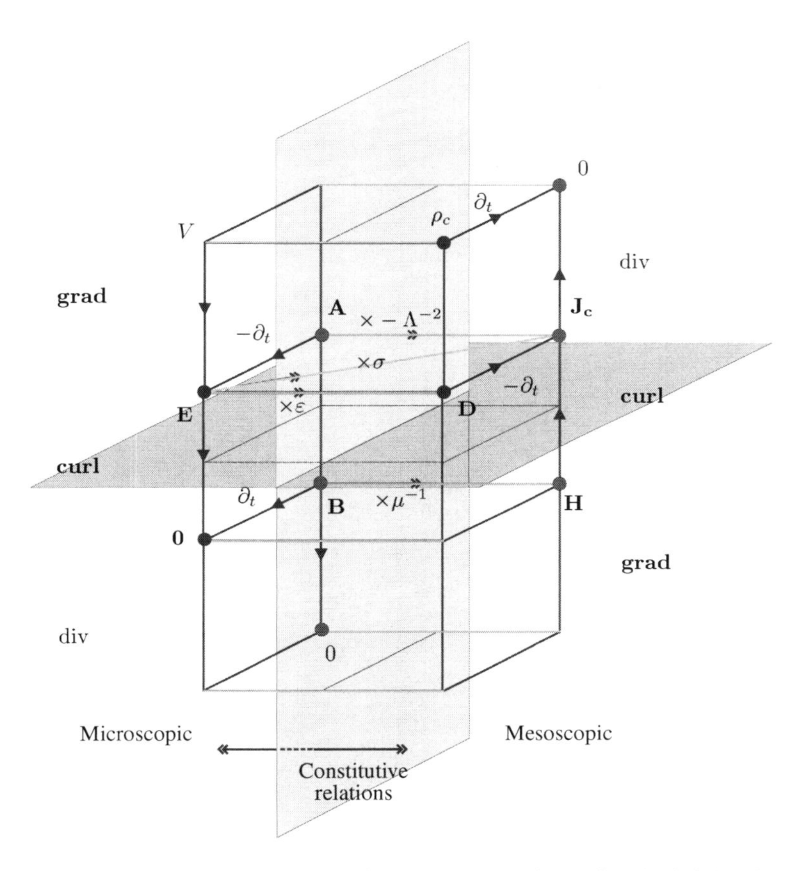

Fig. 2.1 *Tonti's diagram*: Schematic for electromagnetism. On the left-hand side of the figure appear the microscopic quantities $\boldsymbol{E}$ and $\boldsymbol{B}$ whereas on the right-hand side appear the derived mesoscopic quantities $\mathbf{J_c}$, ρ_c, $\boldsymbol{D}$ and $\boldsymbol{H}$. Moreover, Tonti's diagram works as two flow diagrams (the left-hand side and the right-hand side). The value of a node ($\bullet$) equals the sum of the values associated with the incoming arrows ($\longrightarrow$). The value associated with an arrow is given by the action of the operator associated with this arrow on the value of the node at the origin of the aforementioned arrow. *Note that "spatial operators" (*div*,* **grad** *and* **curl***) only act on vertical arrows whereas the "temporal operator" (∂_t) only acts on horizontal arrows.* We find in this way the following relations $\boldsymbol{E} = \mathbf{grad}\, V - \frac{\partial \mathbf{A}}{\partial t}$, $\mathbf{0} = \mathbf{curl}\, \boldsymbol{E} + \frac{\partial \boldsymbol{B}}{\partial t}$, $\boldsymbol{B} = \mathbf{curl}\, \mathbf{A}$ and $0 = \mathrm{div}\, \boldsymbol{B}$ for the right-hand side and $\mathbf{0} = \mathrm{div}\, \mathbf{j} + \frac{\partial \rho}{\partial t}$ and $\mathbf{j} = \mathbf{curl}\, \boldsymbol{H} - \frac{\partial \boldsymbol{D}}{\partial t}$ for the left-hand side, where, of course, we have denoted by $\mathbf{A}$ (resp. V) the vector (resp. scalar) potential. Additionally, the connection between the two sides of the diagram is given by the most classical constitutive relations such as the electric and magnetic constitutive relations (Eqs. 2.18–2.24) and also *Ohm's law* $\mathbf{J_c} = \sigma \boldsymbol{E}$, *London's law* $\mathbf{J_c} = -\frac{1}{\Lambda^2}\mathbf{A}\ldots$

where χ_e is a real number (usually positive in the optical wavelength range) called electric susceptibility. We thus have:

$$\mathbf{D} = \varepsilon_0(1+\chi_e) < \mathbf{E} > := \varepsilon_0\varepsilon_r < \mathbf{E} > \ . \tag{2.11}$$

Analogously, we assume that:

$$\mathbf{M} = \frac{\chi_m}{1+\chi_m} < \mathbf{B} > , \tag{2.12}$$

where χ_m is a real number (positive or negative) called magnetic susceptibility. We thus have:

$$\mathbf{B} = \mu_0(1+\chi_m) < \mathbf{B} > := \mu_0\mu_r < \mathbf{B} > , \tag{2.13}$$

where μ_r and ε_r are functions which can depend upon the space variable $\mathbf{r}$, respectively called relative permeability and relative permittivity (note that they have no physical dimensions). From now on, in order to lighten the notation, we will denote the mesoscopic electromagnetic field ($<\boldsymbol{E}>$, $<\boldsymbol{H}>$) by ($\boldsymbol{E}$, $\boldsymbol{H}$). It is clear that the constitutive relations have been obtained roughly: they cannot convey physical properties as important as dispersion and absorption. We make less restrictive assumptions concerning $\boldsymbol{P}$. We state the existence of an operator L such that

$$\boldsymbol{P}(\boldsymbol{r},t) = L\big(\boldsymbol{E}(\boldsymbol{r},t)\big) . \tag{2.14}$$

where L is a linear and local operator. In that case, it can be proved that L is what is known as an operator of convolution. In other terms, we can find a distribution S such that

$$\boldsymbol{P} = S \star \boldsymbol{E} . \tag{2.15}$$

Moreover, we assume that S is a regular distribution associated with a sufficiently smooth function g such that

$$\boldsymbol{P} = \varepsilon_0 g \star \boldsymbol{E} . \tag{2.16}$$

We have thus

$$\hat{\boldsymbol{P}}(\boldsymbol{r},\nu) = \varepsilon_0\hat{g}(\nu)\hat{\boldsymbol{E}}(\boldsymbol{r},\nu) , \tag{2.17}$$

Consequently, we can find the electric constitutive relation:

$$\hat{\boldsymbol{D}}(\boldsymbol{r},\nu) = \varepsilon_0\varepsilon_r(\boldsymbol{r},\nu)\hat{\boldsymbol{E}}(\boldsymbol{r},\nu) \tag{2.18}$$

with $\varepsilon_r(\nu) = 1 + \chi_e(\nu)$ where $\chi_e(\nu) = \hat{g}(\nu)$. This last relation which may appear innocent at first glance has important consequences. Causality implies that the function g has a positive support. As a consequence:

$$g(t) = \operatorname{sgn}(t)\, g(t) \tag{2.19}$$

where sgn is a function defined as:

$$\operatorname{sgn}(t) = \begin{cases} -1 & \text{if } t < 0 \\ 0 & \text{if } t = 0 \\ 1 & \text{if } t > 0 \end{cases} . \tag{2.20}$$

Therefore

$$\hat{g} = \widehat{\operatorname{sgn}} \star \hat{g} \tag{2.21}$$

with (See Appendix C p. 305)

$$\widehat{\operatorname{sgn}} = \frac{i}{\pi}\operatorname{pv}\left(\frac{1}{\nu}\right) \tag{2.22}$$

which leads after elementary manipulations to the so-called *Kramers-Kronig relation* linking the real part of the relative permittivity ε'_r with its imaginary part ε''_r:

$$\varepsilon'_r(\nu) = 1 + \frac{2}{\pi}\left[\operatorname{pv}\left(\int_0^{+\infty} \frac{u\varepsilon''_r(u)}{\nu^2 - u^2}\, du\right)\right] . \tag{2.23}$$

As a matter of fact, it appears that some materials have a remarkable property ; they are transparent in a certain range of frequencies! In other words, the permittivity is real for certain frequencies as we can see in the infrared for silicon in Fig. 2.2. These bands of frequencies are often called *bands of transparency* in which we can neglect the absorption (we express this by $\varepsilon''_r = 0$) are, of course, of prime importance. The very subject of this book could not be addressed without the existence of such bands. Throughout this book, we will assume that we will be always in a band of transparency. On the other hand, we will draw special attention to the dispersion curves which will cause problems later on. Resuming our investigation with $\mathbf{M}$ and $\boldsymbol{B}$, we find similarly another new constitutive relation (See Fig. 2.1):

$$\hat{\boldsymbol{B}} = \mu_0 \mu_r(\nu) \hat{\boldsymbol{H}} \tag{2.24}$$

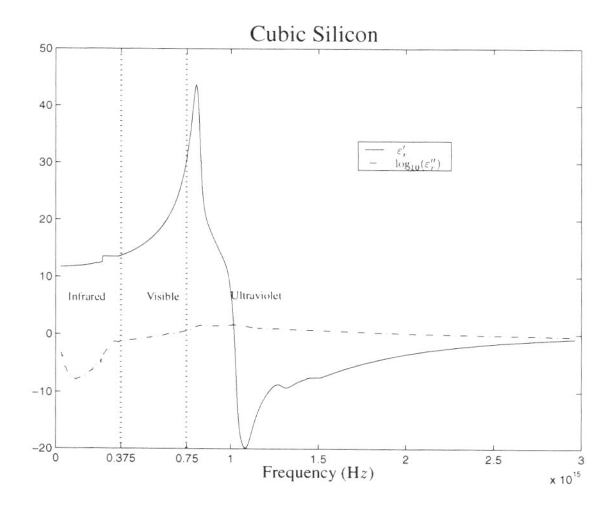

Fig. 2.2 Real and imaginary part of the relative permittivity versus frequency for silicon.

2.2 The Monodimensional Case *(Modes, Dispersion Curves)*

2.2.1 *A first approach*

2.2.1.1 *A special feature of the 1D-case: the decoupling of modes*

In this paragraph, we only deal with *one-dimensional structures*, namely structures being invariant along two directions (x and z)(See Fig. 2.3). Furthermore, we are only concerned with linear, lossless and isotropic materials (cf. section 2.1.2.2). Accordingly, we can write the following relations:

$$\underset{\sim}{\boldsymbol{D}} = \varepsilon(y)\underset{\sim}{\boldsymbol{E}} \quad \text{and} \quad \underset{\sim}{\boldsymbol{B}} = \mu_0 \underset{\sim}{\boldsymbol{H}} \ , \tag{2.25}$$

where $0 < \varepsilon_{min} \leq \varepsilon \leq \varepsilon_{max} < +\infty$ and where $\underset{\sim}{\boldsymbol{E}}$ (resp. $\underset{\sim}{\boldsymbol{D}}$, $\underset{\sim}{\boldsymbol{B}}$ and $\underset{\sim}{\boldsymbol{H}}$) are the complex amplitudes associated with $\boldsymbol{\mathcal{E}}$ (resp. $\boldsymbol{\mathcal{D}}$, $\boldsymbol{\mathcal{B}}$ and $\boldsymbol{\mathcal{H}}$). If we focus our attention only on propagating waves along the $z-$axis, then the field $\underset{\sim}{\boldsymbol{F}} = (\underset{\sim}{\boldsymbol{E}}, \underset{\sim}{\boldsymbol{H}})$ does not depend on the x variable and we have to find solutions of the following form:

$$\underset{\sim}{\boldsymbol{E}} = \boldsymbol{E}_1(y)e^{i\beta z} \quad \text{and} \quad \underset{\sim}{\boldsymbol{H}} = \boldsymbol{H}_1(y)e^{i\beta z} \ , \tag{2.26}$$

where β is the so-called *propagation constant*. In other words, we want to find fields of the following form:

$$\boldsymbol{E}(\mathbf{r},t) = \Re e\{\boldsymbol{E}_1(y)e^{i(\beta z-\omega t)}\} \tag{2.27}$$

and

$$\boldsymbol{H}(\mathbf{r},t) = \Re e\{\boldsymbol{H}_1(y)e^{i(\beta z-\omega t)}\} \tag{2.28}$$

For such fields, we can recast Maxwell equations as:

$$\begin{cases} E_z' - i\beta E_y = i\omega\mu_0 H_x & (2.29\text{a}) \\ i\beta E_x = i\omega\mu_0 H_y & (2.29\text{b}) \\ -E_x' = i\omega\mu_0 H_z & (2.29\text{c}) \end{cases}$$

and

$$\begin{cases} H_z' - i\beta H_y = -i\omega\varepsilon E_x & (2.30\text{a}) \\ i\beta H_x = -i\omega\varepsilon E_y & (2.30\text{b}) \\ -H_x' = -i\omega\varepsilon E_z & (2.30\text{c}) \end{cases}$$

where f' represents the derivative of f with respect to y. These last six equations can be divided into two distinct groups: the first one which is made up of Eqs. 2.29b, 2.29c, 2.30a only involves the components (E_x, H_y, H_z) whereas the second one which is made up of Eqs. 2.29a, 2.30b, 2.30c involves the remaining components, namely (H_x, E_y, E_z). Due to the linearity of the Maxwell equations and the constitutive relations (2.25), any field is a linear combination of two fields: for the first one E_x (*TE field*) is null and for the second one H_x is null (*TM field*). Furthermore, we note that for a TE field ($E_x = 0$), E_y and E_z can be derived from H_x via Eqs. 2.30b and 2.30c: it is therefore the appropriate unknown function, which is a solution of the following equation:

$$(\varepsilon_r^{-1}H_x')' + (k_0^2 - \varepsilon_r^{-1}\beta^2)H_x = 0 \ , \tag{2.31}$$

where $k_0 = \omega\sqrt{\mu_0\varepsilon_0}$.

For a TM field ($H_x = 0$), this time, E_x is the suitable unknown function because H_y and H_z can be derived from Eqs. 2.29b and 2.29c. The function E_x is then a solution of:

$$E_x'' + (k_0^2\varepsilon_r - \beta^2)E_x = 0 \ . \tag{2.32}$$

2.2.1.2 *Physics and functional spaces*

For the problem to be well specified, we still have to set the functional space for the functions E_x and H_x. For *physical reasons*, we assume that a guided mode is associated with a field of finite energy W in any band of finite width l (cf. Fig. 2.3):

$$W(l) = \int_a^{a+l} \int_{-\infty}^{+\infty} w(y)\, dxdy = l \int_{\mathbb{R}} w(y)\, dy\ , \tag{2.33}$$

where w is the total harmonic energy density ($w = \frac{1}{2}\boldsymbol{\mathcal{E}} \cdot \bar{\boldsymbol{\mathcal{D}}} + \frac{1}{2}\boldsymbol{\mathcal{B}} \cdot \bar{\boldsymbol{\mathcal{H}}}$).

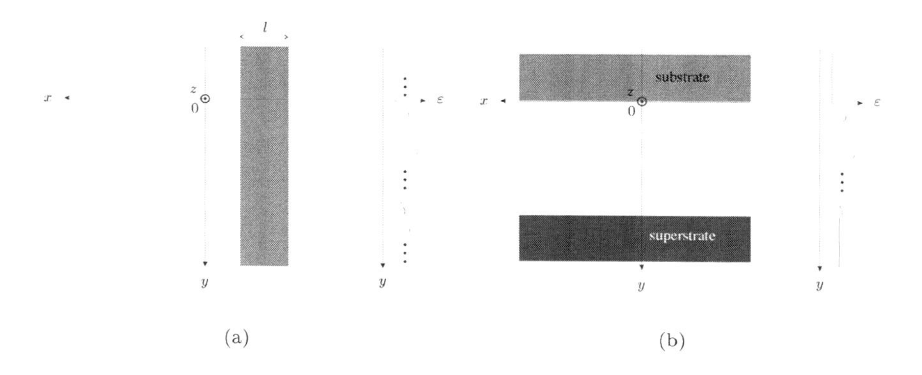

Fig. 2.3 Schematics for propagation in z direction : (a) an arbitrary structure (b) ε is constant for $y < 0$ (the substrate) and $y > h$ (the superstrate).

Theorem 2.1 *If the field is of finite energy in any band of width l (cf. Fig. 2.3) then E_x and H_x are in* [4] $H^1(\mathbb{R})$.

Proof. In the TM case, we can derive $\underset{\sim}{\boldsymbol{H}}$ from $\underset{\sim}{E}_x$

$$\underset{\sim}{\boldsymbol{H}} = \frac{-i}{\omega\mu_0} \mathbf{curl}(\underset{\sim}{E}_x \boldsymbol{e}^x) = \frac{-i}{\omega\mu_0} \mathbf{grad}\, \underset{\sim}{E}_x \times \boldsymbol{e}^x = \frac{-i}{\omega\mu_0}(i\beta E_x \boldsymbol{e}^y - E_x' \boldsymbol{e}^z) e^{i\beta z} \tag{2.34}$$

which leads to

$$\boldsymbol{H} = \frac{-i}{\omega\mu_0}(i\beta E_x \boldsymbol{e}^y - E_x' \boldsymbol{e}^z) \tag{2.35}$$

[4] For the definition of this Sobolev Space, see Math Appendix p. 305.

We can then compute the harmonic magnetic energy density w_m, namely

$$w_m = \frac{1}{4}(\underset{\sim}{\boldsymbol{B}} \cdot \bar{\underset{\sim}{\boldsymbol{H}}}) = \frac{\mu_0}{4}|\boldsymbol{H}|^2 . \tag{2.36}$$

We then obtain

$$w_m = \frac{1}{4}\varepsilon_0 \left[\left(\frac{\beta^2}{k_0^2} \right) |E_x|^2 + \frac{1}{k_0^2}|E_x'|^2 \right] . \tag{2.37}$$

Additionally, the expression for the harmonic electric energy density is:

$$w_e = \frac{1}{4}(\underset{\sim}{\boldsymbol{E}} \cdot \bar{\underset{\sim}{\boldsymbol{D}}}) = \frac{1}{4}\varepsilon_0 |E_x|^2 . \tag{2.38}$$

As a consequence, the total harmonic energy in any band of width l can be expressed as:

$$\mathcal{W}_{TM} = \frac{\varepsilon_0 l}{4} \int_{\mathbb{R}} \left(\frac{\beta^2}{k_0^2} + \varepsilon_r \right) |E_x|^2 + \frac{1}{k_0^2}|E_x'|^2 \, dy . \tag{2.39}$$

For any frequency ω, the quantity $\beta^2 + k_0^2 \varepsilon_r$ is positive and therefore E_x is such that

$$\int_{\mathbb{R}} |E_x|^2 \, dy < +\infty \quad \text{and} \quad \int_{\mathbb{R}} |E_x'|^2 \, dy < +\infty . \tag{2.40}$$

We conclude that the function E_x is in the functional space $H^1(\mathbb{R})$.

In the same way, we can find the expression for the total harmonic energy in the TE case. We find

$$\mathcal{W}_{TE} = \frac{\mu_0 l}{4} \int_{\mathbb{R}} \left(1 + \frac{\beta^2}{k_0^2}\varepsilon_r^{-1} \right) |H_x|^2 + \frac{1}{k_0^2}|H_x'|^2 \, dy , \tag{2.41}$$

which leads to the same conclusion for H_x. □

Corollary 2.1 *The field tends towards zero as y tends to infinity.*

Proof. The function ψ denotes either E_x or H_x. According to the Schwarz inequality:

$$\left| \int_a^b \psi(y)\psi'(y) \, dy \right|^2 \leq \left(\int_a^b |\psi(y)|^2 \, dy \right) \left(\int_a^b |\psi'(y)|^2 \, dy \right) . \tag{2.42}$$

Therefore, if both a and b independently tend to $+\infty$, the right-hand side tends to zero. Moreover, the left-hand side is nothing but $\frac{1}{2}\left(\psi^2(b) - \psi^2(a)\right)$. Therefore, ψ converges to a constant as a and b tend both to $+\infty$. Bearing in mind that ψ is in $L^2(\mathbb{R})$ this constant is bound to be null. The same reasoning can be done with $-\infty$. □

2.2.1.3 *Spectral presentation*

We have just proved that the problem of propagation along a certain direction in a 1-D structure can be reduced to searching for the solutions of Eqs. 2.31 and 2.32. We can summarize these equations as follows:

$$p^{-1}(p\psi')' + (k_0^2\varepsilon_r - \beta^2)\psi = 0 , \quad (\psi \in H^1(\mathbb{R})) , \tag{2.43}$$

with $\psi = E_x$ and $p = 1$ in the TM case and $\psi = H_x$ and $p = \varepsilon_r^{-1}$ in the TE case. In other words, we have to find solutions in $H^1(\mathbb{R})$ of the following eigenproblem:

$$\mathcal{L}_{k_0}\psi = \beta^2\psi , \tag{2.44}$$

where $\mathcal{L}_{k_0}$ is a linear operator depending on k_0 defined by

$$\mathcal{L}_{k_0} = p^{-1}(y)\frac{d}{dy}(p(y)\frac{d\cdot}{dy}) + k_0^2\varepsilon_r(y) . \tag{2.45}$$

In conclusion, it turns out that the function ψ is an eigenfunction of the operator $\mathcal{L}_{k_0}$ and the eigenvalue of the same operator is the square of the propagation constant β. Moreover, these propagation constants β depend on k_0 (and therefore on ω). One of our challenges is thus to obtain the so-called *dispersion curves* $\beta = \beta(\omega)$.

Remark 2.1 *Another spectral representation could be:*

$$\mathcal{M}_\beta\psi = k_0^2\psi , \tag{2.46}$$

where $\mathcal{M}_\beta$ is a linear operator depending on β

$$\mathcal{M}_\beta = \varepsilon^{-1}(y)p^{-1}(y)\frac{d}{dy}(p(y)\frac{d}{dy}) + \beta^2\varepsilon_r^{-1}(y) . \tag{2.47}$$

However, in a band of transparency, the permittivity of the material itself depends on ω. This is the reason why we use the operator $\mathcal{L}_{k_0}$ rather than $\mathcal{M}_\beta$.

Remark 2.2 *The stationary solutions ψ of the Schrödinger equation in the case of a monodimensional potential V are eigenfunctions ψ associated with the eigenvalue E:*

$$(-\frac{2m}{\hbar^2}\frac{d^2}{dy^2} + V(y))\psi = E\psi , \tag{2.48}$$

and therefore the eigenproblem is slightly different from those encountered in electromagnetism.

2.2.1.4 *Orthogonality of modes*

From Eq. 2.43, we derive:

$$(p\,\psi')' + k_0^2 p\,\varepsilon_r \psi = \beta^2 p\,\psi \tag{2.49}$$

Assuming that ψ_1 (resp. ψ_2) is an eigenfunction associated with the eigenvalue β_1 (resp. β_2) with $\beta_1 \neq \beta_2$. We are led to

$$\bar{\psi_2}\frac{d}{dy}\left(p\,\frac{d\psi_1}{dy}\right) + k_0^2 p \varepsilon_r \psi_1 \bar{\psi_2} = p\,\beta_1^2 \psi_1 \bar{\psi_2} \tag{2.50}$$

and

$$\psi_1 \frac{d}{dy}\left(p\,\frac{d\bar{\psi_2}}{dy}\right) + k_0^2 p \varepsilon_r \psi_1 \bar{\psi_2} = p\,\beta_2^2 \psi_1 \bar{\psi_2}\ . \tag{2.51}$$

Accordingly:

$$\int_{\mathbb{R}} \bar{\psi_2}\frac{d}{dy}\left(p\frac{d\psi_1}{dy}\right) - \psi_1\frac{d}{dy}\left(p\frac{d\bar{\psi_2}}{dy}\right) dy = (\beta_1^2 - \beta_2^2)\int_{\mathbb{R}} p\,\psi_1\bar{\psi_2}\,dy\ . \tag{2.52}$$

The integrand of the left-hand side is nothing but

$$\frac{d}{dy}\left[p\left(\bar{\psi_2}\frac{d\psi_1}{dy} - \psi_1\frac{d\bar{\psi_2}}{dy}\right)\right]. \tag{2.53}$$

Consequently:

$$\left[p\left(\bar{\psi_2}\frac{d\psi_1}{dy} - \psi_1\frac{d\bar{\psi_2}}{dy}\right)\right]_{-\infty}^{+\infty} = (\beta_1^2 - \beta_2^2)\int_{\mathbb{R}} p\,\psi_1\bar{\psi_2}\,dy\ . \tag{2.54}$$

Now, we have proved that the functions ψ_1 and ψ_2 vanish at infinity. Therefore, ψ_1 and ψ_2 are orthogonal with respect to the following scalar product.

$$\Big(\psi_1, \psi_2\Big)_p = \int_{\mathbb{R}} p\,\psi_1\bar{\psi_2}\,dy\ . \tag{2.55}$$

2.2.2 *Localisation of constants of propagation*

When dealing with dispersion curves a question arises : can one have some *a priori* information, even if it is rough, about the constants of propagation ? The answer to this question is based on the following pair of theorems.

Theorem 2.2 *If we define the function k as $k(y) = k_0\varepsilon_r(y)$, then:*

$$\beta \leq \max_{y\in\mathbb{R}}(k(y)) \tag{2.56}$$

Proof. From $(p\,\psi')' + p\beta^2(y)\psi = 0$ with $\beta^2(y) = k_0^2\varepsilon_r(y) - \beta^2$, we derive:

$$(p\,\psi')'\,\bar{\psi} + p\beta^2(y)|\psi|^2 = 0\,. \tag{2.57}$$

Besides, ψ is in $L^2(\mathbb{R})$ and for any given pair (k_0, β), β^2 is bounded above and below. Therefore the equation 2.57 implies that $(p\,\psi')'\,\bar{\psi}$ is integrable. As a consequence

$$\int_{\mathbb{R}} (p\,\psi')'\,\bar{\psi}\,dy + \int_{\mathbb{R}} p\beta^2(y)|\psi|^2\,dy = 0\,. \tag{2.58}$$

Let us assume that $(p\,\psi')'\,\bar{\psi}$ has a discontinuity at a point a. When integrating by parts, care must be taken in dealing with this point a:

$$\begin{aligned}\int_{\mathbb{R}} (p\,\psi')'\,\bar{\psi}\,dy &= \int_{-\infty}^{a_-} (p\,\psi')'\,\bar{\psi}\,dy + \int_{a_+}^{+\infty} (p\,\psi')'\,\bar{\psi}\,dy \\ &= \left[p\,\psi'\bar{\psi}\right]_{-\infty}^{a_-} - \int_{-\infty}^{a_-} p\,|\psi'|^2\,dy + \left[p\,\psi'\bar{\psi}\right]_{a_+}^{+\infty} - \int_{a_+}^{+\infty} p|\psi'|^2\,dy \\ &= p(a_-)\psi'(a_-)\bar{\psi}(a_-) - p(a_+)\psi'(a_+)\bar{\psi}(a_+) \\ &\quad -p(-\infty)\psi'(-\infty)\bar{\psi}(-\infty) + p(+\infty)\psi'(+\infty)\bar{\psi}(+\infty) \\ &\quad - \int_{-\infty}^{a_-} p\,|\psi'|^2\,dy - \int_{a_+}^{+\infty} p|\psi'|^2\,dy \end{aligned} \tag{2.59}$$

Now, due to the continuity of both ψ and $p\,\psi'$ and the vanishing behaviour at infinity, we can conclude that

$$\int_{\mathbb{R}} (p\,\psi')'\,\bar{\psi}\,dy = -\int_{\mathbb{R}} p|\psi'|^2\,dy\,. \tag{2.60}$$

It is worth noting that this last equation always holds when dealing with ε_r having a countable set of discontinuities. It follows that

$$-\int_{\mathbb{R}} p|\psi'|^2\,dy + \int_{\mathbb{R}} p\,\beta^2|\psi|^2\,dy = 0 \tag{2.61}$$

If $\beta \geq \max_{y\in\mathbb{R}}(k(y))$ then β^2 is negative for every y and therefore the integral $\int_{\mathbb{R}} p\,\beta^2|\psi|^2\,dy$ is negative as well which is obviously irreconcilable with Eq. 2.61. □

In most cases encountered in electromagnetism the function ε is constant outside a bounded interval, for instance $[0, h]$. We call the lower semi-infinite medium ($y < 0$) the *substrate* for which $\varepsilon_r(y) = \varepsilon_-$ and the upper semi-infinite medium ($y > h$) the *superstrate* for which $\varepsilon_r(y) = \varepsilon_+$. We are now in a position to prove the following inequality:

Theorem 2.3

$$\beta > \max(k_-, k_+) \tag{2.62}$$

with $k_- = k_0\sqrt{\varepsilon_-}$ *and* $k_+ = k_0\sqrt{\varepsilon_+}$.

Proof. *In the substrate* ε_r *equals* ε_-, *therefore* ψ *is a solution of*

$$\psi'' + \beta_-^2 \psi = 0 , \tag{2.63}$$

with $\beta_-^2 = k_0^2\varepsilon_- - \beta^2$ *and* $\beta_- \in \mathbb{R}^+ \cup i\mathbb{R}^+$. *This equation yields:*

$$\psi = B_- e^{-i\beta_- y} + B_+ e^{i\beta_- y} \tag{2.64}$$

Besides this function ψ *vanishes at* $-\infty$, *therefore* β *is a purely imaginary number which leads to* $\beta > k_-$ *(and* $B_+ = 0$*). By the same way, we find* $\beta > k_+$. □

2.2.3 *How can one practically get the dispersion curves and the modes?*

2.2.3.1 *Zerotic approach*

In order to find the modes and the propagation constants let us remark that a solution of Eq. 2.43 is completely characterized by its "initial values" $(\psi(0), \psi'(0))$. It is apropos to introduce a column vector:

$$\Psi(y) = \begin{pmatrix} \psi(y) \\ p(y)\psi'(y) \end{pmatrix} , \tag{2.65}$$

Our aim is to construct a transmission matrix T, that is, the 2×2 matrix such that:

$$\Psi(h) = T\Psi(0) . \tag{2.66}$$

Remark 2.3 *Let us note that the quantity* Ψ *is a continuous function because Eq. 2.43 implies that* $\psi(y)$ *and* $p(y)\psi'(y)$ *are both continuous functions, even for the points of discontinuity of the function* ε. *By a light abuse of notation, we write in the following* $p(0)\psi'(0)$ *(resp.* $p(d)\psi'(d)$*) instead of* $\lim_{y\longrightarrow 0^+} p(y)\psi'(y)$ *(resp.* $\lim_{y\longrightarrow d^-} p(y)\psi'(y)$*).*

In order to obtain the T matrix in closed form, we consider two particular solutions χ_1 and χ_2 defined by their initial conditions as follows:

$$\begin{cases} \chi_1(0) = 1, & p(0)\chi_1'(0) = 0, \quad (2.67a) \\ \chi_2(0) = 0, & p(0)\chi_2'(0) = 1 . \quad (2.67b) \end{cases}$$

Any solution $\psi(y)$ of Eq. 2.43 can be written in terms of $\chi_1(y)$ and $\chi_2(y)$ as:

$$\psi(y) = \psi(0)\chi_1(y) + p(0)\psi'(0)\chi_2'(y), \tag{2.68}$$

and we obtain

$$\begin{cases} \psi(h) = \psi(0)\,\chi_1(h) + p(0)\psi'(0)\,\chi_2(h) \\ \psi'(h) = \psi(0)\,\chi_1'(h) + p(0)\psi'(0)\,\chi_2'(h) \end{cases}. \tag{2.69}$$

These relations are equivalent to the matrix relation (2.66), hence we get the transmission matrix

$$T = \begin{pmatrix} \chi_1(h) & \chi_2(h) \\ p(h)\chi_1'(h) & p(h)\chi_2'(h) \end{pmatrix}. \tag{2.70}$$

The matrix T has the important property that its determinant $\det(T)$ is equal to unity. This result comes by considering the so-called *Wronskian* $W(y)$ of the system:

$$W(y) = p(y)\big(\chi_1(y)\,\chi_2'(y) - \chi_1'(y)\,\chi_2(y)\big). \tag{2.71}$$

A direct computation shows that $\frac{dW}{dy} = 0$. From this property, we obtain that $W(y) = 1$, by considering $W(0)$. As a consequence, we have $W(d) = 1$ and thus $\det(T) = 1$. This important property implies that the eigenvalues λ_1 and λ_2 of T are roots of the so-called *characteristic polynomial* : $X^2 - tr(T)\,X + 1$, where $tr(T)$ denotes the trace of T. As a consequence, three situations may occur.

- If $|tr(T)| < 2$, both eigenvalues of T are complex-valued. They have unit modulus, and they are the conjugates of each other. Therefore they can be written as $e^{i\theta}$ and $e^{-i\theta}$, where θ is a real number.
- If $|tr(T)| > 2$ both eigenvalues are real-valued and the reciprocal of each other. They can be written $\lambda_1 = e^{\theta}$ and $\lambda_1^{-1} = e^{-\theta}$ if $tr(T) > 2$ and $\lambda_1 = -e^{\theta}$ and $\lambda_1^{-1} = -e^{-\theta}$ if $tr(T) < -2$ (by convention $|\lambda_1| > 1$ *i.e.* $\theta > 0$).
- The very particular case $|tr(T)| = 2$ for which the eigenvalues are 1 and -1.

In other terms the eigenvalues can be written as follows:

$$\lambda_1 = \pm e^{u} \quad \text{and} \quad \lambda_2 = \pm e^{-u} \tag{2.72}$$

where u is in $\mathbb{R}^+$ or in $i\mathbb{R}^+$. We have thus obtained a splitting of the set of parameters $(k_0, \beta) \in \mathbb{R}^+ \times \mathbb{R}^+$ as follows:

$$\begin{aligned} \mathcal{G} &= \{(k_0, \beta), |tr(T)| > 2\} \\ \mathcal{B} &= \{(k_0, \beta), |tr(T)| < 2\} \\ \Delta &= \{(k_0, \beta), |tr(T)| = 2\} \end{aligned}$$

Theorem 2.4 *Finding the solution pairs (k_0, β) of Eq. 2.43 involves the identification of pairs which are the zeros of a function $\tilde{\mathcal{T}}_{2,1}$ defined by*

$$\tilde{\mathcal{T}}_{2,1} = (T_{1,1} + -i\beta_- p_- T_{1,2})(-i\beta_+ p_+) + T_{2,1} - i\beta_- p_- T_{2,2} \tag{2.73}$$

with $\beta_-^2 = k_0^2 \varepsilon_-$, $\beta_+^2 = k_0^2 \varepsilon_+$, $p_- = p(0^-)$ and $p_+ = p(h^+)$.

Proof. In the substrate $(y \leq 0)$ ε is constant, therefore ψ is a solution of

$$\psi'' + \beta_-^2 \psi = 0 \quad \text{with } \beta_-^2 = k_0^2 \varepsilon_- - \beta^2 \tag{2.74}$$

This equation yields:

$$\psi = A_- e^{-i\beta_- y} + a_- e^{i\beta_- y} \tag{2.75}$$

with β_- in $i\mathbb{R}^+$. In the same way, we obtain the solution in the superstrate $(y \geq h)$:

$$\psi = A_+ e^{i\beta_+ y} + a_+ e^{-i\beta_+ y} \tag{2.76}$$

We are now in a position to give expressions for $\Psi(0)$ and $\Psi(h)$.

$$\Psi(0) = M_- \begin{pmatrix} A_- \\ a_- \end{pmatrix} \quad \text{and} \quad \Psi(h) = M_+ \begin{pmatrix} A_+ \\ a_+ \end{pmatrix} \tag{2.77}$$

with

$$M_- = \begin{pmatrix} 1 & 1 \\ -i\beta_- p_- & i\beta_- p_- \end{pmatrix} \quad \text{and} \quad M_+ = \begin{pmatrix} e^{i\beta_+ h} & e^{-i\beta_+ h} \\ i\beta_+ p_+ e^{i\beta_+ h} & -i\beta_+ p_+ e^{-i\beta_+ h} \end{pmatrix} \tag{2.78}$$

Finally, we obtain

$$\begin{pmatrix} A_- \\ a_- \end{pmatrix} = \mathcal{T} \begin{pmatrix} A_+ \\ a_+ \end{pmatrix} \tag{2.79}$$

with $\mathcal{T} = M_+^{-1} T M_-$. This last equation leads us to:

$$\begin{pmatrix} 1 & -\mathcal{T}_{1,1} \\ 0 & \mathcal{T}_{2,1} \end{pmatrix} \begin{pmatrix} A_+ \\ A_- \end{pmatrix} = \begin{pmatrix} 0 & -\mathcal{T}_{1,2} \\ 1 & -\mathcal{T}_{2,2} \end{pmatrix} \begin{pmatrix} a_+ \\ a_- \end{pmatrix} . \tag{2.80}$$

If, we now consider only fields without antievanescent ingoing waves[5], we have

$$a_+ = a_- = 0\ . \tag{2.81}$$

The last two equations are consistent if and only if the determinant of the left-hand member matrix is null. Consequently the term corresponding to the first column and the second row vanishes, namely:

$$\mathcal{T}_{2,1}(k_0, \beta) = 0\ . \tag{2.82}$$

After elementary manipulation we find

$$\mathcal{T}_{2,1} = \frac{e^{i\beta_+ h}}{2i\beta_+ p_+} \left[(T_{1,1} - i\beta_- p_- T_{1,2})(-i\beta_+ p_+) + T_{2,1} + i\beta_- p_- T_{2,2} \right] \tag{2.83}$$

which finishes the demonstration. □

Remark 2.4 *This approach allows us to reduce the determination of the dispersion curves to the search for the zeros of a* real function *of two real variables β and k_0. Actually, according to the theorem 2.3 p. 41, β_- and β_+ are both purely imaginary numbers. Consequently, $\beta_- = i\tilde{\beta}_-$ and $\beta_+ = i\tilde{\beta}_+$ with $\tilde{\beta}_-$ and $\tilde{\beta}_+$ real numbers. Bearing in mind that the components of the matrix T are real, it is easy to see that $\tilde{\mathcal{T}}_{2,1}$ is a real function:*

$$\tilde{\mathcal{T}}_{2,1} = T_{2,1} + \tilde{\beta}_- p_- \tilde{\beta}_+ p_+ T_{1,2} + \left(\tilde{\beta}_+ p_+ T_{1,1} + \tilde{\beta}_- p_- T_{2,2} \right) \tag{2.84}$$

If the substrate and the superstrate are identical, we have $\tilde{\beta}_s = \tilde{\beta}_+ p_+ = \tilde{\beta}_- p_-$ and therefore the function $\tilde{\mathcal{T}}_{2,1}$ takes the particularly simple form

$$\tilde{\mathcal{T}}_{2,1} = T_{2,1} + \tilde{\beta}_s^2 T_{1,2} + \tilde{\beta}_s\, tr(T) \tag{2.85}$$

Theorem 2.5 *If the substrate and the superstrate are identical, if (k_0, β) is in $\mathcal{G}$, then (k_0, β) cannot correspond to a mode.*

Proof. To prove this latter theorem, we will adopt a roundabout route by making use of the poles of the reflection coefficient. For a given pair (k_0, β), we artificially introduce the coefficient of reflection r and the coefficient of transmission t. We thus have

$$\psi(y) = e^{i\beta_s y} + r(k_0, \beta) e^{i\beta_s y} \tag{2.86}$$

[5]The field then exponentially decreases at infinity.

in the substrate, and

$$\psi(y) = t(k_0, \beta)e^{-i\beta_s y} \tag{2.87}$$

in the superstrate. The problem is now reduced to finding the poles of r, *i.e.* to finding pairs (k_0, β) in such a way that $r(k_0, \beta) = +\infty$. We suppose that (k_0, β) belongs to $\mathcal{B} \cup \mathcal{G}$. Denoting by $(\mathbf{v}, \mathbf{w})$ a basis of eigenvectors of T we write in the canonical basis of $\mathbb{R}^2$: $\mathbf{v} = (v_1, v_2)$, $\mathbf{w} = (w_1, w_2)$. Eigenvector $\mathbf{v}$ (resp. $\mathbf{w}$) is associated with eigenvalue $\gamma(k_0, \beta)$ (resp. $\gamma^{-1}(k_0, \beta)$). It is of course always possible to choose $(\mathbf{v}, \mathbf{w})$ such that $\det(\mathbf{v}, \mathbf{w}) = 1$. After tedious but easy calculations, we get r in closed form:

$$r(k_0, \beta) = \frac{(\gamma^2 - 1) f(k_0, \beta)}{(\gamma^2 - g^{-1}(k_0, \beta) f(k_0, \beta))} \tag{2.88}$$

in terms of functions f and g have yet to be defined. Denoting

$$q(x_1, x_2) = \frac{i\beta_s x_2 - x_1}{i\beta_s x_2 + x_1} \tag{2.89}$$

functions f and g are defined by

- If $(k_0, \beta) \in \mathcal{G}$

$$\begin{cases} g(k_0, \beta) = q(v_1, v_2) & (2.90\text{a}) \\ f(k_0, \beta) = q(w_1, w_2) & (2.90\text{b}) \end{cases}$$

- If $(k_0, \beta) \in \mathcal{B}$
  - If $|q(v_1, v_2)| < |q(w_1, w_2)|$

$$\begin{cases} g(k_0, \beta) = q(v_1, v_2) & (2.91\text{a}) \\ f(k_0, \beta) = q(w_1, w_2) & (2.91\text{b}) \end{cases}$$

  - If $|q(v_1, v_2)| > |q(w_1, w_2)|$

$$\begin{cases} g(k_0, \beta) = q(w_1, w_2) & (2.92\text{a}) \\ f(k_0, \beta) = q(v_1, v_2) & (2.92\text{b}) \end{cases}$$

Now, we have to notice that the functions f and g are chosen in such a way that for any pair $(k_0, \beta) \in \mathcal{G}$, $|g| < |f|$. Bearing in mind that γ (and hence γ^2) is chosen in such a way that $|\gamma| < 1$ in $\mathcal{G}$, the denominator of r can never vanish. Consequently, the coefficient of reflection cannot possess any pole in $\mathcal{G}$. □

Remark 2.5 *It is worth noting that for a pair* (k_0, β) *in* $\mathcal{B}$ *the situation is completly different since* $|f| = |g|$ *and* γ *is of modulus* 1. *Therefore the equation* $\gamma = \frac{f}{g}$ *is plausible for some pairs* (k_0, β) *which represent the modes. This latter method offers the great advantage that it may be generalized to the case of leaky modes.*

The very heart of this book is devoted to periodic structures: we will establish some properties specific to that case. If the waveguide is made of N identical layers, the matrix $T^{(N)}$ associated with the whole crystal is therefore linked to the matrix T_l associated with *one single layer* by the relation:

$$T^{(N)} = T_l^N \tag{2.93}$$

We can prove the following theorem:

Theorem 2.6 *If we define the following two sets* $\mathcal{G}_N$ *and* $\mathcal{G}_l$ *as*

$$\mathcal{G}_N = \left\{ (k_0, \beta) \, , \left| tr(T^{(N)}) \right| > 2 \right\} \tag{2.94}$$

and

$$\mathcal{G}_l = \{ (k_0, \beta) \, , |tr(T_l)| > 2 \} \tag{2.95}$$

then we have $\mathcal{G}_N = \mathcal{G}_l = \mathcal{G}_1$.

Proof. If we denote by λ_1 and λ_2 the two eigenvalues associated with the matrix T_l, we saw (cf. p. 42) that $\lambda_1 = \pm e^{\theta}$ and $\lambda_2 = \pm e^{-\theta}$ with θ being a real number. Moreover, let us recall that $tr(T_l^N) = \lambda_1^N + \lambda_2^N$. We have thus:

$$\frac{1}{2} |tr(T_l)| = \cosh(\theta) \tag{2.96}$$

and

$$\frac{1}{2} \left| tr(T_l^N) \right| = \cosh(N\theta) \tag{2.97}$$

Now if θ is real we have therefore $\cosh(N\theta) > 1$ which proves the inclusion $\mathcal{G}_l \subset \mathcal{G}_N$. Conversely if $\cosh(N\theta) > 1$, θ is real and therefore $\cosh(\theta) > 1$ which proves the inclusion $\mathcal{G}_N \subset \mathcal{G}_l$. □

In other words, the problem of propagation in the whole crystal amounts to computing the trace of a matrix associated with one layer. The region defined by $\mathcal{G}_l$ is called a *rigorously forbidden region*: whatever the number of layers no propagation is allowed!

(a) Modes in a *simple slab*

As an example, let us compute the matrix T for a multilayered structure. The permittivity is therefore described by a piecewise constant function. To do that, let us begin with a simple homogeneous layer, characterized by its permittivity ε_l and its depth d. In this layer ψ is a solution of:

$$\psi'' + \beta_l^2 \psi = 0 \quad , \text{ with } \beta_l^2 = k_0^2 \varepsilon_l - \beta^2 \ . \tag{2.98}$$

Thus, in this layer, the function ψ can be written in the form

$$\psi(y) = A e^{i\beta_l y} + B e^{-i\beta_l y} \tag{2.99}$$

And consequently, we can express $\Psi(0)$ and $\Psi(d)$ in the following manner:

$$\Psi(0) = M_1 \begin{pmatrix} A \\ B \end{pmatrix} \quad \text{and} \quad \Psi(d) = M_2 \begin{pmatrix} A \\ B \end{pmatrix} \tag{2.100}$$

with

$$M_1 = \begin{pmatrix} 1 & 1 \\ ip_l\,\beta_l & -ip_l\,\beta_l \end{pmatrix} \quad \text{and} \quad M_2 = \begin{pmatrix} e^{i\beta_l d} & e^{-i\beta_l d} \\ ip_l\,\beta_l e^{i\beta_l d} & -ip_l\,\beta_l e^{-i\beta_l d} \end{pmatrix} \tag{2.101}$$

Finally, we find the matrix T associated with the slab as

$$T = M_2 M_1^{-1} \ , \tag{2.102}$$

which leads to

$$T = \begin{pmatrix} \cos(\beta_l d) & \frac{1}{p_l \beta_l} \sin(\beta_l d) \\ -p_l \beta_l \sin(\beta_l d) & \cos(\beta_l d) \end{pmatrix} . [6] \tag{2.103}$$

By making use of this last expression, we can easily derive the function $\tilde{\mathcal{T}}_{2,1}$ that appears in Eq. 2.85.

$$\begin{aligned} \tilde{\mathcal{T}}_{2,1} &= \frac{\tilde{\beta}_s^{\,2}}{p_l \beta_l} \sin(\beta_l d) + 2\tilde{\beta}_s \cos(\beta_l d) - p_l \beta_l \sin(\beta_l d) \\ &= 2\left(\frac{\tilde{\beta}_s^{\,2}}{p_l \beta_l} - p_l \beta_l \right) \sin\left(\frac{\beta_l d}{2}\right) \cos\left(\frac{\beta_l d}{2}\right) \\ &\quad + 2\tilde{\beta}_s \left(\cos^2\left(\frac{\beta_l d}{2}\right) - \sin^2\left(\frac{\beta_l d}{2}\right) \right) . \end{aligned} \tag{2.104}$$

[6] Note that this matrix is in $SL_2(\mathbb{R})$ for every β_l in $\mathbb{R}_+ \cup i\mathbb{R}_+$ *i.e.* all components of this matrix are real and the determinant equals 1.

If we let $X = \frac{\beta_l d}{2}$ and $Y = \frac{\tilde{\beta}_s d}{2}$, we get:

$$\begin{aligned}\tilde{\mathcal{T}}_{2,1} &= 2\left[\left(\frac{Y^2}{pX} - pX\right)\sin X\cos X + Y\left(\cos^2 X - \sin^2 X\right)\right] \\ &= \frac{\sin(2X)}{pX}\left(Y - pX\tan X\right)\left(Y + pX\cot X\right) . \end{aligned} \tag{2.105}$$

The solutions $X = j\pi/2$, $(j \in \mathbb{N})$ lead to $\tan X$ or $\cot X$ infinite and therefore are inconsistent. Hence, the only solutions remaining are very classical (see Ref. [Mar91])

$$Y = pX\tan X \tag{2.106}$$

or

$$Y = -pX\cot X . \tag{2.107}$$

It remains to remark that the quantity $X^2 + \frac{Y^2}{p_s^2}$ is independent of β.

$$X^2 + \frac{Y^2}{p_s^2} = k_0^2\left(\varepsilon_l - \varepsilon_s\right)\frac{d^2}{4} . \tag{2.108}$$

The problem of modes in a simple slab amounts to finding the intersection of the curve described by $Y = pX\tan X$ for the symmetric modes or the curve described by $Y = -pX\cot X$ for the antisymmetric modes and the family of ellipses given by the above equation.

(b) Modes in a *binary multilayered structure*

We saw that the determination of modes can be reduced to the search for the zeros of a real function deduced from the so-called matrix T. In this paragraph, let us call by $T^{(N)}$ the matrix associated with N layers and let us use T_l for the matrix associated with *one* of the layers. With such a notation, we have the following relations:

$$T = T_{(1)}T_{(2)} \tag{2.109}$$

with

$$T_{(j)} = \begin{pmatrix} \cos(\beta_j d_j) & \frac{1}{p_j\beta_j}\sin(\beta_j d_j) \\ -p_j\beta_j\sin(\beta_j d_j) & \cos(\beta_j d_j) \end{pmatrix} , \; j \in \{1,2\} \tag{2.110}$$

and

$$T^{(N)} = T_l^N . \tag{2.111}$$

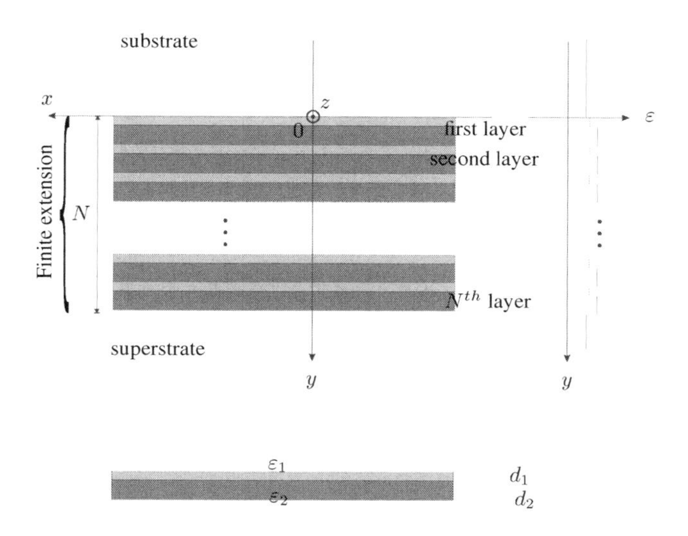

Fig. 2.4 Schematic for a binary multilayered structure.

(c) Some numerical results

In this paragraph, we give some examples of binary multilayered structures (cf. Fig. 2.4) in order to illustrate the theoretical results previously established. We consider a structure consisting of N identical layers, each layer being made up of two sublayers of thicknesses $d_1^{(N)}$ and $d_2^{(N)}$ and with $\varepsilon_{r,1} = 1$ and $\varepsilon_{r,2} = 2.1$. The thicknesses $d_1^{(N)}$ and $d_2^{(N)}$ are chosen in such a way that the total thickness $d_t = N(d_1^{(N)} + d_2^{(N)})$ does not depend on N, namely $d_t = 1$ with an arbitrary unit. We represent in Figs. 2.5– 2.6 the function $\tilde{\mathcal{T}}_{2,1}$ (as a matter of fact $\log(|\tilde{\mathcal{T}}_{2,1}|)$) against k_0 and β. The dispersion curves are therefore depicted by the *dark grooves*.

2.2.4 *Spectral approach*

2.2.4.1 *Strengths and weaknesses*

We have just seen that the spectral problem (See Eq. 2.44 p. 38) is in practice solved by means of the determination of the zeros of an element of the matrix T. We are therefore "doomed" to build the dispersion curves in a point-by-point fashion. Besides, the very computation of this matrix T can turn out to be a delicate task when ε is not a piecewise constant function: the study of solutions of Eq. 2.43 from which one can derive the matrix T follows from the so-called *Sturm-Liouville Problem* (See Refs [GWH02;

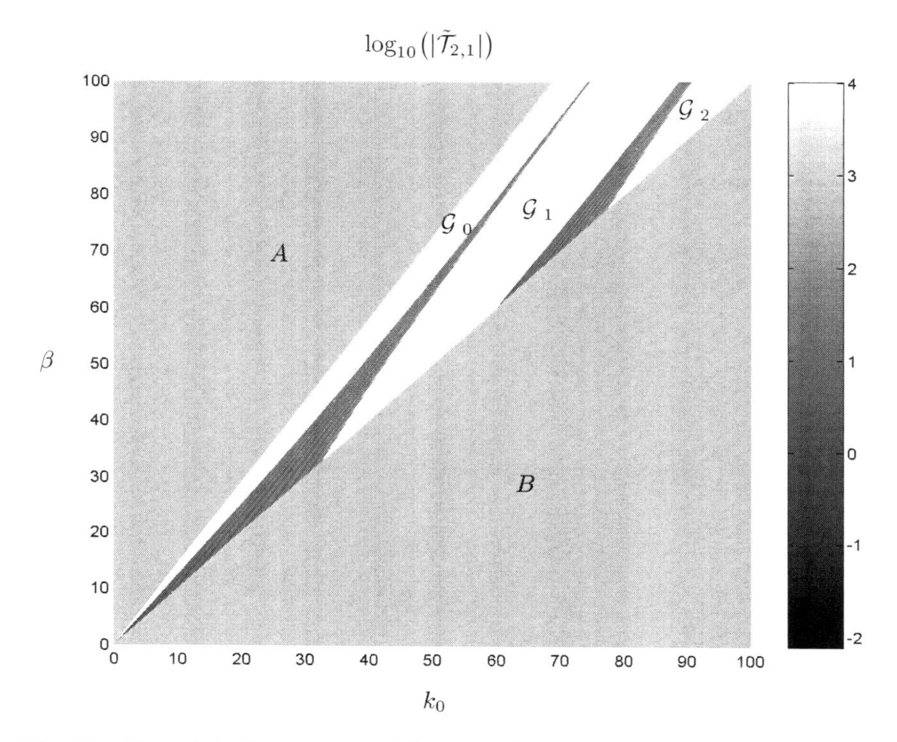

Fig. 2.5 Map of the function $\log_{10}(|\tilde{\mathcal{T}}_{2,1}(k_0,\beta)|)$ in the allowed regions for a binary multilayered structure described in Fig. 2.4 with 10 layers and with $d_1 = d_2 = 0.05$, $\varepsilon_{r,1} = 1$ and $\varepsilon_{r,2} = 2.1$. The dispersion curves are depicted by the dark grooves. The forbidden regions can be divided into three parts: A ($\beta < \max_{\mathbb{R}}(k(y))$, see Theorem 2.2), B ($\beta > \max(k_-, k_+)$, see Theorem 2.3) and the *rigorously forbiden region* ($\mathcal{G}_l = \mathcal{G}_0 \cup \mathcal{G}_1 \cup \mathcal{G}_2 \ldots$, see Theorems 2.5–2.6).

Sch80], for instance). The reader is thus entitled to ask himself the following question:

Can one find a way to approach the operator $\mathcal{L}_{k_0}$ by a matrix $L_{k_0}^{(l)}$, where l is the dimension of the matrix for which the eigenvalues; (resp. eigenfunctions) would be the approximation of the eigenvalues; (resp. eigenfunctions) of the operator $\mathcal{L}_{k_0}$? In other terms, if $\Lambda_{k_0}^{(l)}$ (resp. $\psi_{k_0}^{(l)}$) represents an eigenvalue (resp. eigenfunction) of $L_{k_0}^{(l)}$ and Λ_{k_0} (resp. ψ_{k_0}) represents an eigenvalue (resp. eigenfunction) of $\mathcal{L}_{k_0}$, we expect:

$$\Lambda_{k_0}^{(l)} \stackrel{l \to +\infty}{\longrightarrow} \Lambda_{k_0} \tag{2.112}$$

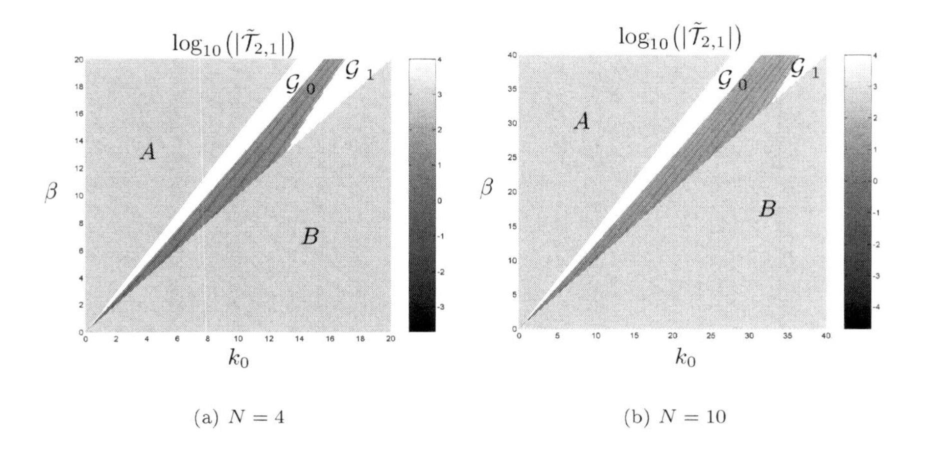

(a) $N = 4$ (b) $N = 10$

Fig. 2.6 Map of the function $\log_{10}(|\tilde{\mathcal{T}}_{2,1}(k_0, \beta)|)$ for a binary multilayered structure described in Fig. 2.4 with $\varepsilon_{r,1} = 1$ and $\varepsilon_{r,2} = 2.1$ and for N layers: (a) $N = 4$, $d_1^{(4)} = d_2^{(4)} = 0.125$, (b) $N = 10$, $d_1^{(10)} = d_2^{(10)} = 0.05$. *Note that the number of dispersion curves (dark grooves) per admissible region equals the number of layers.*

and

$$\psi_{k_0}^{(l)} \overset{l \to +\infty}{\longrightarrow} \psi_{k_0} \quad \text{, in a sense to be made more precise later.} \tag{2.113}$$

As we shall see, the task will not be easy. The reason is that for a given k_0, we expect a certain number of modes, for instance, one for a certain range of low frequencies whereas the matrix $L_{k_0}^{(l)}$ has, of course, l eigenvalues and l eigenfunctions! Thus we will have to sort out the true modes among all the *spurious modes* "given" by the matrix $L_{k_0}^{(l)}$. In order to have a more definite idea about these spurious modes let us consider the following theorem.

Theorem 2.7 *One can find a sequence of functions $\psi_{k_0}^{(l)}$ defined in the following manner:*

(1) $\|\psi_{k_0}^{(l)}\|_{L^2(\mathbb{R})} = 1$
(2) $\lim_{l \longrightarrow +\infty} \|\mathcal{L}_{k_0}(\psi_{k_0}^{(l)}) - K_0^2 \psi_{k_0}^{(l)}\|_{L^2(\mathbb{R})} = 0$
(3) $\psi_{k_0}^{(l)} \rightharpoonup 0$, *weakly in* $L^2(\mathbb{R})$, *i.e.*
$(\forall \Phi \in L^2(\mathbb{R}), \lim_{l \longrightarrow +\infty} \left(\psi_{k_0}^{(l)}, \Phi\right)_{L^2(\mathbb{R})} = 0)$

This sequence of functions is called a sequence of approximated eigenfunctions or a Weyl sequence (see Ref [Ric78]).

Proof. First of all, we introduce a function ξ defined as:

$$\xi(y) = \frac{2}{\sqrt{3}}(y-h)^2 e^{-(y-h)} \chi_{[h,+\infty]}(y) \tag{2.114}$$

where χ_C is the characteristic function of the set C. The reader can verify that this function has the following properties which are used in what follows:

(1) $\|\xi\|_{L^2(\mathbb{R})} = 1$
(2) ξ' and ξ'' are both in $L^2(\mathbb{R})$. (Note that ξ and ξ' are both continuous in h and therefore everywhere. Accordingly, $\xi' = \{\xi\}'$ and $\xi'' = \{\xi\}''$).[7]

We are now in a position to prove that the sequence:

$$\psi_{k_0}^{(l)}(y) = \frac{1}{\sqrt{l}}\xi\left(\frac{y}{l}\right)\phi(y) \quad , \tag{2.115}$$

where ϕ is a solution of:

$$\phi'' + K_0^2\phi = 0 \quad , \tag{2.116}$$

with $K_0^2 = \Lambda - k_0^2\varepsilon_{sup}$ is a Weyl sequence. Let us take e^{iK_0y} as solution of Eq. 2.116. For this choice of function ϕ, we have

$$\int_{\mathbb{R}} |\psi_{k_0}^{(l)}|^2\, dy = \frac{1}{l}\int_{\mathbb{R}} \xi^2\left(\frac{y}{l}\right)\, dy = 1 \,, \tag{2.117}$$

which proves the first statement. For the second one, let us remark that, due to the support of the function $\psi_{k_0}^{(l)}$:

$$\mathcal{L}_{k_0}(\psi_{k_0}^{(l)}) = \psi_l'' + k_0^2\varepsilon_r\psi_{k_0}^{(l)} = \psi_l'' + k_0^2\varepsilon_{sup}\psi_{k_0}^{(l)} \,. \tag{2.118}$$

Additionally,

$$\psi_l'' = l^{-5/2}\xi''\left(\frac{y}{l}\right)\phi(y) + l^{-3/2}\xi'\left(\frac{y}{l}\right) + \phi'(y) + l^{-1/2}\xi\left(\frac{y}{l}\right)\phi''(y) \,, \tag{2.119}$$

which leads to:

$$\mathcal{L}_{k_0}(\psi_{k_0}^{(l)}) - K_0^2\psi_{k_0}^{(l)} = l^{-5/2}\xi''\left(\frac{y}{l}\right)\phi(y) + l^{-3/2}\xi'\left(\frac{y}{l}\right)\phi'(y) + l^{-1/2}\xi\left(\frac{y}{l}\right)\underbrace{\left(\phi''(y) + (\Lambda + k_0^2\varepsilon_{sup})\phi\right)}_{=0} \,. \tag{2.120}$$

[7] ξ' represents the derivative in the sense of distributions whereas $\{\xi\}'$ represents the ordinary derivative, see for instance, math appendix p. 304.

As a consequence, we have:

$$|\mathcal{L}_{k_0}(\psi_{k_0}^{(l)}) - K_0^2\psi_{k_0}^{(l)}|^2 \leq l^{-5}\left[\xi''\left(\frac{y}{l}\right)\right]^2 + K_0^2 l^{-5}\left[\xi'\left(\frac{y}{l}\right)\right]^2 . \quad (2.121)$$

After the change of variable $u = \frac{y}{l}$, we get the following inequality:

$$\left\|\mathcal{L}_{k_0}(\psi_{k_0}^{(l)}) - K_0^2\psi_{k_0}^{(l)}\right\|_{L^2(\mathbb{R})} \leq l^{-4}\int_{\mathbb{R}} [\xi''(u)]^2 \, du + K_0^2\, l^{-2}\int_{\mathbb{R}} [\xi'(u)]^2 \, du \quad (2.122)$$

which concludes the second point. The last point requires the Lebesgue dominated convergence theorem (see, for instance Ref. [CBWMDB82]). It is easily seen that the function ξ is bounded and therefore we have the following inequalities:

$$\frac{1}{\sqrt{l}}\xi\left(\frac{y}{l}\right) \leq \frac{M}{\sqrt{l}} \leq M . \quad (2.123)$$

Therefore, we can find for the function $\psi_{k_0}^{(l)}\,\Phi$ an upper bound independent of l:

$$\left|\frac{1}{\sqrt{l}}\xi\left(\frac{y}{l}\right)\Phi(y)e^{iK_0 y}\right| \leq M|\Phi(y)| \quad (2.124)$$

Furthermore, for almost every y, the sequence $\psi_{k_0}^{(l)}\,\Phi$ converges towards zero. We then apply the Lebesgue dominated convergence theorem which proves the third and last point of the theorem. □

At first sight, one could hastily conclude that for every sequence satisfying the conditions 1–2 the function $\psi_{k_0}^{(l)}$ for a sufficiently large l is a good candidate for representing a propagating mode. The example given in Eq. 2.115 shows that the function $\psi_{k_0}^{(l)}$ is far from looking like a propagating mode (the support of this function is $[lh, +\infty]$!). The third condition which appears innocent, is therefore crucial. If $\psi_{k_0}^{(l)}$ converges to $\psi_{k_0}^{\mathrm{TRUE}}$, where $\psi_{k_0}^{\mathrm{TRUE}}$ is supposed to represent a *true mode*, the third condition obviously fails!

2.2.4.2 *The variational formulation (weak formulation)*

Starting from Eq. 2.43, we multiply each member by $w\,\bar{\psi}_2$, where w is a strictly positive bounded function that is sufficiently smooth ($\mathcal{C}^1(\mathbb{R})$, for instance). We then obtain by integrating over $\mathbb{R}$:

$$\int_{\mathbb{R}} p^{-1}(p\psi_1)'\bar{\psi}_2 w \, dy + k_0^2\int_{\mathbb{R}} \varepsilon_r w\psi_1\bar{\psi}_2 \, dy - \beta^2\int_{\mathbb{R}} w\psi_1\bar{\psi}_2 \, dy = 0 . \quad (2.125)$$

Let us focus our attention on the first term. We begin by making the change of variables $d\xi = w(y)\,dy$ and by integrating by parts, resulting in

$$\int_{\mathbb{R}} p^{-1}(p\psi_1')'\bar{\psi_2}\,d\xi = \left[\psi_1'\bar{\psi_2}\right]_{-\infty}^{+\infty} - \int_{\mathbb{R}} (p^{-1}\bar{\psi_2})' p\,\psi_1\,d\xi\ . \tag{2.126}$$

Recalling that ψ vanishes at infinity, the term between brackets does this also. We are thus led to introduce a bilinear form a:[8]

$$a(\psi_1,\psi_2) = \int_{\mathbb{R}} p^{-1}\bar{\psi_2}(p\psi_1')'\,d\xi = -\int_{\mathbb{R}} (p^{-1}\bar{\psi_2})'(p\,\psi_1')\,d\xi\ . \tag{2.127}$$

This bilinear form is symmetric for T.M. fields

$$a_{\text{T.M.}}(\psi_1,\psi_2) = -\int_{\mathbb{R}} \psi_1'\bar{\psi_2}'\,d\xi \tag{2.128}$$

but not for T.E. fields, which could lead to undesired consequencies. In order to avoid this difficulty, we take up Eq. 2.43 again, but this time, we multiply each side by $p\,w\,\bar{\psi_2}$. After a change of variables and an integration by parts, we obtain a symmetric bilinear form:

$$a_{\text{T.E.}}(\psi_1,\psi_2) = -\int_{\mathbb{R}} p\,\psi_1'\bar{\psi_2}'\,d\xi\ . \tag{2.129}$$

In brief, introducing the optical index $\nu = \sqrt{\varepsilon_r}$, we obtain the two following powerful formulations:

$$a_{\text{T.M.}}(\psi_1,\psi_2) + k_0^2\Big(\nu\psi_1,\nu\psi_2\Big)_w - \beta^2\Big(\psi_1,\psi_2\Big)_w = 0 \tag{2.130}$$

in the T.M. case and

$$a_{\text{T.E.}}(\psi_1,\psi_2) + k_0^2\Big(\psi_1,\psi_2\Big)_w - \beta^2\Big(\nu\psi_1,\nu\psi_2\Big)_w = 0 \tag{2.131}$$

in the T.E. case. Let us assume, now that $(e_n)_{n\in\mathbb{N}}$ is a Hilbert-basis [9] associated with the above-mentioned scalar product $\Big(\cdot,\cdot\Big)_w$ ($\Big(e_n,e_m\Big)_w = \delta_{n,m}$) and $\psi = \sum_{n\in\mathbb{N}}\psi_n e_n$, where ψ_n are complex numbers such as $\psi_n = \Big(\psi,e_n\Big)_w$. We return to Eqs. 2.131–2.130 with $\psi_1 = \psi$, where ψ is a

[8] See for instance, p. 296.

[9] We give an example of a Hilbert-basis in the following paragraph. *Note that the notion of a Hilbertian basis in not absolutely necessary and can be relaxed: if the orthogonality between the functions e_n is not required, the family of functions is called a total family. A practical example is given in the section devoted to finite elements.*

solution of Eq. 2.43 and with any admissible function $\psi_2 = \tilde{\psi}$.

$$\sum_{(n,m)\in\mathbb{N}^2} \bar{\tilde{\psi}}_m \big[M_\beta^{n,m} - k_0^2 \delta_{n,m}\big] \psi_n = 0 \tag{2.132}$$

and

$$\sum_{(n,m)\in\mathbb{N}^2} \bar{\tilde{\psi}}_m \big[L_{k_0}^{n,m} - \beta^2 \delta_{n,m}\big] \psi_n = 0\,. \tag{2.133}$$

with $M_\beta^{n,m} = a_{\text{T.M.}}(e_n, e_m) - \beta^2 \Big(\nu e_n, \nu e_m\Big)_w$ and $L_{k_0}^{n,m} = -a_{\text{T.E.}}(e_n, e_m) + k_0^2 \Big(\nu e_n, \nu e_m\Big)_w$. We admit that the fulfilment of the last two equations for any admissible functions $\tilde{\psi}$ implies that for any $n \in \mathbb{N}$, $\big[M_\beta^{n,m} - k_0^2 \delta_{n,m}\big]\psi_n = 0$ and $\big[L_{k_0}^{n,m} - \beta^2 \delta_{n,m}\big]\psi_n = 0$. In other words, in the T.E. case k_0^2 appears as one of the eigenvalues of the infinite matrix $L_{k_0} = (L_{k_0}^{n,m})_{n,m}$ and, in the T.M. case, β^2 appears as one of the eigenvalues of $M_\beta = (M_\beta^{n,m})_{n,m}$.

2.2.4.3 *An example of a Hilbert-basis in $L^2(\mathbb{R})$: the Hermite polynomials*

Let e_n be a sequence of functions (polynomials) defined as:

$$e_n(y) = (-1)^n \pi^{-1/4} 2^{-n/2} (n!)^{1/2} e^{y^2} \frac{d^n}{dy^n} e^{-y^2} \tag{2.134}$$

then this sequence constitutes a Hilbert-basis with respect to the following scalar product

$$\Big(\psi_1, \psi_2\Big) = \int_{\mathbb{R}} e^{-y^2}\, \psi_1 \bar{\psi}_2 \, dy\,. \tag{2.135}$$

2.3 The Two-Dimensional Vectorial Case *(general case)*

In this paragraph, we tackle the very heart of this book in resuming the study of the propagation of the so-called modes but this time the function ε depends on both x and y. We are looking for the electromagnetic fields $(\underset{\sim}{\boldsymbol{E}}, \underset{\sim}{\boldsymbol{H}})$, solutions of the following harmonic Maxwell equations, a time dependance of the form $e^{-i\omega t}$ being assumed:

$$\begin{cases} \mathbf{curl}\, \underset{\sim}{\boldsymbol{H}} = -i\omega \underset{\sim}{\boldsymbol{D}} & (2.136a) \\ \mathbf{curl}\, \underset{\sim}{\boldsymbol{E}} = i\omega\ \underset{\sim}{\boldsymbol{B}}\,. & (2.136b) \end{cases}$$

At this stage, the materials can be supposed to be anisotropic. The quantities $\underline{\underline{\varepsilon}}$ and $\underline{\underline{\mu}}$ can thus be represented by symmetric matrices:

$$\underline{\underline{\varepsilon}} = \left(\varepsilon_{i,j}\right)_{i,j=\{x,y,z\}} \quad \text{and} \quad \underline{\underline{\mu}} = \left(\mu_{i,j}\right)_{i,j=\{x,y,z\}} \tag{2.137}$$

and the constitutive relations are therefore

$$\underset{\sim}{\boldsymbol{D}} = \underline{\underline{\varepsilon}}\underset{\sim}{\boldsymbol{E}} \quad \text{and} \quad \underset{\sim}{\boldsymbol{B}} = \underline{\underline{\mu}}\underset{\sim}{\boldsymbol{H}}\,. \tag{2.138}$$

Furthermore, taking into account the invariance of the guide along its $z-$axis, we define time-harmonic two-dimensional electric and magnetic fields $\boldsymbol{E}$ and $\boldsymbol{H}$ such that

$$\underset{\sim}{\boldsymbol{E}}(x,y,z) = \boldsymbol{E}(x,y)e^{i\beta z} \quad \text{and} \quad \underset{\sim}{\boldsymbol{H}}(x,y,z) = \boldsymbol{H}(x,y)e^{i\beta z}\,. \tag{2.139}$$

For $(\boldsymbol{E},\boldsymbol{H})$ satisfying Eqs. 2.136–2.139 can be written as:

$$\begin{cases} \mathbf{curl}_\beta\, \boldsymbol{H} = -i\omega\underline{\underline{\varepsilon}}(x,y)\boldsymbol{E} & (2.140a) \\ \mathbf{curl}_\beta\, \boldsymbol{E} = i\omega\underline{\underline{\mu}}(x,y)\boldsymbol{H} & (2.140b) \end{cases}$$

where the quantities $\mathbf{curl}_\beta\, \boldsymbol{H}$ and $\mathbf{curl}_\beta\, \boldsymbol{E}$ are defined by:

$$\begin{cases} \mathbf{curl}_\beta\, \boldsymbol{H}(x,y) = \mathbf{curl}\Big(\boldsymbol{H}(x,y)e^{i\beta z}\Big)e^{-i\beta z} & (2.141a) \\ \mathbf{curl}_\beta\, \boldsymbol{E}(x,y) = \mathbf{curl}\Big(\boldsymbol{E}(x,y)e^{i\beta z}\Big)e^{-i\beta z} & (2.141b) \end{cases}$$

In the literature, three kinds of modes are studied; the guided modes, the leaky modes and the radiative modes.[10] For this purpose we have to define accurately these modes.

(1) **We say that $(\boldsymbol{E},\ \boldsymbol{H})$ is a *guided mode* if the following three conditions are fulfilled:**

   (a) $(\beta,\omega) \in (\mathbb{R}^+)^2$
   (b) $(\boldsymbol{E},\boldsymbol{H}) \neq (\mathbf{0},\mathbf{0})$
   (c) $(\boldsymbol{E},\boldsymbol{H}) \in \left[L^2(\mathbb{R}^2)\right]^3$**.**

[10] In this book, only the former two modes are tackled.

(2) **We say that $(\boldsymbol{E}, \boldsymbol{H})$ is a *leaky mode* if we can find a strictly positive real number Γ in order to fulfil the following three conditions:**

(a) $(\beta, \omega) \in \mathbb{C}^+ \times \mathbb{R}^+$, **where** $\mathbb{C}^+ = \{z \in \mathbb{C}; \Im m\{z\} > 0\}$
(b) $(\boldsymbol{E}, \boldsymbol{H}) \neq (\mathbf{0}, \mathbf{0})$
(c) $(e^{-\Gamma R}\boldsymbol{E}, e^{-\Gamma R}\boldsymbol{H}) \in [L^2(\mathbb{R}^2)]^3$, **where** $R = \sqrt{x^2 + y^2}$.

(3) **We say that $(\boldsymbol{E}, \boldsymbol{H})$ is a *radiative mode* if, for any strictly positive real number Γ the following four conditions are fulfilled:**

(a) $(\beta, \omega) \in (\mathbb{R}^+)^2$
(b) $(\boldsymbol{E}, \boldsymbol{H}) \neq (\mathbf{0}, \mathbf{0})$
(c) $(\boldsymbol{E}, \boldsymbol{H}) \notin [L^2(\mathbb{R}^2)]^3$
(d) $(e^{-\Gamma R}\boldsymbol{E}, e^{-\Gamma R}\boldsymbol{H}) \in [L^2(\mathbb{R}^2)]^3$, **where** $R = \sqrt{x^2 + y^2}$.

The study of the problem of guided modes and also leaky modes is the main topic of this book. To do this, it is sometimes[11] useful to obtain some relations between the transverse components ($\boldsymbol{E}_t$ and $\boldsymbol{H}_t$) and the axial components (E_z and H_z). This is the main subject of what follows.

2.3.1 *Some useful relations between transverse and axial components*

We separate $\boldsymbol{H}(x, y)$ into its *transverse* and *axial components* $\boldsymbol{H}_t(x, y)$ and $H_z(x, y)$:

$$\begin{aligned} \boldsymbol{H}(x, y) &= H_{t,1}(x, y)\mathbf{e}^x + H_{t,2}(x, y)\mathbf{e}^y + H_z(x, y)\mathbf{e}^z \\ &= \boldsymbol{H}_t(x, y) + H_z(x, y)\mathbf{e}^z \end{aligned} \tag{2.142}$$

We define the tranverse gradient, divergence and curl as follows:

$$\left\{ \begin{aligned} &\mathbf{grad}_t\, H_z(x, y) = \frac{\partial H_z}{\partial x}\mathbf{e}^x + \frac{\partial H_z}{\partial y}\mathbf{e}^y && (2.143a) \\ &\mathrm{div}_t(H_{t,1}\mathbf{e}^x + H_{t,2}\mathbf{e}^y) = \frac{\partial H_{t,1}}{\partial x} + \frac{\partial H_{t,2}}{\partial y} && (2.143b) \\ &\mathbf{curl}_t(H_{t,1}\mathbf{e}^x + H_{t,2}\mathbf{e}^y) = (\frac{\partial H_{t,2}}{\partial x} - \frac{\partial H_{t,1}}{\partial y})\mathbf{e}^z\,. && (2.143c) \end{aligned} \right.$$

[11] This study is not absolutely necessary for all methods. For instance, when dealing with one of the variants of the Finite Element Methods it may appear judicious to take $\boldsymbol{E}$ or $\boldsymbol{H}$ as the unknown function (See chapter 2.6.5). On the contrary, for the Multipole Method, this preliminary study is a prerequisite.

We separate $\mathbf{curl}_\beta$ and div_β into their transverse and axial components as follows:

$$\mathrm{div}_\beta\, \boldsymbol{H} = \frac{\partial H_{t,1}}{\partial x} + \frac{\partial H_{t,2}}{\partial y} + i\,\beta H_z = \mathrm{div}_t\, \boldsymbol{H}_t + i\,\beta H_z \tag{2.144}$$

and

$$\begin{aligned}\mathbf{curl}_\beta\, \boldsymbol{H} &= \Big(\frac{\partial H_l}{\partial y} - i\,\beta H_{t,2}\Big)\mathbf{e}^x \\ &+ \Big(i\,\beta H_{t,1} - \frac{\partial H_l}{\partial x}\Big)\mathbf{e}^y + \Big(\frac{\partial H_{t,2}}{\partial x} - \frac{\partial H_{t,1}}{\partial y}\Big)\mathbf{e}^z \\ &= \mathbf{curl}_t\, \boldsymbol{H}_t + (\mathbf{grad}_t\, H_z - i\,\beta \boldsymbol{H}_t) \times \mathbf{e}^z\ .\end{aligned} \tag{2.145}$$

In the same way, we obtain

$$\mathbf{curl}_\beta\, \boldsymbol{E} = \mathbf{curl}_t\, \boldsymbol{E}_t + (\mathbf{grad}_t\, E_z - i\,\beta \boldsymbol{E}_t) \times \mathbf{e}^z\ . \tag{2.146}$$

The first term on the right hand side being along the $z-$axis and the last two terms being in the $xy-$plane we are led to:

$$\begin{cases} i\omega B_z \mathbf{e}^z = \mathbf{curl}_t\, \boldsymbol{E}_t & (2.147\text{a}) \\ i\omega \boldsymbol{B}_t = (\mathbf{grad}_t\, E_z - i\,\beta \boldsymbol{E}_t) \times \mathbf{e}^z & (2.147\text{b}) \\ i\omega D_z \mathbf{e}^z = -\,\mathbf{curl}_t\, \boldsymbol{H}_t & (2.147\text{c}) \\ i\omega \boldsymbol{D}_t = (-\,\mathbf{grad}_t\, H_z + i\,\beta \boldsymbol{H}_t) \times \mathbf{e}^z & (2.147\text{d}) \end{cases}$$

We now introduce the following two notations. We introduce $\underline{\underline{m}}_T$, the 2×2 matrix built from the first two rows and columns of the 3×3 matrix $\underline{\underline{m}}$:

$$\underline{\underline{m}}_T = \begin{pmatrix} m_{xx} & m_{xy} \\ m_{yx} & m_{yy} \end{pmatrix} \tag{2.148}$$

We also introduce the vector $\mathbf{m}_\perp = m_{xz}\mathbf{e}^x + m_{yz}\mathbf{e}^y$. Using this notation we get:

$$\boldsymbol{B}_t = \underline{\underline{\mu}}_T \boldsymbol{H}_t + H_z \boldsymbol{\mu}_\perp \tag{2.149}$$

and

$$\boldsymbol{D}_t = \underline{\underline{\varepsilon}}_T \boldsymbol{E}_t + E_z \boldsymbol{\varepsilon}_\perp \tag{2.150}$$

From Eqs. 2.150 and 2.147d we obtain

$$i\omega \boldsymbol{E}_t = \underline{\underline{\varepsilon}}_T^{-1} \Big[(-\,\mathbf{grad}_t\, H_z + i\beta \boldsymbol{H}_t) \times \mathbf{e}_z - i\omega E_z \boldsymbol{\varepsilon}_\perp\Big] \tag{2.151}$$

Introducing the matrix of rotation $R_{\frac{\pi}{2}}$:

$$R_{\frac{\pi}{2}} = \begin{pmatrix} 0 & 1 \\ -1 & 0 \end{pmatrix} \tag{2.152}$$

we have the following identity

$$\mathbf{u}_T \times \mathbf{e}_z = R_{\frac{\pi}{2}} \mathbf{u}_T \tag{2.153}$$

for any transverse vector. We have therefore

$$i\omega \boldsymbol{E}_t = -\underline{\underline{\varepsilon}}_T^{-1} R_{\frac{\pi}{2}} \,\mathbf{grad}_t\, H_z + i\beta \underline{\underline{\varepsilon}}_T^{-1} R_{\frac{\pi}{2}} \boldsymbol{H}_t - i\omega E_z \underline{\underline{\varepsilon}}_T^{-1} \boldsymbol{\varepsilon}_\perp \tag{2.154}$$

Using now Eqs 2.154, 2.149 and 2.147b we obtain, as expected, the equation giving $\boldsymbol{H}_t$ as a function of E_z and H_z:

$$\begin{aligned} \boldsymbol{H}_t = (-\omega^2 \underline{\underline{\mu}}_T - \beta^2 \underline{\underline{\tilde{K}}}(\underline{\underline{\varepsilon}}_T^{-1}))^{-1} \Big[& i\omega R_{\frac{\pi}{2}} \,\mathbf{grad}_t\, E_z \\ & + i\beta \underline{\underline{\tilde{K}}}(\underline{\underline{\varepsilon}}_T^{-1}) \,\mathbf{grad}_t\, H_z - \beta\omega \underline{\underline{\varepsilon}}_T^{-1} E_z \boldsymbol{\varepsilon}_\perp + \omega^2 H_z \boldsymbol{\mu}_\perp \Big] \end{aligned} \tag{2.155}$$

with $\underline{\underline{\tilde{K}}}(\underline{\underline{\varepsilon}}_T^{-1}) = R_{\frac{\pi}{2}} \underline{\underline{\varepsilon}}_T^{-1} R_{\frac{\pi}{2}}$. Similarly, we obtain

$$\begin{aligned} \boldsymbol{E}_t = (\omega^2 \underline{\underline{\varepsilon}}_T + \beta^2 \underline{\underline{\tilde{K}}}(\underline{\underline{\mu}}_T^{-1}))^{-1} \Big[& i\omega R_{\frac{\pi}{2}} \,\mathbf{grad}_t\, H_z \\ & - i\beta \underline{\underline{\tilde{K}}}(\underline{\underline{\mu}}_T^{-1}) \,\mathbf{grad}_t\, E_z - \beta\omega \underline{\underline{\mu}}_T^{-1} H_z \boldsymbol{\mu}_\perp - \omega^2 E_z \boldsymbol{\varepsilon}_\perp \Big] \end{aligned} \tag{2.156}$$

Remark 2.6 *An interesting relation can be easily derived:*

$$\underline{\underline{\tilde{K}}}(\underline{\underline{m}}_T^{-1}) = -\frac{\underline{\underline{m}}_T}{\det(\underline{\underline{m}}_T)} \,. \tag{2.157}$$

Consequently if $\underline{\underline{m}}_T = m_T Id_2$,[12] *we have* $\underline{\underline{\tilde{K}}}(\underline{\underline{m}}_T^{-1}) = -\frac{Id_2}{m_T}$. *As a result, when dealing with isotropic materials, Eqs. 2.155–2.156 can be written in a significantly more convenient manner as follows*[13]

$$\boldsymbol{H}_t = \frac{1}{k_T^2 - \beta^2} \Big[-i\omega\varepsilon_T R_{\frac{\pi}{2}} \,\mathbf{grad}_t\, E_z + i\beta \,\mathbf{grad}_t\, H_z \Big] \tag{2.158}$$

and

$$\boldsymbol{E}_t = \frac{1}{k_T^2 - \beta^2} \Big[i\omega\mu_T R_{\frac{\pi}{2}} \,\mathbf{grad}_t\, H_z + i\beta \,\mathbf{grad}_t\, E_z \Big] \tag{2.159}$$

with $k_T = \omega\sqrt{\varepsilon_T \mu_T}$.

[12] Id_2 denotes the bidimensional identity matrix.

[13] Note that Eqs. 2.158–2.159 hold true irrespective of whether the materials are homogeneous.

2.3.2 *Equations of propagation involving only the axial components*

The basic problem that we now have to solve is to look for the equations for which the unknowns are the axial components E_z and H_z. For this purpose, B_z and D_z have to be expressed as functions of $\boldsymbol{E}$ and $\boldsymbol{H}$. Due to the symmetry of $\underline{\underline{\varepsilon}}$ and $\underline{\underline{\mu}}$, we have:

$$\begin{cases} B_z = \mu_{zz} H_z + \boldsymbol{\mu}_\perp \cdot \boldsymbol{H}_t & (2.160a) \\ D_z = \varepsilon_{zz} E_z + \boldsymbol{\varepsilon}_\perp \cdot \boldsymbol{E}_t & (2.160b) \end{cases}$$

Besides Eqs. 2.155–2.156 can be summarized formally as follows:

$$\begin{pmatrix} \boldsymbol{E}_t \\ \boldsymbol{H}_t \end{pmatrix} = \mathcal{V} \begin{pmatrix} E_z \\ H_z \end{pmatrix} \tag{2.161}$$

where $\mathcal{V}$ is a differential operator. It is possible to proceed further by means of Eqs. 2.147a–2.147c. We now pursue our task with materials for which one of the principal axes is the z–axis.[14] For such materials, which we call z–anisotropic $\underline{\underline{\varepsilon}}$ and $\underline{\underline{\mu}}$ are such that $\boldsymbol{\varepsilon}_\perp$ and $\boldsymbol{\mu}_\perp$ vanish and the above relation becomes simpler:

$$\begin{pmatrix} \boldsymbol{E}_t \\ \boldsymbol{H}_t \end{pmatrix} = \begin{pmatrix} \underline{\underline{W}}_{EE} & \underline{\underline{W}}_{EH} \\ \underline{\underline{W}}_{HE} & \underline{\underline{W}}_{HH} \end{pmatrix} \begin{pmatrix} \mathbf{grad}_t\, E_z \\ \mathbf{grad}_t\, H_z \end{pmatrix} \tag{2.162}$$

where $\underline{\underline{W}}_{EE}$, $\underline{\underline{W}}_{EH}$, $\underline{\underline{W}}_{HE}$ and $\underline{\underline{W}}_{HH}$ are quite simply 2×2 symmetric matrices depending upon $\underline{\underline{\varepsilon}}_T$, $\underline{\underline{\mu}}_T$, β and ω. Taking into account that $\boldsymbol{\varepsilon}_\perp$ and $\boldsymbol{\mu}_\perp$ both vanish and using Eqs. 2.147a–2.147c, we are led to:

$$\begin{cases} E_z = \dfrac{i}{\omega \varepsilon_{zz}} \mathbf{curl}_t\, \boldsymbol{E}_t \cdot \mathbf{e}^z & (2.163a) \\[2ex] H_z = \dfrac{-i}{\omega \mu_{zz}} \mathbf{curl}_t\, \boldsymbol{H}_t \cdot \mathbf{e}^z & (2.163b) \end{cases}$$

The last step can be done via the following relation:

$$\mathbf{curl}_t \big(\underline{\underline{M}}_T\, \mathbf{grad}_t\, U_z \big) \cdot \mathbf{e}^z = \mathrm{div}_t \big(R_{\frac{\pi}{2}} \underline{\underline{M}}_T\, \mathbf{grad}_t\, U_z \big) \tag{2.164}$$

[14] $\underline{\underline{\varepsilon}}$ and $\underline{\underline{\mu}}$ are therefore of the form $\begin{pmatrix} m_{xx} & m_{xy} & 0 \\ m_{yx} & m_{yy} & 0 \\ 0 & 0 & m_{zz} \end{pmatrix}$, for $m \in \{\varepsilon, \mu\}$. This kind of matrix will be revisited in chapter 2.6.5.

Finally, letting $k_z = \omega\sqrt{\varepsilon_z \mu_z}$ we obtain the expected relations:

$$k_z^2 E_z = i\omega\mu_{zz}\big(\mathrm{div}_t(R_{\frac{\pi}{2}}\underline{\underline{W}}_{HE}\,\mathbf{grad}_t\,E_z) + \mathrm{div}_t(R_{\frac{\pi}{2}}\underline{\underline{W}}_{HH}\,\mathbf{grad}_t\,H_z)\big) \tag{2.165}$$

and

$$k_z^2 H_z = -i\omega\varepsilon_{zz}\big(\mathrm{div}_t(R_{\frac{\pi}{2}}\underline{\underline{W}}_{EE}\,\mathbf{grad}_t\,E_z) + \mathrm{div}_t(R_{\frac{\pi}{2}}\underline{\underline{W}}_{EH}\,\mathbf{grad}_t\,H_z)\big) \tag{2.166}$$

For most microstructured fibers the materials are isotropic. In that case, Eqs. 2.155–2.156 show that $\underline{\underline{W}}_{EE}$, $\underline{\underline{W}}_{EH}$, $\underline{\underline{W}}_{HE}$ and $\underline{\underline{W}}_{HH}$ are particularly simple:

$$\begin{cases} \underline{\underline{W}}_{EE} = \underline{\underline{W}}_{HH} = \dfrac{i\beta}{k_T^2 - \beta^2} Id_2 & \text{(2.167a)} \\ \underline{\underline{W}}_{EH} = \dfrac{i\omega\mu_T}{k_T^2 - \beta^2} R_{\frac{\pi}{2}} & \text{(2.167b)} \\ \underline{\underline{W}}_{HE} = \dfrac{-i\omega\varepsilon_T}{k_T^2 - \beta^2} R_{\frac{\pi}{2}} & \text{(2.167c)} \end{cases}$$

It is important to note that the last equations show that E_z and H_z are coupled even in this simple case:[15] this is the main difficulty that we have to overcome (See chapter 7.6).

2.3.3 *What are the special features of isotropic microstructured fibers?*

In most applications fibers are such that the permittivity and the permeability can be represented by piecewise constant functions. In each medium, ε_T and μ_T are independent of x and y, therefore using the relations $\mathrm{div}_t(Id_2\,\mathbf{grad}_t\,U_z) = \triangle_t U_z$ and $\mathrm{div}_t(R_{\frac{\pi}{2}}\,\mathbf{grad}_t\,U_z) = 0$, we obtain in each medium

$$\begin{cases} \triangle_t E_z + (k_T^2 - \beta^2)\dfrac{\varepsilon_{zz}}{\varepsilon_T} E_z = 0 & \text{(2.168a)} \\ \triangle_t H_z + (k_T^2 - \beta^2)\dfrac{\mu_{zz}}{\mu_T} H_z = 0 & \text{(2.168b)} \end{cases}$$

One should be careful, of course, not to hastily conclude that E_z and H_z are decoupled; the coupling lies hidden in the conditions of continuity deduced from Eqs. 2.165–2.166. The problem of the conditions of continuity is a

[15] Except, of course, for $\beta = 0$. We have indeed $\underline{\underline{W}}_{EE} = \underline{\underline{W}}_{HH} = 0$. However the reader can persuade himself that this last property does not hold when dealing with non $z-$anisotropic materials!

somewhat subtle matter. In order to convince himself of this fact let the reader now suppose that Σ represents the set of discontinuities of either ε or μ and $\mathbf{n}$ the normal vector to Σ. We have then the fundamental relation

$$\mathbf{grad}_t\, U_z = \{\mathbf{grad}_t\, U_z\} + [U_z]_\Sigma\, \mathbf{n}\, \delta_\Sigma \tag{2.169}$$

where $\mathbf{grad}\, U_z$ (resp. $\{\mathbf{grad}\, U_z\}$) represents the gradient in the sense of distribution (resp. ordinary gradient) and where $[U_z]_\Sigma$ represents the jump of the function U_z when crossing Σ. Besides, $E_z\mathbf{e}^z$ and $H_z\mathbf{e}^z$ represent fields which are tangential to the surface spanned by Σ namely $\Sigma \times \mathbb{R}$. Therefore the axial components E_z and H_z are continuous at Σ and consequently in the whole space. It is worth noting that Eqs. 2.165–2.166 contain quantities of the kind $\mathrm{div}_t(A\, \mathbf{grad}_t\, U_z)$ and if U_z were discontinuous when crossing Σ, we would be led to unsound quantities such as "$\mathrm{div}_t(A\, [U_z]_\Sigma\, \mathbf{n}\, \delta_\Sigma)$" because of A which is precisely discontinuous at Σ[16]. Accordingly, axial components are continuous:

$$\begin{cases} [E_z]_\Sigma = 0 & (2.170\text{a}) \\ [H_z]_\Sigma = 0 & (2.170\text{b}) \end{cases}$$

Remark 2.7 *We have supposed up to now that we are only dealing with dielectric materials. What happens in the case of an infinitely conducting metal ? We know that the electromagnetic field vanishes inside the metal. Accordingly E_z and H_z are null inside the metal. Now, if we denote by Σ_m the boundary of the metal it can be verified that E_z satisfies a Dirichlet condition whereas H_z satisfies a Neuman condition:*

$$\begin{cases} E_{z\,|\Sigma_m} = 0 & (2.171\text{a}) \\ \dfrac{dH_z}{dn}_{\,|\Sigma_m} = 0 & (2.171\text{b}) \end{cases}$$

where $\frac{dH_z}{dn} = \mathbf{grad}_t\, H_z \cdot \mathbf{n}$.

We now return to our problem and write the following identity

$$\mathrm{div}\, \mathbf{U} = \{\mathrm{div}\, \mathbf{U}\} + [\mathbf{U}]_\Sigma \cdot \mathbf{n}\, \delta_\Sigma \tag{2.172}$$

where $\mathrm{div}\, \mathbf{U}$ (resp. $\{\mathrm{div}\, \mathbf{U}\}$) represents the divergence in the sense of distribution (resp. ordinary divergence) and where $[\mathbf{U}]_\Sigma$ represents the jump of the vector $\mathbf{U}$ when crossing Σ. Consequently, since U_z is continuous

[16]See the remark concerning the Dirac distribution in the Annex of mathematics p. 304.

($\mathbf{grad}_t\, U_z = \{\mathbf{grad}_t\, U_z\}$), for any function χ we have the following relation

$$\begin{aligned}\mathrm{div}_t(\chi Id_2\, \mathbf{grad}_t\, U_z) &= \{\mathrm{div}_t(\chi Id_2\{\mathbf{grad}_t\, U_z\})\} + \left[\chi \frac{dU_z}{dn}\right]_\Sigma \delta_\Sigma \\ &= \chi\{\triangle_t U_z\} + \left[\chi \frac{dU_z}{dn}\right]_\Sigma \delta_\Sigma \end{aligned} \tag{2.173}$$

where $\frac{dU_z}{dn}$ denotes the normal derivative given by $\frac{dU_z}{dn} = \mathbf{grad}_t\, E_z \cdot \mathbf{n}$. Bearing in mind that the tangent vector $\mathbf{t}$ to Σ is nothing but $R_{\frac{\pi}{2}}\mathbf{n}$, we get

$$\begin{aligned}\mathrm{div}_t(\chi R_{\frac{\pi}{2}}\, \mathbf{grad}_t\, U_z) &= \{\mathrm{div}_t(\chi R_{\frac{\pi}{2}}\{\mathbf{grad}_t\, U_z\})\} + \left[\chi \frac{dU_z}{dt}\right]_\Sigma \delta_\Sigma \\ &= \chi \underbrace{\{\mathrm{div}_t(R_{\frac{\pi}{2}}\{\mathbf{grad}_t\, U_z\})\}}_{=0} + \left[\chi \frac{dU_z}{dt}\right]_\Sigma \delta_\Sigma \\ &= \left[\chi \frac{dU_z}{dt}\right]_\Sigma \delta_\Sigma \end{aligned} \tag{2.174}$$

where $\frac{dU_z}{dt}$ denotes the tangential derivative given by $\frac{dU_z}{dt} = \mathbf{grad}_t\, E_z \cdot \mathbf{t}$. Finally, from Eqs. 2.165–2.166, 2.167, 2.173 and 2.174 and remarking that $[A]_\Sigma + [B]_\Sigma = [A+B]_\Sigma$ we derive, as expected, the so-called relations of continuity:

$$\left\{\begin{aligned} &\left[\frac{\omega\varepsilon_T}{k_T^2-\beta^2}\frac{dE_z}{dn} + \frac{\beta}{k_T^2-\beta^2}\frac{dH_z}{dt}\right]_\Sigma = 0 && (2.175\text{a}) \\ &\left[\frac{-\omega\mu_T}{k_T^2-\beta^2}\frac{dH_z}{dn} + \frac{\beta}{k_T^2-\beta^2}\frac{dE_z}{dt}\right]_\Sigma = 0 && (2.175\text{b}) \end{aligned}\right.$$

2.4 The Two-Dimensional Scalar Case *(weak guidance)*

In this paragraph, we will see that under certain hypotheses recalled in the following, the general problem of propagation in fibers can be reduced to two independent scalar problems. We assume that ε_r becomes constant at a finite distance. In other words, the *jacket* is made up of an homogeneous and isotropic material and we call by ε_J the permittivity of the aforementioned jacket. Therefore the support $\mathcal{C}$ of the function $\varepsilon_r - \varepsilon_J$ is bounded and we denote by R the radius of $\mathcal{C}$:

$$R = \frac{\max_{(M_1,M_2)\in\mathcal{C}^2} ||\mathbf{M_1M_2}||}{2} \tag{2.176}$$

For such a guide, we define a normalized frequency V as:[17]

$$V(k_0) = k_0 R(\varepsilon_{\text{Max}} - \varepsilon_{\text{J}})^{1/2} , \tag{2.177}$$

where $\varepsilon_{\text{Max}} = \max_{\mathbf{x}\in\mathcal{C}} \varepsilon_r(\mathbf{x})$. Additionally, we assume that the distribution of ε_r is characterized by:

$$\varepsilon_r(x, y) = \varepsilon_{\text{J}}(1 + \eta f(x, y)) , \tag{2.178}$$

where η is a small parameter and where f is a real function such that $-1 < a < f \leq 1$. Therefore

$$(\varepsilon_{\text{Max}} - \varepsilon_{\text{J}})^{1/2} = \varepsilon_{\text{J}}^{1/2}\eta^{1/2} \tag{2.179}$$

and for $k_\eta = k_0\eta^{-1/2}$ and for a fixed R, we have

$$V(k_\eta) = k_0 R\varepsilon_{\text{J}}^{1/2} \tag{2.180}$$

One can therefore demonstrate (See Ref. [SY78]) that the transverse components of the electrical field, namely E_x and E_y, are completely independent of each other and they satisfy the same scalar eigenproblem:[18]

$$\triangle_T\psi + k_\eta^2\varepsilon_r\psi = \beta^2\psi \tag{2.181}$$

provided that the two following conditions are fulfilled:

(1) $\varepsilon_{\text{Max}} - \varepsilon_{\text{J}} = \mathcal{O}(\eta)$. The guide is then weakly heterogeneous.
(2) k_η is such that $V(k_\eta) = \mathcal{O}(1)$ (*i.e.* $k_\eta = \mathcal{O}(\eta^{-1/2})$). This condition means that $k_\eta R \gg 1$ and we are therefore in the range of optical frequencies.

Unfortunately, the most interesting properties of MOF's are the prerogative of high index fibres: the condition (1) then fails. We are therefore "doomed" to deal with the two-dimensional vectorial problem.

2.5 Spectral Analysis

2.5.1 *Preliminary remarks*

This paragraph is not a *crash course*: our aim here is extremely modest. The topics relating to Operator Theory and Spectral Analysis which are of

[17] We assume, of course, that $\varepsilon_{\text{Max}} > \varepsilon_{\text{J}}$!
[18] $\triangle_T = \frac{\partial^2}{\partial x^2} + \frac{\partial^2}{\partial y^2}$.

foremost importance in Physics, have been worked out in detail in numerous excellent books. Some of these books are very voluminous and comprehensive. Among this abundant literature, let us quote, for instance, two reference books [Kat95; RS78] and a thesis [Bon98] which are intended for readers with a strong mathematical background and who are interested in the more theoretical aspects of Physics. Secondly, let us suggest a few more accessible books for non-theoretical physicists [Ric78] (generalist book), [Sch80] (mainly devoted to Spectral Analysis linked with the Schrödinger Equation) and a remarkable book, recently published, whose main topic is precisely Spectral Analysis relating to Electromagnetism [GWH02].

2.5.2 *A brief vocabulary*

We saw in the chapter devoted to monodimensional waveguides that the problem of propagation could be reduced to looking for both the eigenfunctions and eigenvalues of an operator depending on a parameter k_0, namely $\mathcal{L}_{k_0}$. We remarked that this operator possesses some *pseudo-eigenfunctions* which are not true eigenfunctions in the sense of being associated with a mode. It is now high time to define precisely the necessary notions in order to tackle the mathematical problems associated with 2D-waveguides. It turns out that the definition and the classification of the spectrum of an operator (the generalization of the set of eigenvalues) is based on a derived operator, the so-called *resolvent operator*.

Definition 2.1 Let us consider an operator $A : H_1 \longrightarrow H_2$, where H_1 and H_2 are two Hilbert spaces. The resolvent operator $R_\lambda(A)$ is defined as:

$$R_\lambda(A) = (A - \lambda I)^{-1} \tag{2.182}$$

We give the name *resolvent set*, which we denote $\rho(A)$, to the set of complex numbers which satisfy the following three conditions:

(1) $R_\lambda(A)$ exists
(2) $R_\lambda(A)$ is bounded
(3) $R_\lambda(A)$ is dense in H_2

- If condition 1 is not fulfilled, we say that λ is an eigenvalue or that λ forms the *point spectrum* of A which we denote $\sigma_p(A)$.
- If conditions 1 and 3 but not condition 2 are fulfilled, we say that λ forms the *continuous spectrum* of A which we denote $\sigma_c(A)$.

- If conditions 1 and 2 but not condition 3 are fulfilled, we say that λ forms the *residual spectrum* of A which we denote $\sigma_r(A)$.

The *total spectrum* $\sigma(A)$ is the complementary in $\mathbb{C}$ of the resolvent set $\rho(A)$, we then have:

$$\sigma(A) = \mathbb{C} \setminus \rho(A) = \sigma_p(A) \cup \sigma_c(A) \cup \sigma_r(A) . \tag{2.183}$$

Remark 2.8 *It can be shown that in problems generally encountered in electromagnetism the residual spectrum is, in fact, reduced to the empty set. The reader to which theoretical aspects of spectral analysis appeal can refer to [Kat95; RS78]. As a consequence, one of our tasks is to accurately characterize both the continuous and the discrete spectra.*

In the literature devoted to spectral analysis, the notion of essential spectrum is often used. Here is the definition.

Definition 2.2 Let us consider an operator $A : H_1 \longrightarrow H_2$, where H_1 and H_2 are two Hilbert spaces. The *essential spectrum*,[19] which we denote $\sigma_{ess}(A)$ consists of all points of the spectrum except isolated eigenvalues of finite multiplicity.[20]

In the following, we will see that the point spectrum is the set of isolated eigenvalues of finite multiplicity.[21] Therefore, taking into account the previous remark, the essential spectrum and the continuous spectrum can be taken to be identical in this case.

2.5.3 *Posing of the problem*

We are now in a position to find the fields $(\boldsymbol{\mathcal{E}},\boldsymbol{\mathcal{H}})$ for a z−invariant structure for which $\boldsymbol{\mathcal{E}}(x,y,z) = \boldsymbol{E}(x,y)e^{i\beta z}$ and $\boldsymbol{\mathcal{H}}(x,y,z) = \boldsymbol{H}(x,y)e^{i\beta z}$ (cf. section 2.3 p. 55) and to characterize the dispersion curves i.e. the feasible pairs (β,ω). We are going to see that this problem takes us back to the study of a family of spectral problems. Finally, for the sake of simplicity, we assume that we are only dealing with non-magnetic materials and that the jacket is

[19] From a practical point of view, the "generalized eigenfunctions" corresponding to the values of the essential spectrum are associated with the Weyl sequences introduced p. 51 (See Ref. [Bon98]).

[20] By multiplicity, we imply geometric multiplicity, *i.e.* the dimension of the nullspace of $A - \lambda I$, where, of course, λ designes the eigenvalue at issue.

[21] Care must be taken when dealing with eigenvalues of infinite multiplicity as shown in Sec. 3.1.4.

surrounded by vacuum. In other words, we assume that $\mu_r = 1$ everywhere and that the support of the function $\varepsilon_r - 1$ is bounded.

2.5.4 *Continuous formulation*

The title of this section is somewhat weird and will be clarified in the next section, where we tackle the discretization of the *continuous formulation* by means of the Finite Element Method. We go over the problem that we left in section 2.3. Denoting by k_0 the wave number $\omega\sqrt{\mu_0\varepsilon_0}$, we are led to the following two systems of Maxwell's type:

$$\begin{cases} \mathbf{curl}_\beta\,\mathbf{curl}_\beta\,\boldsymbol{E} = k_0^2\varepsilon\boldsymbol{E} & (2.184a) \\ \mathrm{div}_\beta(\varepsilon\boldsymbol{E}) = 0 & (2.184b) \end{cases}$$

and

$$\begin{cases} \mathbf{curl}_\beta(\varepsilon^{-1}\,\mathbf{curl}_\beta\,\boldsymbol{H}) = k_0^2\boldsymbol{H} & (2.185a) \\ \mathrm{div}_\beta(\mu_0\boldsymbol{H}) = 0 & (2.185b) \end{cases}$$

div_β being an operator analogously defined to $\mathbf{curl}_\beta$ in Eq. 2.141 p. 56. It must be noted that $\boldsymbol{H}$ (resp. $\boldsymbol{E}$) can be deduced from Eq. 2.184 (resp. Eq. 2.185) via Eq. 2.140b (resp. Eq. 2.140a). Mking the obvious remark that thb divergence of $\boldsymbol{H}$ is null, contrary to that of $\boldsymbol{E}$, we choose a magnetic formulation. We thus have the following lemma:

Lemma 2.1 *Let $V(\beta)$ be a Hilbert space (dependent on β):*

$$V(\beta) = \{\mathbf{w} \in [L^2(\mathbb{R}^3)]^3\ ,\ \mathrm{div}_\beta\,\mathbf{w} = 0\} \tag{2.186}$$

Let s be a positive real. Then, the following two systems 2.187 and 2.188 are equivalent in $V(\beta)$:

$$\begin{cases} \mathbf{curl}_\beta(\varepsilon^{-1}\,\mathbf{curl}_\beta\,\boldsymbol{H}) = k_0^2\boldsymbol{H} & (2.187a) \\ \mathrm{div}_\beta(\mu_0\boldsymbol{H}) = 0 & (2.187b) \end{cases}$$

$$\mathbf{curl}_\beta\big(\varepsilon^{-1}\,\mathbf{curl}_\beta\,\boldsymbol{H}\big) - s\,\mathbf{grad}_\beta(\mathrm{div}_\beta\,\boldsymbol{H}) = k_0^2\boldsymbol{H} \tag{2.188}$$

Proof. A solution $\boldsymbol{H}$ of Eq. 2.187 in the Hilbert space $V(\beta)$ clearly satisfies Eq. 2.188. Conversely, suppose that $\boldsymbol{H}$ satisfies 2.188. Taking the divergence of the two members of Eq. 2.188 and letting ϕ be $\mathrm{div}_\beta\,\boldsymbol{H}$, we get:

$$-s\,\mathrm{div}_\beta(\mathbf{grad}_\beta\,\phi) = k_0^2\phi \tag{2.189}$$

We are thus led to the following equation:

$$-s\Delta\phi = (k_0^2 - \beta^2 s)\phi \;, \tag{2.190}$$

where $\phi \in L^2(\mathbb{R}^2)$. Taking the Fourier transform $\hat{\phi}$ of ϕ, we get:

$$(\Lambda^2 - p^2)\hat{\phi} = 0 \;, \quad \text{in } L^2(\mathbb{R}^2) \tag{2.191}$$

where $\Lambda = \sqrt{\dfrac{k_0^2 - \beta^2}{s}}$ and where we have adopted the form e^{-ipx} as a convention in our Fourier basis. The solution of the preceding equation in the sense of tempered distributions is of the form:

$$\hat{\phi} = A\delta(\Lambda - p) + B\delta(\Lambda + p) \tag{2.192}$$

which is not in $L^2(\mathbb{R}^2)$. Hence, Eq. 2.190 admits only trivial solutions, which prove that $\mathrm{div}_\beta \boldsymbol{H} = 0$. □

Noting that *in vacuo* Eq. 2.190 becomes:

$$-\Delta \boldsymbol{H} + \beta^2 \boldsymbol{H} + (1 - s)\,\mathbf{grad}_\beta(\mathrm{div}_\beta \boldsymbol{H}) = k^2 \boldsymbol{H} \tag{2.193}$$

we take $s = 1$ to get the Helmholtz equation *in vacuo*. Our problem reduces to that of finding the pair of real numbers (β, ε) such that there exists $\mathbf{H}$ that is a solution of:

$$\begin{cases} \boldsymbol{H} \in H^1(\mathbb{R}^2)^3 \;, \boldsymbol{H} \neq 0 & \text{(2.194a)} \\ c(\beta; \boldsymbol{H}, \boldsymbol{H}') = k_0^2(\boldsymbol{H}, \boldsymbol{H}') \;, \forall \boldsymbol{H}' \in H^1(\mathbb{R}^2)^3 & \text{(2.194b)} \end{cases}$$

where $c_(\beta; \cdot, \cdot)$ is the sesquilinear form (cf. p. 298) defined by:

$$c_(\beta; \boldsymbol{H}, \boldsymbol{H}') = \int_{\mathbb{R}^2} \Big(\frac{1}{\varepsilon}\, \mathbf{curl}_\beta \boldsymbol{H} \cdot \mathbf{curl}_\beta \overline{\boldsymbol{H}'} + \mathrm{div}_\beta \boldsymbol{H}\, \mathrm{div}_\beta \overline{\boldsymbol{H}'}\Big)\, dxdy \tag{2.195}$$

Noting that for all $\mathbf{U}$ in $H^1(\mathbb{R}^2)^3$, we have:

$$\int_{\mathbb{R}^2} (|\,\mathbf{curl}_\beta \mathbf{U}|^2 + |\,\mathrm{div}_\beta \mathbf{U}|^2)\, dxdy = \int_{\mathbb{R}^2} (|\,\mathbf{grad}\,\mathbf{U}|^2 + \beta^2 |\mathbf{U}|^2)\, dxdy \tag{2.196}$$

where $|\,\mathbf{grad}\,\mathbf{U}|^2 = tr\,((\mathbf{grad}\,\mathbf{U})^t\, \mathbf{grad}\,\mathbf{U}) = \sum_{i=1}^3 |\partial^i U_i|^2$. Thus assuming that $\inf\limits_{(x,y)\in\mathbb{R}^2} \varepsilon(x, y) = 1$, we deduce that $c(\beta; \cdot, \cdot)$ satisfies, for all β in $\mathbb{R}^+$ and for all $\boldsymbol{H}$ in $H^1(\mathbb{R}^2)^3$

$$\begin{cases} c(\beta; \boldsymbol{H}, \boldsymbol{H}) \geq \frac{1}{\varepsilon^+} \int_{\mathbb{R}^2} (|\,\mathbf{grad}\,\boldsymbol{H}|^2 + \beta^2 |\boldsymbol{H}|^2)\, dxdy & \text{(2.197a)} \\ c(\beta; \boldsymbol{H}, \boldsymbol{H}) \leq \int_{\mathbb{R}^2} (|\,\mathbf{grad}\,\boldsymbol{H}|^2 + \beta^2 |\boldsymbol{H}|^2)\, dxdy & \text{(2.197b)} \end{cases}$$

where ε^+ is given by $\varepsilon^+ = \sup_{(x,y)\in\mathbb{R}^2} \varepsilon(x,y)$.

If $\beta \neq 0$, the bilinear form $c(\beta;\cdot,\cdot)$ is thus continuous and coercive in $H^1(\mathbb{R}^2)^3$. From Lax-Milgram lemma, we then deduce that Eq. 2.187 admits a unique solution in $H^1(\mathbb{R}^2)^3$ given by the minimum of the following functional in the Hilbert space $H^1(\mathbb{R}^2)^3$:

$$\begin{aligned}\mathcal{R}(\beta;\boldsymbol{H},\boldsymbol{H}') &= \int_{\mathbb{R}^2} \frac{1}{\varepsilon}\, \mathbf{curl}_\beta\, \boldsymbol{H} \cdot \overline{\mathbf{curl}_\beta\, \boldsymbol{H}'}\, dxdy \\ &+ s \int_{\mathbb{R}^2} \mathrm{div}_\beta\, \boldsymbol{H}\, \overline{\mathrm{div}_\beta\, \boldsymbol{H}'}\, dxdy \\ &- k_0^2 \int_{\mathbb{R}^2} \boldsymbol{H} \cdot \overline{\boldsymbol{H}'}\, dxdy \end{aligned} \tag{2.198}$$

Lemma 2.2

(1) *Let β be in $\mathbb{R}^+$. Then the operator $C(\beta)$ defined for all $\boldsymbol{H}$ and $\boldsymbol{H}'$ in $[L^2(\mathbb{R}^2)]^3$ by $(C(\beta)\boldsymbol{H},\boldsymbol{H}') = c(\beta;\boldsymbol{H},\boldsymbol{H}')$ is a self adjoint operator and we have the following two inclusions:*

$$\sigma(C(\beta)) \subset [\frac{\beta^2}{\varepsilon^+}, +\infty[\tag{2.199}$$

(2) *Besides, if we denote by $\sigma_p(C(\beta))$ the set of eigenvalues of $C(\beta)$, then $\frac{\beta^2}{\varepsilon^+}$ is not in $\sigma_p(C(\beta)$.*

Proof.

(1) From Eq. 2.197a, we deduce that

$$c(\beta;\boldsymbol{H},\boldsymbol{H}) \geq \frac{\beta^2}{\varepsilon^+} \|\boldsymbol{H}\|^2_{L^2(\mathbb{R}^2)} \tag{2.200}$$

for all $\boldsymbol{H}$ in $H^1(\mathbb{R}^2)^3$.

(2) We have to prove that $\frac{\beta^2}{\varepsilon^+}$ is not an eigenvalue of $C(\beta)$. Supposing that $\boldsymbol{H}$ is such that the above inequality is in fact an equality, we deduce that

$$\int_{\mathbb{R}^2} |\,\mathbf{grad}\, \boldsymbol{H}|^2\, dxdy = 0 \tag{2.201}$$

which proves that $\mathbf{grad}\, \boldsymbol{H}$ is null and therefore that $\boldsymbol{H}$ is a constant vector. Bearing in mind that $\boldsymbol{H}$ is in $[L^2(\mathbb{R}^2)]^3$ this constant is bound to be null which leads to a contradiction.

□

As mentioned in section 2.3 we can separate $\boldsymbol{H}(x, y)$ into its transverse and axial components $\boldsymbol{H}_t(x, y)$ and $H_z(x, y)$:

$$\begin{aligned} \boldsymbol{H}(x,y) &= H_{t,1}(x,y)\mathbf{e}^x + H_{t,2}(x,y)\mathbf{e}^y + H_z(x,y)\mathbf{e}^z \\ &= \boldsymbol{H}_t(x,y) + H_z(x,y)\mathbf{e}^z \end{aligned} \tag{2.202}$$

We are now in a position to establish the following lemma

Lemma 2.3 *For all positive real numbers β, and for $\boldsymbol{H}$ and $\boldsymbol{H}'$ in $H^1(\mathbb{R}^2)^3$, we have the following equality:*

$$c(\beta; \boldsymbol{H}, \boldsymbol{H}') = d(\beta; \boldsymbol{H}, \boldsymbol{H}') + \beta^2(\boldsymbol{H}, \boldsymbol{H}') \tag{2.203}$$

with $d(\beta, ., .)$ defined by:

$$d(\beta; \boldsymbol{H}, \boldsymbol{H}') = d^{(0)}(\boldsymbol{H}, \boldsymbol{H}') + \beta d^{(1)}(\boldsymbol{H}, \boldsymbol{H}') + \beta^2 d^{(2)}(\boldsymbol{H}, \boldsymbol{H}') \tag{2.204}$$

where $d^{(0)}$, $d^{(0)}$ and $d^{(1)}$ are given by:

$$\left\{ \begin{aligned} d^{(0)}(\boldsymbol{H}, \boldsymbol{H}') &= \int_{\mathbb{R}^2} \Big(\frac{1}{\varepsilon_r} \, \mathbf{curl}_t \, \boldsymbol{H}_t \cdot \overline{\mathbf{curl}_t \, \boldsymbol{H}'_t} \\ &\qquad + \mathrm{div}_t \, \boldsymbol{H}_t \, \overline{\mathrm{div}_t \, \boldsymbol{H}'_t} + \tfrac{1}{\varepsilon_r} \, \mathbf{grad}_t \, H_z \cdot \overline{\mathbf{grad}_t \, H'_z} \Big) \, dxdy & \text{(2.205a)} \\ d^{(1)}(\boldsymbol{H}, \boldsymbol{H}') &= i \int_{\mathbb{R}^2} \kappa_r \left(\boldsymbol{H}_t \cdot \overline{\mathbf{grad}_t \, H'_z} - \mathbf{grad}_t \, H_z \cdot \overline{\boldsymbol{H}'_t} \right) dxdy & \text{(2.205b)} \\ d^{(2)}(\boldsymbol{H}, \boldsymbol{H}') &= \int_{\mathbb{R}^2} \kappa_r \, \boldsymbol{H}_t \cdot \overline{\boldsymbol{H}'_t} \, dxdy & \text{(2.205c)} \end{aligned} \right.$$

with $\kappa_r = \frac{1}{\varepsilon_r} - 1$.

Proof. We derive the lemma from Eqs. 2.143, 2.144, 2.145 and the following Green formula:

$$\begin{aligned} &\int_{\mathbb{R}^2} \left(\mathbf{grad} \, H_z \cdot \boldsymbol{H}'_t - \boldsymbol{H}_t \cdot \mathbf{grad} \, H'_z \right) dxdy \\ &= \int_{\mathbb{R}^2} \left(\mathrm{div}_t \, \boldsymbol{H}_t \, H'_z - H_z \, \mathrm{div} \, \boldsymbol{H}'_t \right) dxdy \end{aligned} \tag{2.206}$$

□

Remark 2.9 *From Eqs. 2.203–2.204, we see that $C(\beta)$ depends on both β and β^2 i.e. ε being fixed, the calculus of β such that (β, ε) is a solution of Eq. 2.194, is a non linear problem. We thus choose to look for ω as a function of β.*

Lemma 2.4 *For all $\beta \in \mathbb{R}^+$, the essential spectrum $\sigma_{ess}(C(\beta))$ of $C(\beta)$ satisfies:*

$$\sigma_{ess}(C(\beta)) = \left[\beta^2; +\infty\right[\tag{2.207}$$

Proof. $d^{(1)}$ and $d^{(2)}$ are compact perturbations of $d^{(0)}$, thus they do not change its continuous spectrum.

$$\sigma_{ess}(C(\beta)) = \left\{\lambda + \beta^2 \ ; \ \lambda \in \sigma_{ess}(C(0))\right\} \tag{2.208}$$

Furthermore, lemma 2.3 ensures that $d^{(0)}(\boldsymbol{H}, \boldsymbol{H})$ satisfies for all $\boldsymbol{H} \in H^1(\mathbb{R}^2)^3$:

$$d^{(0)}(\boldsymbol{H}, \boldsymbol{H}) \geq \frac{1}{\varepsilon^+} \int_{\mathbb{R}^2} |\ \mathbf{grad}\, \boldsymbol{H}\ |^2 \, dxdy \tag{2.209}$$

Hence, we deduce that:

$$\sigma_{ess}(C(0)) = \mathbb{R}^+ \tag{2.210}$$

The proof of the lemma is thus straightforward. □

Lemma 2.5 *The discrete spectrum of $C(\beta)$ satisfies:*

$$\sigma_p(C(\beta)) \subset] - \infty; \beta^2] \tag{2.211}$$

Proof. *In other words, we must prove that there are no eigenvalues in the continuous spectrum except for β^2. Let $\boldsymbol{H}$ be a mode in $H^1(\mathbb{R}^2)^3$ and let $\lambda > \beta^2$ be an eigenvalue such that:*

$$c(\beta; \boldsymbol{H}, \boldsymbol{V}) = \lambda(\boldsymbol{H}, \boldsymbol{V})\ ,\ \forall\ \boldsymbol{V} \in H^1(\mathbb{R}^2)^3 \tag{2.212}$$

From 2.188, we see that for all $\mathbf{x}$ in vacuo, $\boldsymbol{H}(\mathbf{x})$ satisfies:

$$-\Delta \boldsymbol{H} + (\lambda - \beta^2)\boldsymbol{H} = 0 \tag{2.213}$$

The above equation admits only trivial square integrable solutions since $\lambda - \beta^2$ is positive. Therefore the theory of partial differential equations ensures us of the nullity of $\boldsymbol{H}$ in $\mathbb{R}^2$. □

We deduce from the preceding five *lemmata* that the eigenvalues $\Lambda^{(j)}(\beta)$, $j \in \{1, .., k\}$ of the operator $C(\beta)$ satisfy:

$$\frac{\beta^2}{\sup\limits_{x \in \mathbb{R}^2} \varepsilon_r(x)} < \Lambda^{(1)}(\beta) \leq \Lambda^{(2)}(\beta) \leq ... \leq \Lambda^{(k)}(\beta) \leq \beta^2\ . \tag{2.214}$$

2.5.5 *Discrete finite element formulation*

Up to now we have manipulated only infinite rank operators. For numerical purpose we ought to approximate these operators by finite rank operators namely matrices. To do this, we have to make sure that it is legitimate ; this is the cornerstone of the following chapter.

2.6 Bloch Wave Theory

In this section, we shall elaborate on the fields that can exist in an infinite (photonic) crystal. The cross-section of a microstructured fibre is clearly a finite piece of such an infinite medium. Considering the infinite structure allows us to obtain a very precise and elegant way of characterising the (photonic) band structure and the dispersion curves of the medium. The treatment follows very closely that for quantum waves in non-artificial crystals.

2.6.1 *The crystalline structure*

We shall be very concise here because the subject is academic material [Kit95]. The crystal is defined by repeating periodically an elementary cell Y along its basis vectors $\mathbf{a}_i$, where according to the dimension of the crystal i belongs to $\{1\}, \{1,2\}$ or $\{1,2,3\}$. The underlying structure is thus an integer lattice with basis $\mathbf{a}_i$. In one-dimension, the crystal is simply characterized by its period $[0, d[$. In higher dimensions, the period is made of all the points M such that: $\mathbf{OM} = x^i \mathbf{a}_i, x^i \in [0,1[$ (where a sum is implied over each pair of repeated index). Generically, a vector belonging to the lattice is denoted by $\mathbf{T}$, *i.e.* $\mathbf{T} = n^i \mathbf{a}_i$ with integer coefficients n^i. For later use, we also define the so-called reciprocal lattice, which is a lattice whose basis vectors $\mathbf{a}^i$ are defined by:[22]

$$\mathbf{a}^i \cdot \mathbf{a}_j = 2\pi \delta^i_j.$$

The basic cell of this lattice is denoted by Y^*, and it is the so-called Brillouin zone. Explicitely, it is defined as the set of points P such that $\mathbf{OP} = y_i \mathbf{a}^i, y_i \in [-1/2, 1/2[$. Generically a vector belonging to the reciprocal lattice is denoted by $\mathbf{G}$, i.e. $\mathbf{G} = n_i \mathbf{a}^i$ for some integers n_i.

The interest of these definitions shall become clear in the next section.

[22] δ^j_i is the Kronecker symbol.

2.6.2 *Waves in a homogeneous space*

We want to be able to characterise the waves that can exist in an infinite periodic medium. The dimension N of the medium can be $1, 2$ or 3. Let us first consider the case of a homogeneous medium and the scalar wave equation for harmonic waves: $\Delta u + k^2 u = 0$. We want to find *all the bounded functions* u satisfying this equation, for all values of k. Let us pretend for the moment that we do not know that a basis of solutions is the plane waves of the form $\exp(\mathbf{k} \cdot \mathbf{y})$. If we rewrite the problem in the following form:

$$\text{Find a function } u \text{ and a positive number } E \text{ such that: } -\Delta u = Eu, \tag{2.215}$$

it appears as a spectral one: the point is to determine the eigenvalues and eigenvectors of some linear operator (here the Laplacian). In order to do so, let us Fourier transform the function $u(\mathbf{y})$:

$$u(\mathbf{y}) = (2\pi)^{-N/2} \int_{\mathbb{R}^N} \widehat{u}(\mathbf{k}) e^{i\mathbf{k}\cdot\mathbf{y}} d\mathbf{k}. \tag{2.216}$$

Inserting this decomposition in Eq. 2.215, we find: $(\|\mathbf{k}^2\| - E)\widehat{u}(\mathbf{k}) = 0$. This shows that $\widehat{u}(\mathbf{k})$ is not a function but a Schwartz distribution, in fact: $\widehat{u} = A(\mathbf{k})\delta(\|\mathbf{k}^2\| - E)$, that is, it is proportional to the Dirac distribution whose support is a spherical shell of radius E.[23] We then obtain $u(\mathbf{y}) = (2\pi)^{-N/2} \int_{\mathbb{S}_E^{N-1}} A(\mathbf{k}) e^{i\mathbf{k}\cdot\mathbf{y}} d\mathbf{k}$.[24] A solution to the spectral problem (2.215) is thus a continuous sum of plane waves with some amplitude factor: the spectral problem is parametrized by plane waves. The very reason why this decomposition works is the fact that all translations of space $T_\mathbf{u}, \mathbf{u} \in \mathbb{R}^N$,[25] commute with the Laplacian, and therefore the translations and the Laplacian have a common basis of eigenvectors. This basis is formed with plane waves $(T_\mathbf{u}(e^{i\mathbf{k}\cdot\mathbf{y}}) = e^{-i\mathbf{k}\cdot\mathbf{u}} e^{i\mathbf{k}\cdot\mathbf{y}})$.

If the wavenumber $k = \|\mathbf{k}\|$ is given, the set of parameters is a spherical shell of dimension $N - 1$ (note that a shell of dimension 0 is just a pair

[23] its action on a regular test function ϕ is $\langle \delta(\|\mathbf{k}\|^2 - E), \psi \rangle = \int_{\mathbb{S}_E^{N-1}} \phi(s) ds$, where $\mathbb{S}_E^{N-1}$ is the sphere of radius $\sqrt{E}$ in $\mathbb{R}^N$

[24] This is quite an abstract formula. In $N = 2$ it turns out to read as:

$$u(y_1, y_2) = (2\pi)^{-1} \int_{-\sqrt{E}}^{\sqrt{E}} e^{ik_1 y_1} \left(A^+(k_1) e^{i\sqrt{E - k_1^2} y_2} + A^-(k_1) e^{-i\sqrt{E - k_1^2} y_2} \right) dk_1.$$

In this formula, y_1 and y_2 can be exchanged.

[25] A translation acts on a function f of the variable $\mathbf{y} \in \mathbb{R}^N$ in the following way: $T_\mathbf{u}(f)(\mathbf{y}) = f(\mathbf{y} - \mathbf{u})$.

of points symmetric with respect to the origin). The decomposition can also be taken in reverse order: we can begin by fixing the wavevector $\mathbf{k}$. Then for the plane wave with wavevector $\mathbf{k}$ the eigenvalue is k^2 and the associated frequency is $\omega = ck$. From this point of view, with one frequency is associated only one energy. Still, we can remark that it is possible to decompose any $\mathbf{k} \in \mathbb{R}^N$, in the following form:

$$\mathbf{k} = \mathbf{k}_b + 2\pi\mathbf{p},$$

where $\mathbf{p}$ is a vector with integer components (i.e. $\mathbf{p} \in \mathbb{Z}^N$) and $\mathbf{k}$ belongs to[26] $Y^* = [-\pi, \pi[^N$. If we use only Y^* and not the entire space $\mathbb{R}^N$ to parametrize the spectral problem, then with a wavevector $\mathbf{k}_b \in Y^*$ is now associated an infinite set of frequencies $\omega_p = c|\mathbf{k}_b + 2\pi\mathbf{p}|$ and an infinite set of eigenvectors, the so-called Bloch waves:

$$\psi_{\mathbf{p}}(\mathbf{k}_b, \mathbf{y}) = \exp(i\mathbf{k}_b \cdot \mathbf{y})\phi_{\mathbf{p}}(\mathbf{k}_b, \mathbf{y}), \tag{2.217}$$

where $\phi_{\mathbf{p}}(\mathbf{y}) = \exp(2i\pi\mathbf{p}\cdot\mathbf{y})$ (note that it is a Y−periodic function). Using this formulation, the Fourier integral of $u(\mathbf{y})$ can be written:

$$u(\mathbf{y}) = \int_{Y^*} \sum_{\mathbf{p}} u_{\mathbf{p}}(\mathbf{y}) e^{i\mathbf{k}_b\cdot\mathbf{y}} \phi_{\mathbf{p}}(\mathbf{y}) d\mathbf{y}, \tag{2.218}$$

where $u_{\mathbf{p}}(\mathbf{y}) = (2\pi)^{-N/2}\widehat{u}(\mathbf{k} - 2\pi\mathbf{p})$. We shall see in the following that this expression can be extended so as to deal with non-homogeneous media.

2.6.3 *Bloch modes of a photonic crystal*

Let us now consider a photonic crystal with basic cell Y and a partial differential operator with periodic coefficient $\mathcal{L}$ that describes wave propagation in the crystal. For instance, operator $\mathcal{L}$ can be the Helmholtz-like operator: $-\varepsilon(\mathbf{y})^{-1}\Delta$ or, for the other polarization: $-\operatorname{div}(\varepsilon(\mathbf{y})^{-1}\,\mathbf{grad}(\cdot))$, or else for the full Maxwell system: $\mathbf{curl}(\varepsilon(\mathbf{y})^{-1}\,\mathbf{curl}(\cdot))$. In such a situation, the Fourier transform cannot lead easily to the solution, because of the inhomogeneity of space. The idea behind Bloch waves is to find a way to generalize the Fourier transform. The first point at issue is that the space is now really periodic and not homogeneous. From the point of view of reference frames, this means that the reference frames deduced from one another by an arbitrary translation are no longer equivalent. Only those related by a

[26] The notation is not innocent. What we do amounts to decomposing the space $\mathbb{R}^N$ into cubic boxes of side 1, which endows it with a lattice structure of basic cell $Y = [0, 1[^N$, whose corresponding Brillouin zone is Y^*.

translation that is of the form $n^i \mathbf{a}_i, n^i \in \mathbb{Z}$ are. The consequence is that we can no longer expect plane waves to be solutions of the propagation equation. However, the form of Bloch waves given in Eq. 2.217, where now $\phi_{\mathbf{p}}(\mathbf{k}, \mathbf{y})$ is an unknown Y-periodic function, shows that they transform in the following way under a translation of the direct lattice:

$$\psi_{\mathbf{p}}(\mathbf{k}_b, \mathbf{y} + \mathbf{T}) = \exp(i\mathbf{k}_b \cdot \mathbb{T})\psi_{\mathbf{p}}(\mathbf{k}_b, \mathbf{y}) \tag{2.219}$$

The function is then said to be *pseudo-periodic*. This suggests that we look for a decomposition such as that in Eq. 3.34, where the plane waves of the homogeneous space are now replaced by Bloch waves (i.e. the product of a plane wave by a $Y-$periodic function).

For such a decomposition to hold, we have to show that we can reduce the spectral problem by imposing the quasi-periodicity condition (2.219) and obtain an equivalent problem[27].

The first step consists in associating with *any square integrable function u on $\mathbb{R}^N$* a family of pseudo-periodic functions defined by $\mathbf{k}$, namely $\mathcal{W}(u)$. This is the so-called Wannier transform:

$$\mathcal{W}(u)(\mathbf{k}, \mathbf{y}) = \sum_{\mathbf{T}} u(\mathbf{y} - \mathbf{T}) e^{i\mathbf{k}\cdot\mathbf{T}}, \tag{2.220}$$

where the sum runs over all vectors of the direct lattice, and $\mathbf{k}$ belongs to the Brillouin zone[28] Y^*. It is easy to check that the transformed function is quasi-periodic with respect to $\mathbf{y}$:

$$\mathcal{W}(\mathbf{k}, \mathbf{y} + \mathbf{T}') = \sum_{\mathbf{T}} u(\mathbf{y} + \mathbf{T}' - \mathbf{T}) e^{i\mathbf{k}\cdot\mathbf{T}} \tag{2.221}$$

$$= e^{i\mathbf{k}\cdot\mathbf{T}'} \sum_{\mathbf{T}''} u(\mathbf{y} - \mathbf{T}'') e^{i\mathbf{k}\cdot\mathbf{T}''} \tag{2.222}$$

$$= e^{i\mathbf{k}\cdot\mathbf{T}'} \mathcal{W}(\mathbf{k}, \mathbf{y} + \mathbf{T}'). \tag{2.223}$$

We can get back the original function by applying the inverse transform:

$$\mathcal{W}^*(\psi)(\mathbf{y}) = \frac{1}{|Y^*|} \int_{Y^*} \psi(\mathbf{k}, \mathbf{y}) d\mathbf{k}. \tag{2.224}$$

This last result is easily obtained. Let us apply the inverse Wannier trans-

[27]In other words, we do not want to remove any solution by requesting that they be pseudo-periodic.

[28]instead of $\mathbf{k}_b$, in order to lighten the notation.

form to $\mathcal{W}(u)$:

$$\begin{aligned}\mathcal{W}^* \left(\mathcal{W}(u)(\mathbf{k},\mathbf{y})\right) = & \quad \tfrac{1}{|Y'|}\int_{Y^*} \mathcal{W}(u)(\mathbf{k},\mathbf{y})d\mathbf{k} & (2.225)\\ & = \tfrac{1}{|Y^*|}\int_{Y^*}\sum_{\mathbf{T}} u(\mathbf{y}-\mathbf{T})e^{i\mathbf{k}\cdot\mathbf{T}}d\mathbf{k} & (2.226)\\ & = \sum_{\mathbf{T}} u(\mathbf{y}-\mathbf{T})\tfrac{1}{|Y^*|}\int_{Y^*} e^{i\mathbf{k}\cdot\mathbf{T}}d\mathbf{k}. & (2.227)\end{aligned}$$

The conclusion follows by the identity: $\frac{1}{|Y^*|}\int_{Y^*} e^{i\mathbf{k}\cdot\mathbf{T}}d\mathbf{k} = \delta_{\mathbf{0}}^{\mathbf{T}}$. Conversely, starting with a pseudo-periodic function $\psi(\mathbf{k},\mathbf{y})$ we have:

$$\mathcal{W}\left(\mathcal{W}^*(\psi)\right)(\mathbf{k}',\mathbf{y}) = \frac{1}{|Y^*|}\sum_{\mathbf{T}}\int_{Y^*}\psi(\mathbf{k},\mathbf{y}-\mathbf{T})e^{i\mathbf{k}'\cdot\mathbf{T}}d\mathbf{k}$$

Using the pseudo-periodicity of ψ, we get

$$\mathcal{W}\left(\mathcal{W}^*(\psi)\right)(\mathbf{k}',\mathbf{y}) = \frac{1}{|Y^*|}\int_{Y^*}\psi(\mathbf{k},\mathbf{y})\sum_{\mathbf{T}} e^{i(\mathbf{k}'-\mathbf{k})\cdot\mathbf{T}}d\mathbf{k}$$

where the conclusion follows from the identity:

$$\frac{1}{|Y^*|}\sum_{\mathbf{T}} e^{i(\mathbf{k}'-\mathbf{k})\cdot\mathbf{T}} = \sum_{\mathbf{G}}\delta(\mathbf{k}-\mathbf{k}'-\mathbf{G})\,.$$

In order to be clearer, we state explicitly the spaces involved: the Wannier transform is defined on the space $\mathcal{H} = L^2(\mathbb{R}^N)$ and is into $\mathcal{V}$ which is the set of functions defined on $\mathbb{R}^N \times Y^*$ such that

$$\|\psi\|^2 = \int_{Y^*\times Y}|\psi(\mathbf{y},\mathbf{k})|^2 d\mathbf{k}d\mathbf{y} < +\infty\,,$$

$\mathcal{V}$ is a Hilbert space for the scalar product:

$$(\psi_1,\psi_2) = \int_{Y^*\times Y}\psi_1(\mathbf{k},\mathbf{y})\overline{\psi_2(\mathbf{k},\mathbf{y})}\,d\mathbf{k}d\mathbf{y}.$$

We have $\mathcal{W}\mathcal{W}^* = I_{\mathcal{V}}$ and $\mathcal{W}^*\mathcal{W} = I_{\mathcal{H}}$. We can obtain all functions[29] of $\mathcal{V}$ by fixing first the wavevector $\mathbf{k}$ and then by considering the functions $u_{\mathbf{k}}$ such that:

$$u_{\mathbf{k}}(\mathbf{y}+\mathbf{T}) = e^{i\mathbf{k}\cdot\mathbf{T}}u_{\mathbf{k}}(\mathbf{y}).$$

[29] Mathematically speaking, the space $\mathcal{V}$ can be identified with a direct Hilbertian integral: $\mathcal{V} = \int_{Y^*}^{\oplus}\mathcal{H}_{\mathbf{k}}d\mathbf{k}$, which corresponds to the notion of a "continuous" sum of Hilbert spaces.

This set of functions is denoted by $\mathcal{H}_{\mathbf{k}}$:

$$\mathcal{H}_{\mathbf{k}} = \left\{ u, u(\mathbf{y}+\mathbf{T}) = e^{i\mathbf{k}\cdot\mathbf{T}} u(\mathbf{y}), \|u\|^2 = \frac{1}{|Y|}\int_Y |u(\mathbf{y})|^2 d\mathbf{y} < +\infty \right\}. \tag{2.228}$$

We now have a mathematical set-up that shows that any square integrable function can be considered as a sum of quasi-periodic functions, by using the Wannier transform. In order to be able to use this transform, it should commute with $\mathcal{L}$. Indeed, we want to find u such that : $\mathcal{L}(u) = Eu$. If the commutator $[\mathcal{W}, \mathcal{L}] = \mathcal{W}\mathcal{L} - \mathcal{L}\mathcal{W} = 0$, then we get, by applying $\mathcal{W}$ to this equation

$$\mathcal{L}(\mathcal{W}(u)) = E\mathcal{W}(u)\,.$$

The commutation is due to the invariance of the medium, and hence that of the permittivity, by translation along the vectors $\mathbf{T}$ of the direct lattice. The eigenvalues and eigenvectors of $\mathcal{W}$ can thus be obtained by solving the equation in $\mathcal{H}_{\mathbf{k}}$, then by varying $\mathbf{k}$ in Y^*. For each $\mathbf{k} \in Y^*$, we therefore look for functions $u \in \mathcal{H}_{\mathbf{k}}$ such that $\mathcal{L}(u) = E(\mathbf{k})u$. Once the eigenvalues $E(\mathbf{k})$ are obtained the corresponding frequencies are $\omega/c = \sqrt{E(\mathbf{k})}$. In the Hilbert space $\mathcal{H}_{\mathbf{k}}$, the operator $\mathcal{L}$ has a set of quasi-periodic eigenfunctions that form a Hilbert-basis, the so-called *Bloch waves*. They are numbered by an integer[30] p, and are of the form:

$$\psi_p(\mathbf{k},\mathbf{y}) = e^{i\mathbf{k}\cdot\mathbf{y}}\phi_p(\mathbf{k},\mathbf{y}), \tag{2.229}$$

where ϕ_p is a $Y-$periodic function. They are associated with a set of eigenvalues $E_p(\mathbf{k})$ that are ordered in ascending order: $E_1 < E_2 < \ldots < E_p < \ldots$ By varying $\mathbf{k}$ in Y^*, we obtain all of the eigenvalues as a set of surfaces indexed by p.

We have in fact obtained a new way of decomposing a square integrable function[31] u. Indeed,for a given $\mathbf{k}$, $\mathcal{W}(u)(\cdot,\mathbf{k})$ belongs to $\mathcal{H}_{\mathbf{k}}$, and therefore it can be expanded on the basis $\{\psi_p\}_p$:

$$\mathcal{W}(u)(\mathbf{k},\mathbf{y}) = \sum_p W_p(\mathbf{k})\psi_p(\mathbf{k},\mathbf{y})$$

[30] This is the band index, and it labels the various allowed frequencies for a given wavevector $\mathbf{k}$.

[31] The reader can remark that in fact the decomposition is valid in $\mathbb{R}^N$ for an arbitrary N, not necessarily for $N = 1, 2, 3$.

where $W_p(\mathbf{k}) = \int_Y \mathcal{W}(u)(\mathbf{k},\mathbf{y})\overline{\phi_p(\mathbf{k},\mathbf{y})}e^{-i\mathbf{k}\cdot\mathbf{y}}d\mathbf{y}$, which can be written:

$$W_p(\mathbf{k}) = \int_Y \sum_{\mathbf{T}} u(\mathbf{y}-\mathbf{T})e^{i\mathbf{k}\cdot\mathbf{T}}\overline{\phi_p(\mathbf{k},\mathbf{y})}e^{-i\mathbf{k}\cdot\mathbf{y}}d\mathbf{y} \tag{2.230}$$

$$= \sum_{\mathbf{T}} \int_{Y+\mathbf{T}} u(\mathbf{y})\overline{\phi_p(\mathbf{k},\mathbf{y})}e^{-i\mathbf{k}\cdot\mathbf{y}}d\mathbf{y} \tag{2.231}$$

$$= \int_{\mathbb{R}^N} u(\mathbf{y})\overline{\phi_p(\mathbf{k},\mathbf{y})}e^{-i\mathbf{k}\cdot\mathbf{y}}d\mathbf{y} \tag{2.232}$$

By using the inverse transform $\mathcal{W}^*$, we get:

$$u(\mathbf{y}) = \frac{1}{|Y^*|}\int_{Y^*}\sum_p W_p(\mathbf{k})\phi_p(\mathbf{k},\mathbf{y})e^{i\mathbf{k}\cdot\mathbf{y}}d\mathbf{k}\,.$$

Finally, we can state the *Bloch decomposition theorem*, which is a generalisation of the Fourier transform:

Theorem 2.8 *Let u be a function of $L^2(\mathbb{R}^N)$, where its pth Bloch coefficient is defined by:*

$$\widehat{u}_p(\mathbf{k}) = \int_{\mathbb{R}^N} u(\mathbf{y})\overline{\phi_p(\mathbf{k},\mathbf{y})}e^{-i\mathbf{k}\cdot\mathbf{y}}d\mathbf{y}\ .$$

We then have the Bloch decomposition formula:

$$u(\mathbf{y}) = \frac{1}{|Y^*|}\int_{Y^*}\sum_p \widehat{u}_p(\mathbf{k})\phi_p(\mathbf{k},\mathbf{y})e^{i\mathbf{k}\cdot\mathbf{y}}d\mathbf{k}$$

and the Parseval identity:

$$\int_{\mathbb{R}^N}|u(\mathbf{y})|^2d\mathbf{y} = \frac{1}{|Y^*|}\int_{Y^*}\sum_p|\widehat{u}_p(\mathbf{k})|^2d\mathbf{k}.$$

2.6.4 *Computation of the band structure*

The previous section has shown that the bounded fields that can exist in an infinite crystal could be parametrized by the set Y^*. In order to describe in a concise way these fields, only a reduced part of the Brillouin zone is used. Indeed, the crystal is invariant under some group of symmetries, and hence it is not necessary to use the entire Brillouin zone in order to compute the spectrum. For instance, let us consider a $2D$ photonic crystal with a square lattice of side a. The crystal is made of rods of radius R and relative permittivity ε_2 embedded in a matrix of relative permittivity ε_1 (See Fig.

2.7). The Brillouin zone is a square of side $2\pi/a$ (See Fig. 2.8). It suffices to describe only $1/8^{th}$ of this square (namely, the triangle in bold lines on Fig. 2.8) in order to characterize the spectrum entirely. The description can be further reduced by restricting $\mathbf{k}$ to the lines connecting the points of higher symmetries: Γ, X, M (a more detailed treatment of the symmetries can be found in [Sak01]).

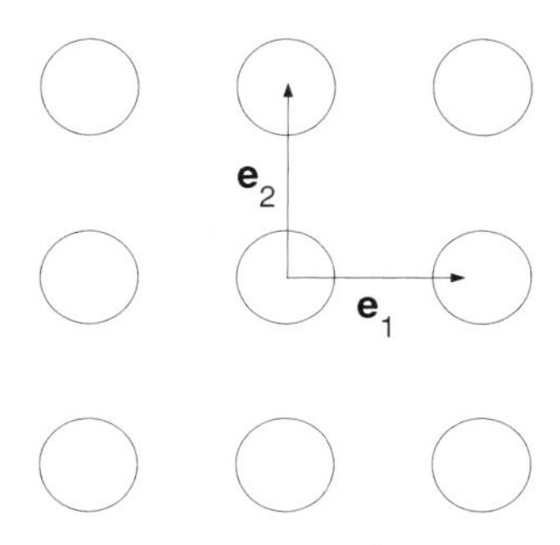

Fig. 2.7 A few cells of a square photonic crystal.

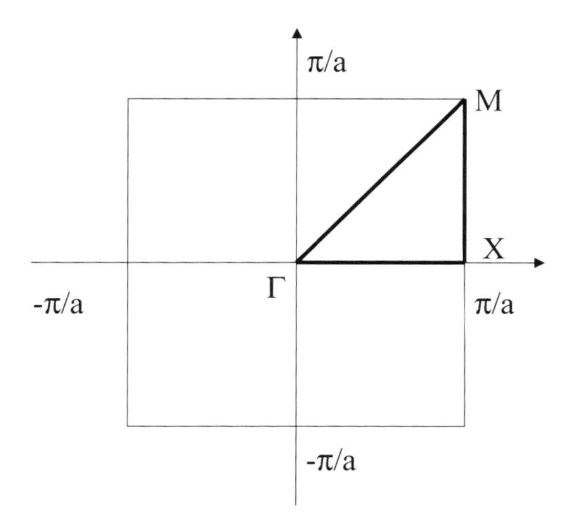

Fig. 2.8 The Brillouin zone Y^* of the crystal.

Let us characterize the $z-$independent fields that can exist in such a structure. The Maxwell system can be reduced to two fundamental cases of polarization:

(1) The TM case, in which the electric field is parallel to the z axis, and its z component E_z satisfies

$$-\varepsilon^{-1}(\mathbf{y})\Delta E_z = \left(\frac{\omega}{c}\right)^2 E_z \tag{2.233}$$

and:

(2) The TE case, in which the magnetic field is parallel to the z axis, and its z component H_z satisfies:

$$-\mathrm{div}(\varepsilon^{-1}(\mathbf{y})\mathbf{grad}H_z) = \left(\frac{\omega}{c}\right)^2 H_z. \tag{2.234}$$

In both cases, we have to compute the Fourier series of $\varepsilon^{-1}(\mathbf{y})$. We write:

$$\varepsilon^{-1}(\mathbf{y}) = \sum_{\mathbf{G}} \widehat{\varepsilon^{-1}}(\mathbf{G})e^{i\mathbf{G}\cdot\mathbf{y}}.$$

For a circular fibre, we have explicitly:

$$\widehat{\varepsilon^{-1}}(\mathbf{G}) = \begin{cases} \frac{1}{\varepsilon_1}f + \frac{1}{\varepsilon_2}(1-f), & \mathbf{G}=0 \\ \left[\frac{1}{\varepsilon_1} - \frac{1}{\varepsilon_2}\right] f \frac{2J_1(\|\mathbf{G}\|R)}{\|\mathbf{G}\|R}, & \mathbf{G}\neq 0 \end{cases} \tag{2.235}$$

where $f = \frac{\pi R^2}{a^2}$ is the filling fraction and J_1 is the Bessel function of order 1.

We now choose a vector $\mathbf{k} \in Y^*$ and look for Bloch waves solving these equations. First, we expand any Bloch wave associated with E_z and H_z in Fourier series:

$$\begin{cases} E_z(\mathbf{k},\mathbf{y}) = \sum_{\mathbf{G}} \widehat{E}(\mathbf{k},\mathbf{G})e^{i(\mathbf{G}+\mathbf{k})\cdot\mathbf{y}} \\ H_z(\mathbf{k},\mathbf{y}) = \sum_{\mathbf{G}} \widehat{H}(\mathbf{k},\mathbf{G})e^{i(\mathbf{G}+\mathbf{k})\cdot\mathbf{y}}. \end{cases} \tag{2.236}$$

It suffices now to insert the expansions into Eqs. 2.233, 2.234, to obtain two eigenvalues problems:

$$\begin{cases} \sum_{\mathbf{G}'}(\mathbf{k}+\mathbf{G})\cdot(\mathbf{k}+\mathbf{G}')\widehat{\varepsilon^{-1}}(\mathbf{G}-\mathbf{G}')\widehat{H}(\mathbf{k},\mathbf{G}') = \left(\frac{\omega}{c}\right)^2 \widehat{H}(\mathbf{k},\mathbf{G}) \\ \sum_{\mathbf{G}'}(\mathbf{k}+\mathbf{G}')^2\widehat{\varepsilon^{-1}}(\mathbf{G}-\mathbf{G}')\widehat{E}(\mathbf{k},\mathbf{G}') = \left(\frac{\omega}{c}\right)^2 \widehat{E}(\mathbf{k},\mathbf{G}) \end{cases} \tag{2.237}$$

Solving these linear systems for a given value of $\mathbf{k}$ and keeping only the positive eigenvalues, we obtain the allowed frequencies $\frac{\omega_p}{c}$. By varying $\mathbf{k}$ along the lines connecting the points of high symmetry, we obtain the curves

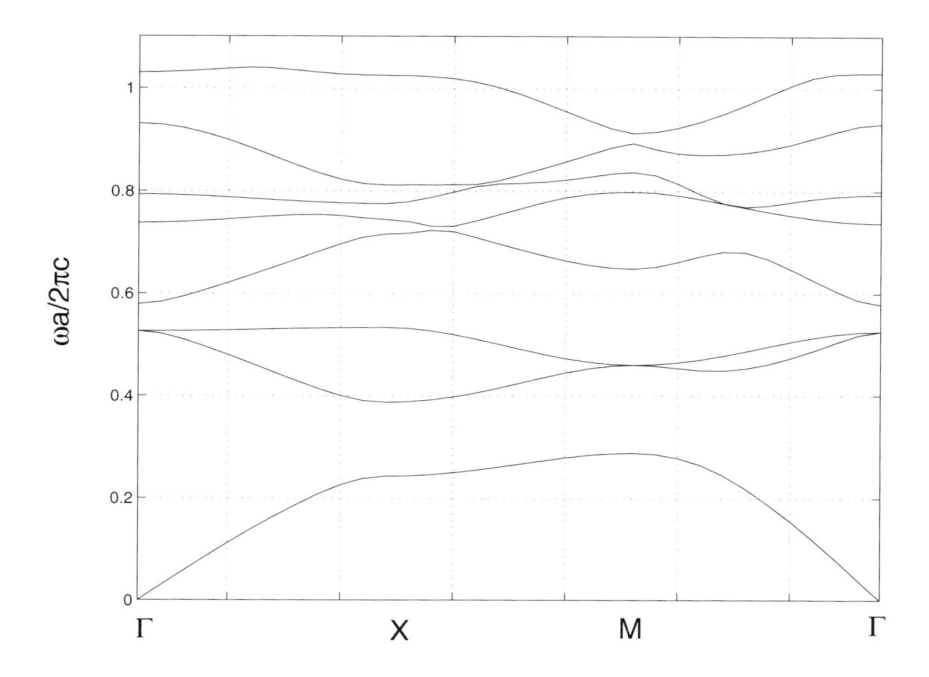

Fig. 2.9 The band structure for the TM case with $\varepsilon_1 = 1, \varepsilon_2 = 9, R/a = 1/4$

in Figs. 2.9 and 2.10 (a complete example of a triangular lattice is given in [PM91]).

2.6.5 *A simple 1D illustrative example: the Kronig-Penney model*

This example is taken from the book by A.B. Movchan. *et al.*[32] For the sake of simplicity we only are interested in the T.E. fields and non-magnetic media are only considered. For this kind of field, we have just seen that ψ (*i.e.* H_z) is a solution of :

$$(\varepsilon_r^{-1}\psi')' + (k_0^2 - \varepsilon_r^{-1}\beta^2)\psi = 0 \tag{2.238}$$

[32]See [MMP02, p. 48] where the authors consider propagation of out-of-plane shear elastic waves through a one-dimensional periodic array composed of binary multilayered structures (See Fig. 2.11) and see [dLKP31] for a historical article by Kronig and Penney themselves.

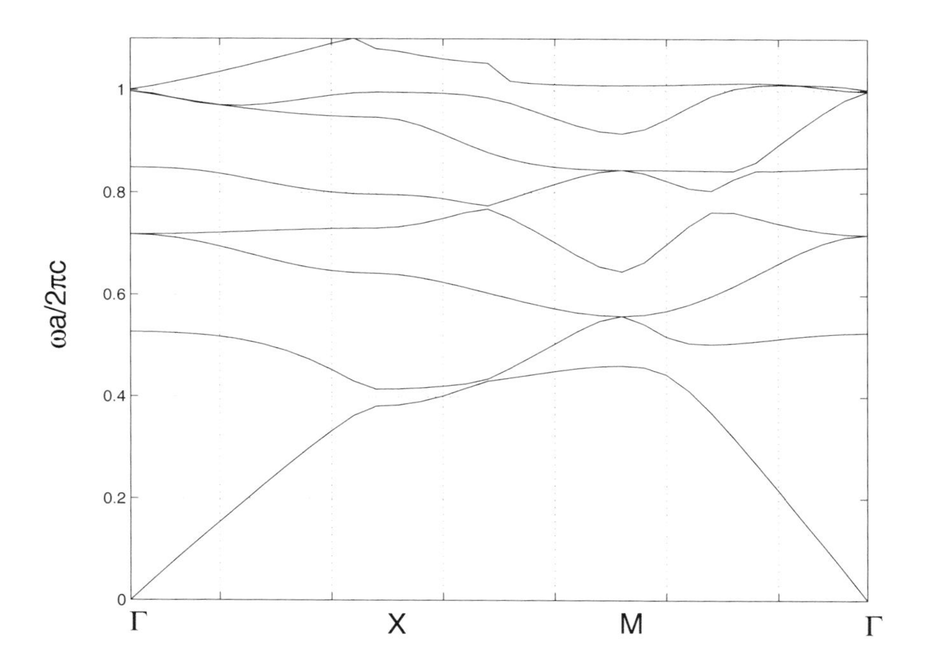

Fig. 2.10 The band structure for the TE case. The parameters are that of Fig. 2.9

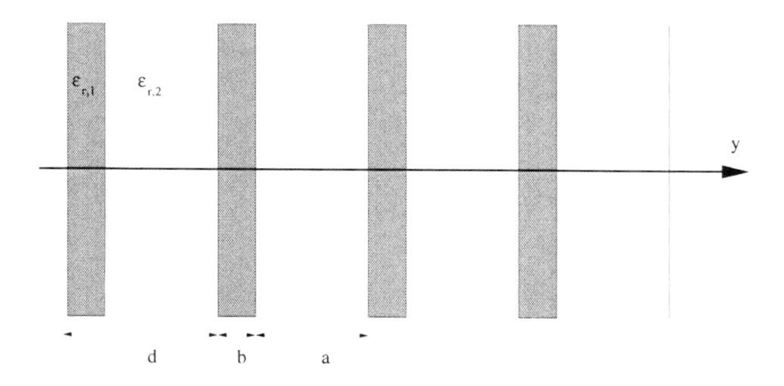

Fig. 2.11 A periodic structure along the y-axis of infinite extent in both directions.

where ε_r is a real $d-$periodic function. We now turn our attention to the fields propagating in the $x-$direction: the constant β is therefore null. It is worth noting that, whatever the number of layers, this situation cannot occur when dealing with structures of finite extent like the ones encountered in Section 2.2 by virtue of Theorem 2.3; it is thus a specificity of the periodic case. Be that as it may Eq. 2.238 can be reduced as follows:

$$(\varepsilon_r^{-1}\psi')' + k_0^2\psi = 0 \,. \tag{2.239}$$

Moreover, we have seen, in a more general context that ψ is a continuous function ; we have thus $\psi' = \{\psi'\}$ since $[\psi]_{x_n} = 0$ for any x_n lying in the set of discontinuities of the function ε_r namely D. Additionally we have:

$$(\varepsilon_r^{-1}\psi')' = \varepsilon_r^{-1}\{\psi''\} + \sum_{x_n \in D} \left[\varepsilon_r^{-1}\psi'\right]_{x_n} \delta_{x_n} \tag{2.240}$$

where D is the set of discontinuities of the function ε_r^{-1}. For any integer n, we are eventually led to:

$$\psi''_{n,j} + k_0^2\varepsilon_{r,j}\psi_{n,j} = 0 \,,\; x \in S_j^{(n)} \,,\; j = 1,\, 2, \tag{2.241}$$

on the one hand where $S_1^{(n)} =]nd, nd+b[$, $S_2^{(n)} =]nd+b, (n+1)d[$, where d is the period ($d = a + b$) and $\varepsilon_{r,j}$, $j = 1,\, 2$, are the relative permittivities of each medium and for each $x_n \in D$:

$$\begin{cases} \psi_{n,1}(x_n) = \psi_{n,2}(x_n) & (2.242\text{a}) \\ \dfrac{1}{\varepsilon_{r,1}}\psi'_{n,1}(x_n) = \dfrac{1}{\varepsilon_{r,2}}\psi'_{n,2}(x_n) \,, & (2.242\text{b}) \end{cases}$$

on the other hand. In order to solve this problem, we have seen previously that the solution amounts to looking for the so-called Bloch waves, namely:

$$\psi_k(x) = e^{ikx}\psi_{k,\sharp}(x) \tag{2.243}$$

where k is in the so-called first Brillouin zone ($k \in [-\frac{\pi}{2d}, \frac{\pi}{2d}[$) and where $\psi_{k,\sharp}$ is a real $d-$periodic function. This last equation implies that the following two relations were fulfilled:

$$\begin{cases} \psi_k(d) = e^{ikd}\psi_k(0) & (2.244\text{a}) \\ \dfrac{1}{\varepsilon_r}\psi'_k(d) = e^{ikd}\dfrac{1}{\varepsilon_r}\psi'_k(0) \,. & (2.244\text{b}) \end{cases}$$

Resuming our previous notation, namely the column vector (See sec. 2.2.3.1), the above relations can be summarised in the following way:

$$\Psi_k(d) = e^{ikd}\Psi_k(0) \,. \tag{2.245}$$

We are now in a position to find the dispersion curves. To do this, we first recall the transfer matrix T_{k_0} associated with a basic cell of thickness d, consisting of two layers of respective thicknesses a and b (note that in the present situation, the order in which we perform the matrix product does not matter, since the basic cell is repeated periodically):

$$T_{k_0} = \begin{pmatrix} \cos(k_2 b) & \frac{\varepsilon_{r,2}}{k_2}\sin(k_2 b) \\ -\frac{k_2}{\varepsilon_{r,2}}\sin(k_2 b) & \cos(k_2 b) \end{pmatrix} \begin{pmatrix} \cos(k_1 a) & \frac{\varepsilon_{r,1}}{k_1}\sin(k_1 a) \\ -\frac{k_1}{\varepsilon_{r,1}}\sin(k_1 a) & \cos(k_1 a) \end{pmatrix}, \tag{2.246}$$

where $k_1 = k_0\sqrt{\varepsilon_{r,1}}$ and $k_2 = k_0\sqrt{\varepsilon_{r,2}}$.

We note that the half trace of T_{k_0} takes the following form:

$$\begin{aligned} \frac{1}{2}tr(T_{k_0}) &= \cos(k_1 a)\cos(k_2 b) \\ &\quad - \frac{1}{2}\Big(\frac{k_1}{k_2} + \frac{k_2}{k_1}\Big)\sin(k_1 a)\sin(k_2 b)\ . \end{aligned} \tag{2.247}$$

From Eq. 2.245, we further deduce that

$$\Psi_k(d) = e^{ikd}\Psi_k(0) = T_{k_0}\Psi_k(0)\ , \tag{2.248}$$

where Ψ_k stands for the column vector defined by Eq. 2.66 in which we take $p(y) = \varepsilon_{r,1}^{-1}$ (resp. $p(y) = \varepsilon_{r,2}^{-1}$) in the layer of thickness a (resp. b). This means that e^{ikd} is an eigenvalue of T_{k_0} associated with the eigenvector $\Psi_k(0)$. Also, if λ is an eigenvalue of T_{k_0}

$$det(T_{k_0} - \lambda I) = 0 \iff \lambda^2 - \lambda\, tr(T_{k_0}) + det(T_{k_0}) = 0\ . \tag{2.249}$$

From Eq. 2.246, we note that $det(T_{k_0}) = 1$, hence the characteristic polynomial in Eq. 2.249 has some complex-valued roots if $\mid tr(T_{k_0}) \mid < 2$ (in other words, the pair $(k_0, 0)$ is in $\mathcal{B}$). In which case, we can be assured that the uni-modular T_{k_0} possesses two eigenvalues of unit modulus that are conjugates of each other. For a given Bloch parameter $k \neq 0$, the eigenvalue e^{ikd} of T_{k_0} belongs to $\mathbb{C} \setminus \mathbb{R}$ and hence

$$\frac{1}{2}tr(T_{k_0}) = \frac{\lambda_1 + \overline{\lambda_1}}{2} = \frac{e^{ikd} + e^{-ikd}}{2} = \cos(kd)\ . \tag{2.250}$$

Let us now introduce the non-dimensional parameter $\xi = \varepsilon_{r,2}/\varepsilon_{r,1}$. As a result of Eq. 2.247 and Eq. 2.250, we obtain a condition which represents the dispersion equation relating the frequency ω (through $k_0 = \omega\sqrt{\varepsilon_0\mu_0}$) and the Bloch parameter k

$$\cos(kd) = \cos(k_1 b)\cos(k_1 a\sqrt{\xi}) - \frac{\xi + 1}{2\sqrt{\xi}}\sin(k_1 b)\sin(k_1 a\sqrt{\xi})\ . \tag{2.251}$$

This analytical formula proves to be really handy since we can now compute the band diagrams associated with our one-dimensional photonic crystal in transverse magnetic polarisation. We illustrate the variation of the widths of the photonic band gaps of such a structure with respect to its opto-geometric parameters (*i.e.* the contrast of the permittivities and the thicknesses of its layers) in Fig. 2.12.

Let us further assume that the layers of width b are relatively thin i.e. $k_1 b \ll 1$ and that the permittivity in layers of width a is much larger than that of layers of width b i.e. $\xi = \varepsilon_{r,2}/\varepsilon_{r,1} \ll 1$. Expanding the corresponding trigonometric parts into power series and retaining the terms up to k_0^4, we obtain the approximate dispersion relation

$$\cos(kd) = k_0^4 p_1 - k_0^2 p_2 + 1 \, , \tag{2.252}$$

where

$$\begin{cases} p_1 = \dfrac{1}{2}\Big[\dfrac{a^2b^2\xi}{4} + \dfrac{b^4 + a^4\xi^2}{12} + \dfrac{a^3b\xi + ab^3}{6}(1+\xi)\Big]\varepsilon_{r,1}^2 & \text{(2.253a)} \\[2ex] p_2 = \dfrac{1}{2}\Big[b^2 + ab(1+\xi) + \xi a^2\Big]\varepsilon_{r,1} \, . & \text{(2.253b)} \end{cases}$$

which is consistent with Eq. 1.171 p. 48 of [MMP02]. One can easily solve Eq. 2.252. This provides us with the dispersion diagram depicted in Fig. 2.13.

There is some physics, of course, running this whole process. This can be loosely classified as perturbation theory, though mathematicians would rather call the previous asymptotic algorithm "asymptotic models for fields in thin-walled composite structures" [MMP02]. The physicist would tell you that it is all about perturbing otherwise degenerate eigenstates. We should perhaps say a few words about this formal approach since the ideas behind this *esoteric* physicist jargon are rather simple.

Let us first assume that our 1D structure is homogeneous with say, $\varepsilon = 1$. This has some planewave eigensolutions $\omega(k) = ck$. We note that ε has trivial periodicity d for any $d > 0$. In particular, the mode corresponding to $k = -\pi/d$ is associated with the same eigenvalue as the mode corresponding to $k = \pi/d$. This accidental degeneracy is an *artefact* of the 'artificial' period we have chosen. We can write down the wave solutions as $u(x) \sim e^{\pm i\pi x/d}$, or equivalently as linear combinations $u_1(x) = \cos(\pi x/d)$ and $u_2(x) = \sin(\pi x/d)$, both at $\omega = c\pi/d$. Let us now suppose that we perturb ε slightly so that it has now a non trivial period d (this amounts to introducing a periodic arrangement of layers with permittivities

ε_1 and $\varepsilon_2 \neq 1$ in the 1D structure as shown in Fig. 2.11). For simplicity we take $\varepsilon(x) = 1 + \chi \cos(2\pi/d)$. In the presence of this oscillating dielectric function, the accidental degeneracy between u_1 and u_2 is broken. If χ is different from 0, then the field $u_1(x)$ is more concentrated in the region of high contrast ε than $u_2(x)$ is. Therefore, u_1 lies at a lower frequency than u_2. This mutual repulsion of the bands creates a band gap.

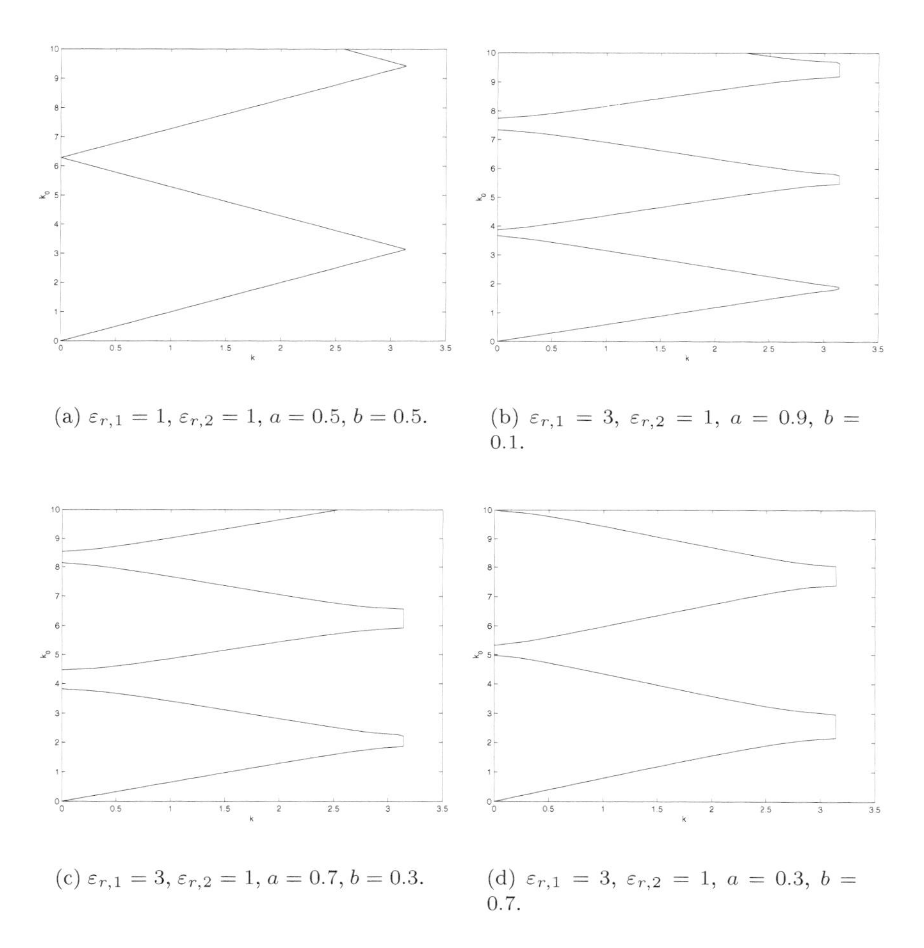

(a) $\varepsilon_{r,1} = 1$, $\varepsilon_{r,2} = 1$, $a = 0.5$, $b = 0.5$.

(b) $\varepsilon_{r,1} = 3$, $\varepsilon_{r,2} = 1$, $a = 0.9$, $b = 0.1$.

(c) $\varepsilon_{r,1} = 3$, $\varepsilon_{r,2} = 1$, $a = 0.7$, $b = 0.3$.

(d) $\varepsilon_{r,1} = 3$, $\varepsilon_{r,2} = 1$, $a = 0.3$, $b = 0.7$.

Fig. 2.12 On diagrams (2.12(a)), (2.12(b)), (2.12(c)), (2.12(d)), the vertical axes correspond to normalised wavenumbers $k_0 d$ running from 0 to 10 and the horizontal axes correspond to normalised Bloch parameters kd running from 0 to π ($d = 1$). The dispersion curves are shaped both by permittivities $\varepsilon_{r,1}$ and $\varepsilon_{r,2}$ and thicknesses a and b of the layers 1 and 2. On diagram (2.12(a)), the dispersion curve is associated with a homogeneous medium with enforced Bloch conditions. The lower dispersion curve is called the light line. On diagram (2.12(b)) we consider a multi-layered structure with a permittivity contrast $\xi = 3$ between layers of radically different ticknesses ($a/b = 9$). On diagrams (2.12(c)) and (2.12(d)) we keep the contrast $\xi = 3$ unchanged but we interchange the (less different) thicknesses of layers (respectively $a/b = 0.7/0.3$ and $a/b = 0.3/0.7$).

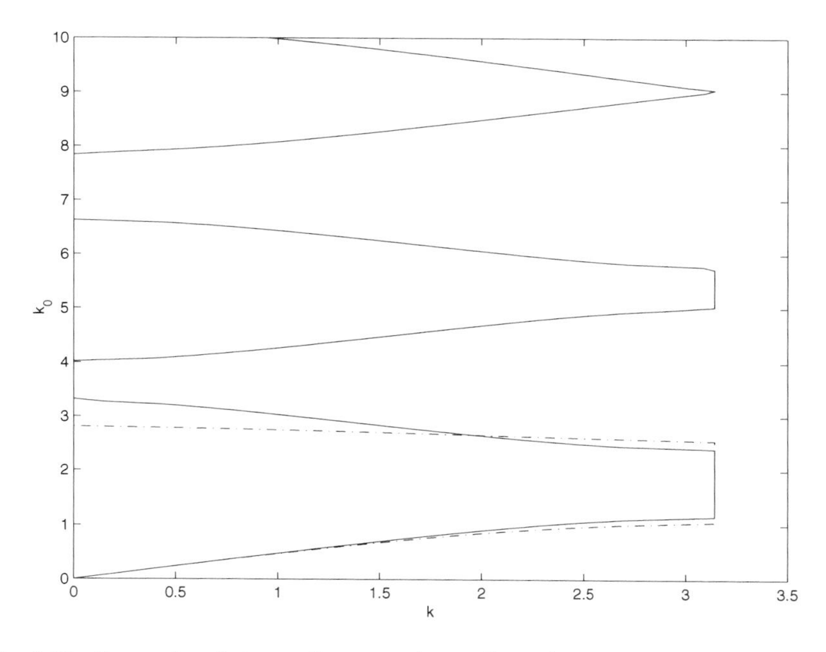

Fig. 2.13 Comparison between the approximate dispersion curves given by Eq. 2.252 and Eq. 2.253 (– · –·) together with the exact dispersion curves given by Eq. 2.251 (—). The vertical axis corresponds to a normalised wavenumber $k_0 d$ running from 0 to 10 and the horizontal axis corresponds to a normalised Bloch parameter kd running from 0 to π ($d = 1$). Here, the opto-geometric parameters are permittivities $\varepsilon_{r,1} = 12$ and $\varepsilon_{r,2} = 1$ and thicknesses $a = 0.3$ and $b = 0.7$ for the layers 1 and 2.

Chapter 3

Finite Element Method

3.1 Finite Elements: Basic Principles

The finite element method is extremely general as it allows the treatment of a large class of partial differential equations, including non-linear problems and without any limitation on the geometry involved. Contrary to the multipole method presented in chapter 4 which is particularly well suited to operators associated with geometries comprising exclusively circular contours, the finite element method is an all-purpose method. The price to pay is, of course, a higher computational cost together with an associated reduction in speed. Nevertheless, the regular increase of computer speed and memory size at constant price together with the improvement of numerical algorithms for sparse matrix algebra make such a general method increasingly promising. In the last twenty years, the finite element method in electromagnetism has not only been improved but also better understood and, nowadays, the reason why some approaches work much better than others is quite clear. In particular, the variant involving Whitney elements presented in this chapter is widely recognized as being superior, for instance, to the scalar approximation of Cartesian components of vector fields.

Among the theoretical progresses, mixed formulations that draw on the whole field of finite element modelling have not only led to new methods but also have shed new light on existing ones, and the inf-sup theorem (also known as the Babuška-Brezzi theorem) has become a fundamental tool in the theoretical analysis of finite element methods, including those having applications in electromagnetism. The geometric approach is more specialised: using the tools of differential geometry instead of coarse vector analysis, the geometric structure of electromagnetism appears clearly as a de Rham complex and experience has shown that it is a fundamental requirement to regard this structure at the discrete level. This is the case for instance with the Yee algorithm that lies at the heart of the FDTD method, probably, up to now, the most widely used method for electromagnetic wave propagation study. In the finite element method, Whitney elements appear to be the right tool for developing the finite element method in a correct geometric setting. Moreover, it appears that this approach also satisfies naturally the Babuška-Brezzi criterion.

In the case of propagation mode computation, the strategy of the finite element method is quite straightforward: a matrix (*i.e.* a finite rank operator) that approximates the partial differential operator of the problem is constructed numerically and a numerical eigenvalue problem is solved using numerical algebra techniques.

3.1.1 *A one-dimensional naive introduction*

The general principles are quite simple and the main ingredients are presented here in the context of a one-dimensional toy problem:

- The boundary value problem is to solve a differential equation:

$$L\,u(x) = -\frac{d}{dx}(p(x)\frac{du(x)}{dx}) + q(x)u(x) = f(x)$$

  where $p(x)$, $q(x)$, and $f(x)$ are given functions and $u(x)$ is the unknown function. The geometrical domain is merely an interval $[a, b]$ and the boundary conditions are taken here to be homogeneous *Dirichlet conditions* $u(a) = 0$ and $u(b) = 0$ which ensure the uniqueness of the solution. The operator L is self-adjoint (See mathematical appendix p. 307) and it is a problem of the Sturm-Liouville type. This is very different to a Cauchy initial value problem, in which two conditions are given on a, *i.e.* the value of u and its derivative.
- The unknown function $u(x)$ is approximated by linear combinations

$$\tilde{u}(x) = \sum u_i \varphi_i(x) \tag{3.1}$$

  of simple functions $\varphi_i(x)$ called *shape functions* that form a basis of a finite dimensional functional space. There are several ways to choose such functions. Choosing, for instance, sine and cosine functions leads to the Fourier method since the approximation is in this case a truncated Fourier series. This is an extremely useful method but this approach has two drawbacks. The first one is that the approximation is smooth ($\in C^\infty[a,b]$) and a large number of terms is necessary if there are discontinuities in the $p(x), q(x)$, and $f(x)$ functions. The second drawback is that the generalization to several dimensions is easy only with simple shapes for the geometrical domain, *e.g.* a rectangle or a disk in two dimensions. In practice, if you decide to adopt this approach, you will get in the end a variant of the plane wave method. In the finite element method, the interval is divided into smaller subintervals on which the approximation has a simple behavior *e.g.* polynomial. A common choice is to use the approximation that the function is a piecewise linear function *i.e.* a linear combination of "hat" functions. Given the *nodes* x_i, *i.e.* the end points of the subintervals such that $x_0 = a < x_1 < \ldots < x_i < \ldots < x_n = b$, the hat function associated with x_i is defined by:

$$\varphi_i(x) = \begin{cases} \frac{x-x_i}{x_{i+1}-x_i} \text{ if } x \in [x_i, x_{i+1}] \\ \frac{x-x_i}{x_{i-1}-x_i} \text{ if } x \in [x_{i-1}, x_i] \\ 0 \text{ elsewhere,} \end{cases} \text{ for } i = 1, \ldots n-1$$

$$\varphi_0(x) = \frac{x-x_1}{x_0-x_1} \text{ if } x \in [x_0, x_1] \text{ and } 0 \text{ elsewhere,}$$

$$\varphi_n(x) = \frac{x-x_{n-1}}{x_n-x_{n-1}} \text{ if } x \in [x_{n-1}, x_n] \text{ and } 0 \text{ elsewhere.}$$

With this choice, the coefficients u_i are the *nodal values* of the approximation.

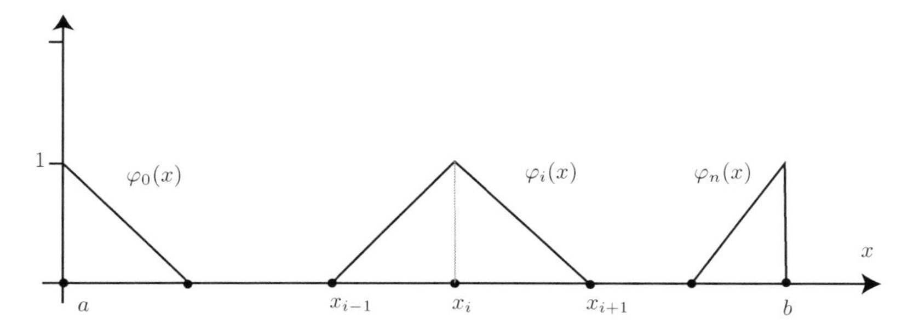

Fig. 3.1 Shape functions in the one-dimensional case.

- In order to determine the coefficients u_i, *weighted residuals* $R(u,w) = \int_a^b (Lu(x) - f(x))w(x)dx$ are built by integrating the product of the equation to be solved $Lu - f = 0$ with a *weight function* $w(x)$, integrating of course on the domain of definition of the problem $[a, b]$. For the solution u, the weighted residuals are equal to zero for any weight function since the integrand is equal to zero: $R(u, w) = 0$. A numerical solution is obtained by introducing the approximation $\tilde{u}(x) = \sum_{i=1}^{n-1} u_i \varphi_i(x)$ in the weighted residuals: $R(\tilde{u}, w) = \sum u_i \int_a^b (L\varphi_i(x))w(x)dx - \int_a^b f(x)w(x)dx = 0$, and choosing a suitable set of weight functions. The most common choice for the weight functions is to take the shape functions themselves. This leads to the *Galerkin method* where the equations to solve are: $R(\sum u_i \varphi_i, \varphi_j) = \sum u_i \int_a^b (L\varphi_i(x))\varphi_j(x)dx - \int_a^b f(x)\varphi_j(x)dx = 0$.

- This is extremely simple but there is a pitfall: the evaluation of the residue involves a second derivative with respect to x which makes no sense for the piecewise linear approximation since it is equal to zero (the first derivative is piecewise constant and discontinuous). Rather than taking distributional derivatives, the common solution is to introduce the *weak formulation* obtained via an integration by parts:

$$\int_a^b \frac{d}{dx}(p(x)\frac{du(x)}{dx})w(x)dx = \\ -\int_a^b p(x)\frac{du(x)}{dx}\frac{dw(x)}{dx}dx + [p(x)\frac{du(x)}{dx}w(x)]_a^b .$$

A term appears that involves the boundary conditions. In the present example, homogeneous Dirichlet conditions have to be imposed. This is easily done by taking an approximation satisfying these *essential conditions*: the nodal values corresponding to $x = a$ and $x = b$ are equal to zero, the weight functions are the shape functions of the internal nodes only so that $w(a) = w(b) = 0$ and the boundary term cancels out.

Note that keeping the nodal values on the boundary as unknowns so that the approximation is now $\tilde{u}(x) = \sum_{i=0}^{n} u_i \varphi_i(x)$ (take note of the limits of the summation) and introducing the corresponding shape functions φ_0 and φ_n in the set of weight functions corresponds to a problem with homogeneous *Neumann conditions* or *natural conditions* $du/dx|_{w=a} = 0$ and $du/dx|_{w=b} = 0$ on the boundary.

The residual can be expressed as:

$$\int_a^b (p(x)\frac{du(x)}{dx}\frac{dw(x)}{dx} + q(x)u(x)w(x))dx - \int_a^b f(x)w(w)dx = 0 \quad (3.2)$$

Introducing the symmetric bilinear form:

$$a(u, w) = \int_a^b (p(x)\frac{du(x)}{dx}\frac{dw(x)}{dx} + q(x)u(x)w(x))dx$$

and writing $\int_a^b f(x)w(w)dx =< f, w >$ as a duality product, Eq. 3.2 can be expressed in a very concise abstract form:

$$a(u, w) =< f, w > .$$

It appears that the constraints on the differentiability of u are now weaker, satisfied by piecewise linear functions, and have been transferred partially on w. The two functions play a symmetrical role which suggests that the Galerkin approach is quite natural.
In terms of functional spaces, the weak solution is in $H^1([a, b])$, though the differential problem involves a second order operator, and is defined by:

$$a(u, w) =< f, w >, \forall w \in H^1([a, b])$$

The numerical solution is in the finite dimensional subspace $V_h \subset H^1([a, b])$ spanned by the φ_i. The approximation $\tilde{u}$ given in Eq. 3.1 is defined by:

$$a(\tilde{u}, \varphi) =< f, \varphi > , \forall \varphi \in V_h$$

- The equality above is satisfied if and only if it holds for all the basis functions φ_j and this gives the following linear system of equations:

$$\sum_i u_i \; a(\varphi_i, \varphi_j) =< f, \varphi_j > , \text{for } j = 1, \cdots, n$$

This linear system is sparse since only two nodes which are the end points of the same subinterval give a nonzero coefficient in the system matrix.

3.1.2 *Multi-dimensional scalar elliptic problems*

3.1.2.1 *Weak formulation of problems involving a Laplacian*

Consider now a two-dimensional or three-dimensional scalar Helmholtz problem with homogeneous Dirichlet conditions:

$$\begin{aligned} -\Delta u + u &= f \text{ in } \Omega \\ u|_{\partial\Omega} &= 0 \end{aligned} \tag{3.3}$$

where Ω is an open set of $\mathbb{R}^2$ or $\mathbb{R}^3$ with a smooth boundary $\partial\Omega$, $u \in H^1(\Omega)$ is the unknown function (alternatively the boundary condition can be replaced by saying that $u \in H_0^1(\Omega)$), and $f \in H^{-1}(\Omega)$ is a given function (and may be a distribution). The Laplacian is responsible for the difference of 2 in the Sobolev space indices.

The trace theorem gives the regularity required by functions involved in the generalisation of the problem to inhomogeneous boundary conditions, *e.g.* $u|_{\partial\Omega} = g$ with $g \in H^{1/2}(\partial\Omega)$.

It seems sensible to choose $L^2(\Omega)$ as a pivot space and indeed this way of posing the problem is very convenient in setting up the classical finite element method. It is therefore possible to write for any $\varphi \in H_0^1(\Omega)$ that $< -\Delta u + u, \varphi > = < f, \varphi >$ (note that the duality pairing is used here but it is in fact an L^2 scalar product if f and Δu are functions) and integration by parts gives the weak form (where $u \in H^1$ is obviously allowed) of Eq. 3.3:

$$a(u,\varphi) = (\mathbf{grad}\, u, \mathbf{grad}\, \varphi)_{L^2} + (u,\varphi)_{L^2} = (u,\varphi)_{H^1} = < f, \varphi > , \forall \varphi \in H_0^1(\Omega) \tag{3.4}$$

Alternatively, a more explicit form is obtained by expressing the abstract products as integrals:

$$a(u,\varphi) = \int_\Omega (\mathbf{grad}\, u \cdot \mathbf{grad}\, \varphi + u\varphi) d\mathbf{x} = \int_\Omega f\varphi d\mathbf{x}$$

where $a(u,\varphi)$ is a symmetric bilinear form associated with the problem.

3.1.2.2 *Generalizations*

It appears that weak formulations of elliptic problems with self-adjoint operators lead to a symmetric form and in this framework, the *Lax-Milgram theorem* is a fundamental tool:

Theorem 3.1 *Let H be a Hilbert space, $a : H \times H \to \mathbb{R}$ a bilinear form that is:*

- *continuous: there exists $C > 0$ such that* $|a(u,v)| \leq C \parallel u \parallel \parallel v \parallel, \forall u, v \in H$
- *H-elliptic or coercive: for some $c > 0$, $a(v,v) \geq c \parallel v \parallel^2, \forall v \in H$*

Then, for every $f \in H'$, the problem

$$a(u,v) = < f, v > , \forall v \in H$$

has a unique solution in H. Moreover if a is symmetric, i.e. $a(u,v) = a(v,u)$, u is the unique solution of the minimization problem

$$J(u) = \min_{v \in H} J(v)$$

where the functional J is defined by:

$$J(v) = \frac{1}{2}a(v,v) - < f, v > .$$

This theorem reduces the demonstration that a problem is well-posed to merely checking that the associated bilinear form is continuous and coercive.

In the case where a is symmetric, the conditions imposed on the bilinear form a in the Lax-Milgram theorem are such that it defines, on H, a new scalar product $(u,v)_a = a(u,v)$ and its associated norm $\|u\|_a^2 = (u,u)_a$, called the *energy scalar product* and the *energy norm* respectively because they are often associated with the expression of an energy in a physical context.

An operator $A : H \to H'$ such that $a(u,v) =< Au, v >$ is associated with the bilinear form a. If a is symmetric then A is symmetric *i.e.* $< Au, v >=< Av, u >$.

The theorem provides also a quadratic functional

$$J(v) = \frac{1}{2}a(v,v) - < f, v >= \frac{1}{2} < Av, v > - < f, v >,$$

called a Lagrangian, that is minimized by the solution (and the Euler-Lagrange equations associated with this functional regurgitate the original differential problem). The stationarity condition is that the variation of the functional defined by $\delta J = J(v + \delta v) - J(v)$ is zero to first order for a variation δv of v. As $< A(v + \delta v), v + \delta v >=< Av, v > + < A\delta v, v > + < Av, \delta v > + < A\delta v, \delta v >=< Av, v > +2 < Av, \delta v > +O(\delta v^2)$, the variation to first order gives the condition:

$$\frac{1}{2} < A(v + \delta v), v + \delta v > - < f, v + \delta v > -(\frac{1}{2} < Av, v > - < f, v >) = \\ < Av, \delta v > - < f, \delta v >= 0$$

holding for any δv respecting the boundary conditions so that the equivalent differential problem is $Au = f$ (with the associated boundary conditions).

A common generalization of problem 3.3 is to replace the partial differential equation with:

$$-\operatorname{div}[\alpha(\mathbf{x}) \,\mathbf{grad}\; u(\mathbf{x})] + \beta(\mathbf{x})u(\mathbf{x}) = f(\mathbf{x})$$

where $\alpha(\mathbf{x})$ and $\beta(\mathbf{x})$ are given scalar functions of the coordinates. Some conditions on $\alpha(\mathbf{x})$ and $\beta(\mathbf{x})$ are necessary to ensure continuity and coercivity. In physics, these functions represents some material properties and a common situation corresponds to $\alpha_* < \alpha(\mathbf{x}) < \alpha^*$, $\forall \mathbf{x} \in \Omega$ and $\beta_* < \beta(\mathbf{x}) < \beta^*$, $\forall \mathbf{x} \in \Omega$ where $\alpha_*, \alpha^*, \beta_*, \beta^*$ are strictly positive constants which guarantees that the problem is well-posed (however, the new metamaterials appearing in electromagnetism such as lefthanded materials with negative permeabilities and permittivities will raise new questions about the well-posedness of problems in which they appear).

With this operator, the associated bilinear form is: $a(u,v) = \int_\Omega (\alpha\, \mathbf{grad}\, u \cdot \mathbf{grad}\, v + \beta uv) d\mathbf{x}$.

This may be generalised further so that $\underline{\underline{\alpha}}$ is a rank 2 tensor field and sufficient conditions correspond often to its symmetry and positive definition: $\alpha^{ij}(\mathbf{x}) = \alpha^{ji}(\mathbf{x}) \in L^\infty(\Omega)$ for $i, j = 1, 2, 3$ and $\exists \alpha^* > 0$ such that $\alpha^{ij}(\mathbf{x}) v_i v_j \leq \alpha^* \|\mathbf{v}\|^2$, $\forall \mathbf{v} \in \mathbb{R}^3$.

The finite element method can also deal with complex valued functions. In this case, $a(u,v)$ is a sesqui-linear form and the coercivity condition in the Lax-Milgram theorem has to be replaced by $\Re e\ a(u,u) \geq c\|\mathbf{u}\|^2$ and the solution corresponds to the minimum of a functional if a is Hermitian: $a(u,v) = \overline{a(v,u)}$ [N9́1].

Of course, other conditions than the Dirichlet homogeneous conditions can be dealt with but we direct the attention of the reader to the classical literature [Bra97] for further details.

3.1.2.3 *The finite element method*

The numerical solution of this kind of problem is found by choosing a finite dimensional subspace $V_h \subset H_0^1(\Omega)$ and looking for an approximation $\tilde{u}$ that satisfies $a(\tilde{u}, \varphi) =< f, \varphi >$, $\forall \varphi \in V_h$, *i.e.* exactly the same process as in the one dimensional case from a formal point of view. The difference lies in the construction of V_h. For instance, the domain Ω is now divided into a set of triangles which pairwise have in common either a full edge, or a node (vertex), or nothing in the two-dimensional case (see Fig. 3.10) and into a set of tetrahedra which pairwise have in common either a full triangular facet, or a full edge, or a node, or nothing in the three-dimensional case (see Fig. 3.2). Triangles in the two-dimensional case and tetrahedra in the three-dimensional case are called the *elements* and their union is the *mesh* $\mathcal{T}_h$. A shape function $\varphi_i(\mathbf{x})$ associated with a node i is a piecewise linear function on the elements having the value 1 on node i and the value

0 on all the other nodes. Its support is therefore the set of elements having the node i among their vertices. This leads again to a sparse linear system. Elements with such shape functions that provide a continuous interpolation of scalar functions, are called *Lagrange elements* (or *nodal elements* since the degrees of freedom are associated to the nodes) of the first order and the corresponding discrete space is denoted $P_1(\mathcal{T}_h)$.

The h in V_h indicates that this discrete space is constructed with the help of a mesh $\mathcal{T}_h$ such that the largest distance inside an element is smaller than h. The approximation of u in V_h is denoted u_h.

An important property of a family of spaces V_h is the *consistency*, *i.e.* the ability of the elements of V_h to converge to an element of $H_0^1(\Omega)$ (and more generally to an element of the considered functional space $V \supset V_h$ depending on the problem) when the mesh is progressively refined. Suppose that $\mathcal{T}_{h_1}$ and $\mathcal{T}_{h_2}$ are two meshes, then there always exists a mesh $\mathcal{T}_h$ which is a refinement of both $\mathcal{T}_{h_1}$ and $\mathcal{T}_{h_2}$ (*e.g.* one can always build a mesh such that every node of $\mathcal{T}_{h_1}$ and $\mathcal{T}_{h_2}$ is a node of $\mathcal{T}_h$). This makes the set of meshes on Ω an (upward filtering) ordered set (where "finer" plays the role of "greater"), which allows the definition of the convergence of an approximation. A sequence of $u_h \in V_h$ converges to $u \in V$, when the mesh is refined, if $\|u_h - u\|_V < \varepsilon$ for all meshes finer than some $\mathcal{T}_{h(\varepsilon)}$.

The numerical procedure that involves the minimization of the quadratic functional associated with the problem (by annulling the partial derivatives with respect to the coefficients of the linear approximation), if it exists, is called the *Rayleigh-Ritz* method but if we use directly the annulation of the weighted residuals with the shape and weight functions in the same space it is called the Galerkin method. When the Lax-Milgram theorem applies, the two methods are of course strictly equivalent. If a different set of functions to the shape functions is chosen for the weight functions, the method is called a *Petrov-Galerkin* method and is sometimes useful for problems with singularities.

As for errors and convergence issues in the Galerkin method, it can be proved easily that the error $e = u - u_h$ is orthogonal to the space V_h with respect to the energy scalar product (*i.e.* $(e, v_h)_a = 0$, $\forall v_h \in V_h$) and that u_h is the best estimate in terms of energy norm:

$$\|u - u_h\|_a \leq \|u - v_h\|_a \,, \forall v_h \in V_h.$$

Moreover, suppose that a is an H-elliptic bilinear form and f provides a bounded linear form, and consider the variational problem stated in the Lax-Milgram theorem using these forms. In this case, *Céa's lemma* [Bra97;

Red98] states that if u and u_h are solutions of this variational problem in spaces H and V_h respectively, then there exists a constant $C(\Omega)$, independent of h but depending on Ω, such that:

$$\|u - u_h\|_a \leq C(\Omega) \inf_{v_h \in V_h} \|u - v_h\|_a.$$

3.1.3 *Mixed formulations*

A variant approach called the *mixed formulation* is now presented that leads naturally to useful functional spaces and justifies our introduction of a new kind of finite element.

The equation $\operatorname{div} \alpha\, \mathbf{grad}\, u = f$ (still assuming homogeneous Dirichlet conditions) can be decomposed into the equivalent first order system:

$$\begin{aligned} \mathbf{grad}\, u &= \alpha^{-1}\mathbf{v} \\ \operatorname{div} \mathbf{v} &= -f. \end{aligned} \tag{3.5}$$

The scalar product of the first equation with a vector weight function $\mathbf{p}$ is integrated over the whole domain Ω, which gives, after integration by parts: $\int \alpha^{-1}\mathbf{v}\cdot\mathbf{p}d\mathbf{x} = -\int u \operatorname{div} \mathbf{p}d\mathbf{x}$. As for the second equation, it is multiplied by a scalar weight function q and integrated over the whole domain to obtain: $\int q \operatorname{div} \mathbf{v}d\mathbf{x} = -\int fqd\mathbf{x}$. The examination of these expressions indicates that q, u, $\operatorname{div} \mathbf{v}$, and $\operatorname{div} \mathbf{p}$ must be in $L^2(\Omega)$. A new space $H(\operatorname{div}, \Omega) = \{\mathbf{p} : \mathbf{p} \in [L^2(\Omega)]^3, \operatorname{div} \mathbf{p} \in L^2(\Omega)\}$ is introduced for $\mathbf{p}$ and $\mathbf{v}$. Therefore, the *mixed formulation* is:

$$\begin{aligned} &\text{Find } \mathbf{v} \in H(\operatorname{div}, \Omega) \text{ and } u \in L^2(\Omega) \text{ such that :} \\ &\textstyle\int \alpha^{-1}\mathbf{v} \cdot \mathbf{p}d\mathbf{x} + \int u \operatorname{div} \mathbf{p}d\mathbf{x} = 0 \qquad , \forall \mathbf{p} \in H(\operatorname{div}, \Omega) \\ &\textstyle\int q \operatorname{div} \mathbf{v}d\mathbf{x} \qquad\qquad\qquad = -\int fqd\mathbf{x} \ \ , \forall q \in L^2(\Omega). \end{aligned} \tag{3.6}$$

Note that this formulation can lead to a very different discrete approximation since u can now be approximated by piecewise constant functions that are in $L^2(\Omega)$ and not in $H^1(\Omega)$.

The space $H(\operatorname{div}, \Omega)$ is a proper subset of $[H^1(\Omega)]^3$. Take a vector field $\mathbf{v}$ that is differentiable in the complement $\Omega \setminus \Sigma$ of a surface Σ and suffers a discontinuity across this surface Σ but in a way such that only the tangential component is discontinuous while the normal one is continuous. This vector field is in $H(\operatorname{div}, \Omega)$ but not in $[H^1(\Omega)]^3$. As an example, take Σ to be the plane $z = 0$ and $\mathbf{v}$ continuous with respect to x, y but with components v_x and v_y discontinuous and component v_z continuous with respect to z on Σ. The distributional divergence $\operatorname{div}\mathbf{v} = \frac{\partial v_x}{\partial x} + \frac{\partial v_y}{\partial y} + \frac{\partial v_z}{\partial z}$ is well defined as a function and can be in $L^2(\Omega)$.

The Lagrangian $J^m(u, \mathbf{v}) = \int_\Omega (\frac{1}{2}\alpha^{-1}\mathbf{v} \cdot \mathbf{v} + u \operatorname{div}\mathbf{v} + fu) d\mathbf{x}$ over $H(\operatorname{div}, \Omega) \times L^2(\Omega)$ is associated with the mixed formulation but the critical point is no longer a minimum but a *saddle point* such that $J^m(u, \mathbf{s}) \leq J^m(u, \mathbf{v}) \leq J^m(w, \mathbf{v})$, $\forall \mathbf{s} \in H(\operatorname{div}, \Omega), \forall w \in L^2(\Omega)$, *i.e.*

$$\inf_{w \in L^2(\Omega)} \sup_{\mathbf{s} \in H(\operatorname{div},\Omega)} J^m(w, \mathbf{s}) = J^m(u, \mathbf{v})$$

The Euler-Lagrange equations for $J^m(u, \mathbf{v})$ corresponding to variations $\delta\mathbf{v}$ and δu regurgitate the first and second equations of 3.5 respectively. In this context, u can be interpreted as a *Lagrange multiplier* so that the system appears as a minimization of $J'(\mathbf{v}) = \int_\Omega \frac{1}{2}\alpha^{-1}\mathbf{v} \cdot \mathbf{v} d\mathbf{x}$ with the constraint $\operatorname{div}\mathbf{v} + f = 0$.

The well-posedness of mixed formulations relies on the satisfaction of the inf-sup (LBB) criterion as explained in the annex A. In the case of problem 3.6, the two spaces to be discretised are $U = L^2(\Omega)$ and $\Sigma = H(\operatorname{div}, \Omega)$ and an example of a good choice of discrete spaces associated with a simplicial mesh $\mathcal{T}_h$ of Ω are the space $\mathcal{P}_0(\mathcal{T}_h)$ of piecewise constant functions on the element for $U_h \subset L^2(\Omega)$ and the space $\mathcal{RT}_0(\mathcal{T}_h)$ of *Raviart-Thomas elements* [RT77] (or *face elements*) of the lowest order for $\Sigma_h \subset H(\operatorname{div}, \Omega)$. The space $\mathcal{RT}_0(\mathcal{T}_h)$ is the space of vector fields $\mathbf{v}$ whose restriction on an element T is $\mathbf{v}|_\mathrm{T} = \mathbf{a}_\mathrm{T} + b_\mathrm{T}\mathbf{r}$ where $\mathbf{r}$ is the position vector, $\mathbf{a}_\mathrm{T}$ is a constant vector and b_T is a constant scalar. The normal component of $\mathbf{v}|_\mathrm{T}$ is constant on each face of the element T and, in dimension n, the number of degrees of freedom is equal to $n + 1$, the number of faces of the simplicial element, so that the normal components of $\mathbf{v}$ on these faces can be chosen as the degrees of freedom. A fundamental property of the approximations built with these elements is that they have a continuous normal component across the boundaries of the elements but a discontinuous tangential one. Such elements whose basis functions span a discrete subspace of $H(\operatorname{div}, \Omega)$ are called *div-conforming*.

3.1.4 *Vector problems*

A further generalization of problem 3.3 is to solve vector problems such as:

$$\begin{aligned} &\mathbf{curl}[\alpha\, \mathbf{curl}\, \mathbf{u}] = \mathbf{f} \text{ in } \Omega \\ &\operatorname{div}\mathbf{u} = 0 \qquad\qquad \text{in } \Omega \\ &\mathbf{u} \times \mathbf{n}|_{\partial\Omega} = 0 \end{aligned} \tag{3.7}$$

The second equation is called a *gauge* and is a constraint added to guarantee the uniqueness of the solution. Without such a constraint, the gradient of an arbitrary function (in $H_0^1(\Omega)$) can be added to $\mathbf{u}$ to obtain a new solution. Physically, if the vector field $\mathbf{u}$ corresponds to a vector potential, the gauge condition is quite arbitrary provided that it guarantees the uniqueness. The local numerical value of the vector potential has no physical meaning and the relevant object is rather the equivalence class of the vector field modulo gradients. If the vector field $\mathbf{u}$ corresponds to a field such as the electric or magnetic field, an equation of the form $\mathbf{curl}(\alpha\,\mathbf{curl}\,\mathbf{u}) + \beta\mathbf{u} = \mathbf{f}$ is instead encountered where the addition of a gradient is no longer neutral. Moreover, taking the divergence of this equation gives $\mathrm{div}(\beta\mathbf{u}) = \mathrm{div}\,\mathbf{f}$ and shows that there is no longer any freedom in the choice of the divergence.

The weak formulation for the first equation of 3.7 is:

$$\int_\Omega \alpha\,\mathbf{curl}\,\mathbf{u}\cdot\mathbf{curl}\,\mathbf{v}d\mathbf{x} = \int \mathbf{v}\cdot\mathbf{f}d\mathbf{x}\ ,\forall\mathbf{v}\in H(\mathbf{curl},\Omega)$$

where the space $H(\mathbf{curl},\Omega) = \{\mathbf{v} : \mathbf{v}\in[L^2(\Omega)]^3, \mathbf{curl}\,\mathbf{v}\in[L^2(\Omega)]^3\}$ enters quite naturally into play. In order to include the boundary conditions in this setting, the space $H_0(\mathbf{curl},\Omega) = \{\mathbf{v} : \mathbf{v}\in H(\mathbf{curl},\Omega), \mathbf{v}\times\mathbf{n}|_{\partial\Omega} = 0\}$ is introduced. Just like $H(\mathrm{div},\Omega)$, the space $H(\mathbf{curl},\Omega)$ is a proper subset of $[H^1(\Omega)]^3$. Take a vector field $\mathbf{v}$ that is differentiable in the complement $\Omega\setminus\Sigma$ of a surface Σ and suffers a discontinuity across this surface Σ but in a way such that only the normal component is discontinuous while the tangential one is continuous. This vector field is in $H(\mathbf{curl},\Omega)$ but not in $[H^1(\Omega)]^3$. As an example, take Σ to be the plane $z=0$ and $\mathbf{v}$ with v_x and v_y continuous and v_z discontinuous with respect to z on Σ. The distributional $\mathbf{curl}\,\mathbf{v} = (\frac{\partial v_z}{\partial y} - \frac{\partial v_y}{\partial z})\mathbf{e}^x + (\frac{\partial v_x}{\partial z} - \frac{\partial v_z}{\partial x})\mathbf{e}^y + (\frac{\partial v_y}{\partial x} - \frac{\partial v_x}{\partial y})\mathbf{e}^z$ is well defined as a function and can be in $[L^2(\Omega)]^3$.

In practice, specific gauges specially adapted to discrete problems have been designed that use a tree (in the sense of graph theory) on the edges of the mesh to impose uniqueness directly on the discrete space. If the gauge condition $\mathrm{div}\,\mathbf{u} = 0$ has to be added as a constraint, the weak form of the system to solve becomes:

$$\begin{array}{lll}\text{Find } \mathbf{u}\in H_0(\mathbf{curl},\Omega) \text{ and } p\in H_0^1(\Omega) \text{ such that :} & & \\ \int_\Omega \alpha\,\mathbf{curl}\,\mathbf{u}\cdot\mathbf{curl}\,\mathbf{v}d\mathbf{x} + \int_\Omega \mathbf{v}\cdot\mathbf{grad}\,pd\mathbf{x} & = \int\mathbf{v}\cdot\mathbf{f}d\mathbf{x} & ,\forall\mathbf{v}\in H_0(\mathbf{curl},\Omega) \\ \int_\Omega \mathbf{u}\cdot\mathbf{grad}\,qd\mathbf{x} & = 0 & ,\forall q\in H_0^1(\Omega)\end{array} \tag{3.8}$$

where the last equation is a weak form of the null divergence condition.

If the equation is decomposed into the following first order system,

$$\begin{aligned} \mathbf{curl}\,\mathbf{u} &= \alpha^{-1}\mathbf{p} \\ \mathbf{curl}\,\mathbf{p} &= \mathbf{f} \end{aligned} \tag{3.9}$$

the corresponding mixed formulation is:

$$\begin{array}{lll} \text{Find } \mathbf{u} \in H_0(\mathbf{curl},\Omega) \text{ and } \mathbf{p} \in H_0(\mathrm{div}^0,\Omega) \text{ such that :} & & \\ \int \alpha^{-1}\mathbf{u}\cdot\mathbf{v}d\mathbf{x} \ + \int \mathbf{p}\cdot\mathbf{curl}\,\mathbf{v}d\mathbf{x} = 0 & , \forall\mathbf{v} \in H_0(\mathbf{curl},\Omega) & \\ \int \mathbf{curl}\,\mathbf{u}\cdot\mathbf{q}d\mathbf{x} \qquad\qquad = \int \mathbf{f}\cdot\mathbf{q}d\mathbf{x} & , \forall\mathbf{q} \in H_0(\mathrm{div}^0,\Omega) & \end{array} \tag{3.10}$$

where $H_0(\mathrm{div}^0,\Omega) = \{\mathbf{v} \in [L^2(\Omega)]^3, \mathrm{div}\,\mathbf{v} = 0, \mathbf{v}\cdot\mathbf{n}|_{\partial\Omega} = 0\}$. Note that $\mathbf{curl}\,H_0(\mathbf{curl},\Omega) \subset H_0(\mathrm{div}^0,\Omega)$.

The first idea that came historically to solve vector problems of electromagnetism by means of the finite element method was to consider each Cartesian component of the vector field as a scalar function that can be discretized with the classical elements where degrees of freedom are nodal values. This leads to some interesting results but it soon became apparent that this approach is plagued by fundamental flaws.

A better approach is to use *curl-conforming* elements that are basis functions for finite dimensional subspaces of $H(\mathbf{curl},\Omega)$. One of their fundamental properties is that the corresponding approximation has tangential components continuous across the element boundaries but a discontinuous normal one. They also provide families of elements that match the inf-sup criterion in the case of the mixed formulation (3.10).

Such elements have been built by Nédélec [N8́0; N8́6; N9́1]. In two dimensions, the lowest order space $\mathcal{N}_0(\mathcal{T}_h)$ of *Nédélec elements* (or *edge elements*) associated with the triangles are just the Raviart-Thomas elements rotated by $\pi/2$: $\mathbf{v}|_{\mathrm{T}} = \mathbf{a}_{\mathrm{T}} + b_{\mathrm{T}}\mathbf{r}^*$ where $\mathbf{r}^* = y\mathbf{e}_x - x\mathbf{e}_y$. In the three dimensional case, the space $\mathcal{N}_0(\mathcal{T}_h)$ is the space of vector fields whose restriction to the tetrahedron T is $\mathbf{v}|_{\mathrm{T}} = \mathbf{a}_{\mathrm{T}} + \mathbf{b}_{\mathrm{T}} \times \mathbf{r}$ where $\mathbf{a}_{\mathrm{T}}$ and $\mathbf{b}_{\mathrm{T}}$ are constant three-dimensional vectors. There are 6 degrees of freedom that are the line integrals along the 6 edges of the tetrahedra. We postpone a more detailed discussion of these elements to Sec. 3.3.4.

3.1.5 *Eigenvalue problems*

Up to now, the finite element method has been applied to elliptic problems to obtain an algebraic system whose solution gives the approximation. The method can also deal with dynamical problems involving parabolic or hyperbolic operators. In this case, by far the most common approach is the semi-discrete Galerkin method: the space undergoes the same treatment as before, involving a mesh and shape functions, but now the degrees of freedom are no longer scalar constants but functions of time. As the operators involve time derivatives, what we get in the end is a system of differential (and sometimes algebraico-differential) equations instead of algebraic ones. This system is then solved using traditional tools for the numerical solution of ordinary differential equations (see for example the description of the leapfrog scheme in Sec. 3.2.4 below).

In the case of a time harmonic problem with pulsation ω, the time derivative amounts to a multiplication by $-i\omega$ and the problem reverts back to a (complex) algebraic one. This is in fact the most common origin for Helmholtz problems that we have already introduced. But it is also possible to look for non zero solutions that can exist for a given frequency without any source. The frequency is in this case an unknown to be determined by solving what is called an eigenvalue problem. This is the kind of problem that will be encountered when trying to compute the propagation modes.

In this section, the general philosophy of the numerical treatment of an eigenvalue problem is to build a matrix system whose eigenvalues approximate correctly those of the original operator. This is not always easy depending on the kind of operators we are dealing with. For instance, compact operators seem easier to approximate (do you prefer to estimate numerically a derivative or an integral?) since an alternate definition is:

Definition 3.1 An operator $A \in \mathcal{L}(H)$ on an Hilbert space H is said to be compact if it can be approximated in norm by finite rank operators in the sense that for all $\varepsilon > 0$, there exists a finite dimensional subspace $E \subset H$ such that[1] $\|A|_{E^\perp}\| < \varepsilon$.

Moreover the spectrum of a compact operator is a point spectrum, either finite or countably infinite, and its only possible limit point is zero.

As an example, consider the determination of eigenvalues associated with the Laplacian on a domain Ω defined by:

$$\begin{aligned} &\text{Find } \lambda \in \mathbb{R} \text{ and } 0 \neq u \in H^1(\Omega) \text{ such that :} \\ &-\Delta u = \lambda u \text{ in } \Omega \\ &u|_{\partial\Omega} = 0 \end{aligned} \tag{3.11}$$

The value λ is called the eigenvalue and the function u the eigenvector. This function is of course determined up to a scalar factor and can be fixed

[1]The definition of the norm of an operator acting on a vector space is given in the mathematical pot-pourri.

by a normalization imposing $\|u\|_{H^1} = 1$. In some cases, several linearly independent eigenvectors can be associated with the same eigenvalue and the dimension of the eigenspace, *i.e.* the vector space spanned by these eigenvectors, is called the (geometric) multiplicity.

The eigenvalue problem can be set up in an equivalent weak form:

$$\int_\Omega \mathbf{grad}\, u \cdot \mathbf{grad}\, v \, d\mathbf{x} = \lambda \int_\Omega u \, v \, d\mathbf{x} \ , \forall v \in H_0^1(\Omega)$$

and the corresponding discrete finite element formulation is straightforwardly

$$\int_\Omega \mathbf{grad}\, u_h \cdot \mathbf{grad}\, v \, d\mathbf{x} = \lambda_h \int_\Omega u_h \, v \, d\mathbf{x} \ , \forall v_h \in V_h \subset H_0^1(\Omega)$$

for which it can be proved that there exists a constant c such that:

$$\|u - u_h\|_{H^1} \leq c \inf_{v \in V_h} \|u - v\|_{H^1} \ \text{ and } \ |\lambda - \lambda_h| \leq c \|u - u_h\|_{H^1}^2$$

where u and λ are the solutions of the previous corresponding continuous problem.

It is not at all obvious at first sight to see where compactness is involved since the Laplace operator here is not compact. To explain that, consider again the abstract problem associated with the operator $A : H \to H'$: find $u \in V$ such that $Au = f$ (for a given $f \in H' \supset H$) and the corresponding weak form associated with the bilinear form a: find $u \in H$ such that $a(u, v) =< f, v >$ for all $v \in H$. If a is coercive (H-elliptic) and continuous, according to the Lax-Milgram theorem, a unique solution u exists that can be associated with f so that an operator $T : H' \to H$ can be defined by $u = Tf$ such that $ATf = f$ and $a(Tf, v) =< f, v >, \forall v \in H$. This operator T is called the resolvent operator and is the inverse [2] of A: $T = A^{-1}$. More generally, the resolvent $R_\lambda(A)$ of an operator A is the one parameter family of operators defined by $R_\lambda(A) = (A - \lambda I)^{-1}$ where λ are complex numbers (such that the corresponding $R_\lambda(A)$ operator exists) and I is the identity operator, and where $T = R_0(A)$. The resolvent is a fundamental tool to define the spectrum of A (See definition p. 2.5.2).

An important class of operators are the *operators with compact resolvent*. If A is a closed operator such that $R_\lambda(A)$ exists and is compact for

[2]It appears here that the domain of definition of an operator is a fundamental part of its definition. Here A is defined on H, a Hilbert space of functions with a given regularity, defined themselves on a given domain $\Omega \subset \mathbb{R}^n$ with given boundary conditions and all these elements are involved in the inversion of A.

at least some $\lambda = \lambda_0 \in \mathbb{C}$, then the spectrum of A is a point spectrum that consists entirely of isolated eigenvalues with finite multiplicities and $R_\lambda(A)$ is compact for every λ in the resolvent set [Kat95]. Therefore, operators with a compact resolvent share along with compact operators the nice property that the spectrum reduces to the point spectrum.

Consider now the eigenvalue problem: find $(\lambda, u) \in \mathbb{R} \times H$ such that $Au = \lambda u$ and its weak form $a(u,v) = \lambda < u, v >, \forall v \in H$. There exists $w \in H'$ such that $u = Tw$ and therefore $a(Tw, v) =< w, v >= \lambda < Tw, v >$ $, \forall v \in H$. If $\lambda^{-1} = \mu$, the eigenvalue problem $Tw = \mu w$ is obtained where T is a compact operator.

It can be proved that if the canonical embedding of H in H' is compact and if the bilinear form is H-elliptic then the operator T is compact from H to H (*i.e.* $T|_H$). In this case, A is an operator with compact resolvent.

Moreover, if a is symmetric and positive ($a(u,v) > 0 \ , \forall 0 \neq v \in H$) then the eigenvalues are an increasing sequence tending to $+\infty$: $0 < \lambda_1 \leq \lambda_2 \leq \cdots \lambda_m \leq \cdots$ [RT83].

In problem 3.11, $H = H_0^1(\Omega)$ and $H' = H^{-1}(\Omega)$ and it can be proved that the canonical embedding of $H_0^1(\Omega)$ in $H^{-1}(\Omega)$ is compact as an extension of the Rellich-Kondrachov theorem[RT83].

Note that the eigenvalue problem can also be directly set up in some kind of variational disguise by means of the *min-max theorem* [CPV95]:

Theorem 3.2 *(Courant-Fisher) Let $A \in \mathcal{L}(H)$ be a compact self-adjoint operator on a Hilbert space H. For all $0 \neq u \in H$, the associated Rayleigh quotient is defined by:*

$$R_A(u) = \frac{(Au, u)_H}{\|u\|_H^2}$$

Then, for all $m \geq 1$, one has:

$$\lambda_m = \max_{U_m \in H_m} \ \min_{0 \neq u \in U_m} R_A(u)$$

$$\lambda_m = \min_{U_{m-1} \in H_{m-1}} \ \max_{0 \neq u \in U_{m-1}^{\perp}} R_A(u)$$

where H_m is the set of subspaces of H of dimension m and $\lambda_1 \geq \cdots \geq \lambda_m \geq \cdots \to 0$ is the decreasing set of eigenvalues of A.

As a simple example, the largest eigenvalue is given by:

$$\lambda_1 = \max_{0 \neq u \in H} R_A(u).$$

Note that the weak form of the Rayleigh quotient is

$$R(u) = \frac{a(u,u)}{\|u\|_H^2}$$

where $a(u,u)$ is the bilinear form associated with operator A. For instance, the weak form of the Rayleigh quotient corresponding to problem 3.11 is:

$$R(u) = \frac{\int_\Omega \mathbf{grad}\, u \cdot \mathbf{grad}\, u \; d\mathbf{x}}{\int_\Omega u \; v \; d\mathbf{x}}$$

As for the vector analogue, we are led to consider the following eigenvalue problem:

$$\begin{array}{ll} \text{Find } \lambda \in \mathbb{R} \text{ and } 0 \neq \mathbf{u} \in H(\mathbf{curl}, \Omega) \text{ such that :} & \\ \mathbf{curl}\,\mathbf{curl}\,\mathbf{u} = \lambda \mathbf{u} \quad \text{in } \Omega & \\ \operatorname{div} \mathbf{u} = 0 \qquad\qquad\; \text{in } \Omega & \\ \mathbf{u} \times \mathbf{n}|_{\partial\Omega} = 0 & \end{array} \qquad (3.12)$$

The second equation imposing the condition on the divergence is nearly redundant since taking the divergence of the first equation gives $\lambda \operatorname{div} \mathbf{u} = 0$. For the non-zero eigenvalues this is equivalent to the null divergence condition but not for the zero eigenvalue. In this case, $\lambda = 0$ would be associated with the infinite dimensional eigenspace made of gradients: $\{\mathbf{grad}\, \phi,\ \phi \in H^1(\Omega)\}$. The null divergence condition eliminates such eigenvectors since it leads to $\Delta\phi = 0$ in Ω and the boundary condition is that the tangential part of the gradient is equal to zero, which implies that ϕ is constant on the boundary $\partial\Omega$ so that the solutions are such that ϕ is a constant in Ω and their gradients are all zero.

We now turn to the numerical solution of such a problem. Much of the initial work has been performed in the context of nodal elements for vector fields. A numerical approach could be to work with system 3.12 and to impose the divergence condition by means of a penalty term. Another way would be to disregard the divergence condition in the initial computation. One could hope that the eigenvalues associated with the non-zero divergence modes are small enough to be easily discarded. This is not the case and these eigenvalues have an unfortunate tendency to spread out among the whole set of numerical values. Therefore, an idea that has been proposed is to compute afterwards some norm of the divergence of the numerical solutions to sort out the wrong modes. The undesirable modes arising in the numerical computations were called the *spurious modes* and have led to a vast literature. None of the previous solutions have given really good results whatever the effort to improve the computation. It was practically impossible to guarantee the quality of the computation and often the numerical eigenvalues were not a mixture of clearly separated true and spurious values but rather a completely smeared out set of values having often nothing to do with the true eigenvalues [Arn02]. It appeared that the numerical determination of the eigenvalues using nodal elements is intrinsically very unstable.

An alternative is to use curl-conforming elements. The weak discrete formulation can be expressed as:

$$\begin{array}{l}\text{Find } \lambda_h \in \mathbb{R} \text{ and } 0 \neq \mathbf{u} \in V_h \subset H_0(\mathbf{curl}, \Omega) \text{ such that:} \\ \int_\Omega \mathbf{curl}\, \mathbf{u}_h \cdot \mathbf{curl}\, \mathbf{v}\, d\mathbf{x} = \lambda_h \int \mathbf{u}_h \cdot \mathbf{v}\, d\mathbf{x} \quad , \forall \mathbf{v} \in V_h \subset H_0(\mathbf{curl}, \Omega)\end{array} \tag{3.13}$$

V_h is taken for instance to be the space of edge elements. The zero divergence condition is not included in the formulation but in practice the zero divergence modes appear to be well associated with eigenvalues equal to zero within numerical round-off. The other numerical eigenvalues provide good approximations of the true ones. This good behaviour can be explained, for instance, by means of an equivalent mixed problem:[3]

$$\begin{array}{lll}\text{Find } \lambda_h \in \mathbb{R} \text{ and } 0 \neq \mathbf{u}_h \in V_h \subset H_0(\mathbf{curl}, \Omega) \text{ and} & & \\ \mathbf{p}_h \in U_h = \mathbf{curl}\, V_h \subset H_0(\mathrm{div}^0, \Omega) \text{ such that:} & & \\ \int \alpha^{-1} \mathbf{u}_h \cdot \mathbf{v} d\mathbf{x} \; + \int \mathbf{p}_h \cdot \mathbf{curl}\, \mathbf{v} d\mathbf{x} = 0 & & , \forall \mathbf{v} \in V_h \\ \int \mathbf{curl}\, \mathbf{u}_h \cdot \mathbf{q} d\mathbf{x} & = -\lambda_h \int \mathbf{p}_h \cdot \mathbf{q} d\mathbf{x} & , \forall \mathbf{q} \in U_h\end{array} \tag{3.14}$$

If $(\lambda_h, \mathbf{u}_h)$ is a solution of 3.13 with $\lambda_h > 0$ then $(\lambda_h, \mathbf{u}_h, \mathbf{p}_h = -\lambda_h^{-1} \mathbf{curl}\, \mathbf{u}_h)$ is a solution of 3.14. The two formulations are equivalent except that 3.13 admits zero eigenvalues that the mixed formulation suppresses. As explained in [BFGP99], the accuracy of the mixed eigenvalue problem hinges on the stability of the corresponding mixed source problem, hence the stability depends on the Babuška-Brezzi conditions and also on a property called the discrete compactness [Bof01] that are satisfied by the edge elements and whose satisfaction follows from properties encoded in a commutative diagram (See the annex A and Sec. 3.2.3 below).

A more direct approach [Bos90] is to remark that $\mathbf{grad}\, H^1(\Omega) = \{\mathbf{grad}\, \varphi : \varphi \in H^1(\Omega)\} \subset H(\mathbf{curl}, \Omega)$ since, by definition of $H^1(\Omega)$, $\mathbf{grad}\, \phi$

[3] With the notation introduced below in Sec. 3.3.1, $V_h = \mathbf{W}_1^{\mathbf{T}} \mathbf{e}$ and $U_h = \mathbf{curl}(\mathbf{W}_1^{\mathbf{T}} \mathbf{e}) = \mathbf{W}_2^{\mathbf{T}} C \mathbf{e}$.

is in $L^2(\Omega)$ and $\mathbf{curl}\,\mathbf{grad}\,\varphi = 0$ is certainly in $L^2(\Omega)$. Now, suppose that $P_h \in H^1(\Omega)$ is a consistent family of discrete spaces built on meshes $\mathcal{T}_h$ such that $\mathbf{grad}\,P_h \subset V_h$. This is the case for $P_h = P_1(\mathcal{T}_h)$, the Lagrange elements, and $V_h = \mathcal{N}_0(\mathcal{T}_h)$, the Nédélec elements. This property is fundamental and can be encoded in a commutative diagram (see Sec. 3.2.3). In Eq. 3.13, it is therefore possible to choose $\mathbf{v} = \mathbf{grad}\,\varphi$ with φ in P_h . In this case $\int_\Omega \mathbf{curl}\,\mathbf{u}_h \cdot \mathbf{curl}\,\mathbf{grad}\,\varphi d\mathbf{x} = \lambda_h \int \mathbf{u}_h \cdot \mathbf{grad}\,\varphi d\mathbf{x} = 0\ , \forall \varphi \in P_h$ and for non-zero eigenvalues:

$$\int \mathbf{u}_h \cdot \mathbf{grad}\,\varphi d\mathbf{x} = 0\ , \forall \varphi \in P_h\ .$$

This expresses the fact that the divergence of $\mathbf{u}_h$ is weakly equal to zero. When the mesh is refined, $\mathbf{u}_h$ converges to $\mathbf{u}$ and $\operatorname{div}\mathbf{u}_h$ converges to zero. This means that the $\mathbf{u}_h$ associated with non-zero λ_h (within numerical round-off) converges to divergence-free modes.

3.2 The Geometric Structure of Electromagnetism and Its Discrete Analog

In this section, we restart from the very beginning in order to expose the deep geometrical structure of the classical electromagnetic theory. This is not only for aesthetic pleasure but also because it brings some fundamental information on the desirable properties of numerical approximations. Not surprisingly, the conclusions will be in perfect agreement with the previous section.

A remarkable fact is that most of the elements of the geometric structure of electromagnetism (whose natural framework is differential geometry), can be transposed *exactly* at the discrete level, *i.e.* on an irregular tessellation of the space made of tetrahedra, similar to those used to form a mesh for the finite element methods, and therefore involving a denumerable (and even finite for bounded domains) number of geometrical elements and degrees of freedom [Ton01b; Ton95]. The word "exactly" means here that most of the equations at the continuous level correspond to true equalities at the discrete level, and only the Hodge star operator has to be approximated! Experience over these this last few years shows that it is a rather good idea for discrete methods to respect this geometrical structure.

In order to establish a direct connection between the concepts, the continuous case and its discrete analogue are presented in parallel, the first paragraph concerning the continuous case followed by a corresponding paragraph on the analogous concept in the discrete case.

3.2.1 *Topology*

The basic geometric framework of the models is a topological structure, *i.e.* a set of points with no reference to either distance or angle but only relying on neighbourhood and continuity.

• The topological structure required for electromagnetism is the classical $\mathbb{R}^3$ *topology* given by open sets. In fact, a little bit more is necessary and the differentiable structure of $\mathbb{R}^3$ with suitable regularity is usually considered (differentiable manifold structure). Special subsets are useful such as volumes (3-dimensional submanifolds), surfaces (2-dimensional submanifolds), curves (1-dimensional submanifolds) and isolated points (0-dimensional submanifolds). Roughly speaking, a p-dimensional submanifold is a map from a domain Ω of $\mathbb{R}^p$ (parameter space, $0 \leq p \leq 3$) to $\mathbb{R}^3$.

It may be useful to consider *oriented* elements in order to determine the sign of physical quantities associated with geometrical elements. Each element has two possible orientations: 3-dimensional elements are oriented by choosing a righthanded or lefthanded frame, 2-dimensional elements by choosing a rotation direction, 1-dimensional elements by choosing a direction of running and 0-dimensional elements by choosing if they are incoming or outgoing [Ton95].

• Discrete models require an irregular tessellation of the geometrical domain (called a *mesh* or *meshing* in the context of the finite element method but rather a *complex* in the context of algebraic topology).

Consider a simplicial mesh K on a three-dimensional domain Ω of $\mathbb{R}^3$, that is, a set of tetrahedra which pairwise have in common either a full facet, or a full edge, or a node (vertex), or nothing, and whose set union is Ω. We also assume a numbering of the nodes n, so that edge e or facet f can alternatively be described by a list of node numbers. We call $\mathcal{T}$, $\mathcal{F}$, $\mathcal{E}$, $\mathcal{N}$ respectively, the sets of *tetrahedra* (or volumes), (triangular) *facets*, *edges* and *nodes* which constitute the mesh. The number of elements in $\mathcal{T}$, $\mathcal{F}$, $\mathcal{E}$, $\mathcal{N}$ are denoted by $\sharp\mathcal{T}$, $\sharp\mathcal{F}$, $\sharp\mathcal{E}$, $\sharp\mathcal{N}$ respectively.

Hence a particular node n is an element of $\mathcal{N}$ and it will be denoted by $\{i\}$ ($i \in \mathbb{N}$ is the number of the particular node n, an integer with a value between 1 and $\sharp\mathcal{N}$). An edge e is an element of $\mathcal{E}$ and it will be denoted by

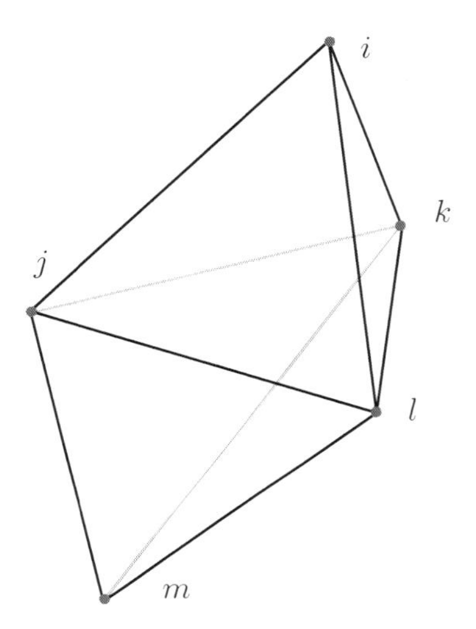

Fig. 3.2 Two tetrahedra $\{i,\ j,\ k,\ l\}$ and $\{m,\ l,\ k,\ j\}$ with a common facet $\{j,\ k,\ l\}$.

the ordered set $\{i,\ j\}$ of the node numbers of its two extremities. A facet f is an element of $\mathcal{F}$ and it will be denoted by $\{i,\ j,\ k\}$. A tetrahedron is an element of $\mathcal{T}$ and it will be denoted by $\{i,\ j,\ k,\ l\}$ (see Fig. 3.2). The generic term of *p-simplices* will be adopted from now on to refer to nodes ($p = 0$), edges ($p = 1$), facets ($p = 2$) and volumes ($p = 3$): more precisely, a p-simplex denotes the convex envelope of $p + 1$ nodes, *i.e.* if $\mathbf{x}_i$ denotes the Cartesian coordinates of the i-th node, the p-simplex is the set $\left\{\sum_{i=1}^{p+1} \lambda_i\, \mathbf{x}_i \ ,\ \lambda_i \geq 0\ ,\ \sum_{i=1}^{p+1} \lambda_i = 1\right\}$. The λ_i are called the *barycentric coordinates* of the simplex. Let us emphasize that if edge $\{i,\ j\}$ belongs to $\mathcal{E}$, or if facet $\{i,\ j,\ k\}$ belongs to $\mathcal{F}$, then $\{j,\ i\}$ does not belong to $\mathcal{E}$, and neither do $\{k,\ i,\ j\}$, $\{j,\ k,\ i\}$, etc., belong to $\mathcal{F}$: each p-simplex appears only once, with a definite orientation given by the ordering of the nodes.

Note that only connectivity is important here and that there is no reference to any sizes of the elements.

3.2.2 *Physical quantities*

Quite surprisingly the physical quantities may already be defined by duality with the topological elements. A *physical quantity* is a map from a given class of (oriented) geometrical objects to (real or complex) numbers.

• In the continuous theory, the language of *differential forms* is adopted here (see the mathematical pot-pourri). 0-forms such as the electric scalar potential $\mathcal{V}$ are maps from points to numbers, *i.e.* the usual definition of a function on $\mathbb{R}^3$. 1-forms such as the electric field $\boldsymbol{\mathcal{E}}$, the magnetic field $\boldsymbol{\mathcal{H}}$, and the vector potential $\boldsymbol{\mathcal{A}}$ are maps from curves to numbers. 2-forms such as the electric displacement $\boldsymbol{\mathcal{D}}$, the magnetic flux density $\boldsymbol{\mathcal{B}}$, and the current density $\boldsymbol{\mathcal{J}}$ are maps from surfaces to numbers. 3-forms such as the charge density ρ are maps from volumes to numbers. To emphasize the duality, we write $< \Omega, \alpha >= \int_\Omega \alpha$ for the number associated with the p-submanifold Ω (a p-dimensional geometric object) and with the p-form α, *i.e.* the integral of the form on the submanifold and the very nature of a p-form may indeed be considered as the integrand on a p-dimensional submanifold (see the mathematical pot-pourri). For instance, $< S, \boldsymbol{\mathcal{B}} >$ is an abstract notation for the surface integral of flux density $\boldsymbol{\mathcal{B}}$ across the surface S. This taxonomy is obviously more accurate than that of vector analysis that merges 0-forms and 3-forms into scalar fields and 1-forms and 2-forms into vector fields. If such definitions seem abstract at first sight, they are in fact deeply related to real life experimental processes [Ton95]. A magnetic flux density is never directly accessible but is only known via measurements on magnetic fluxes through finite-size loops in various positions.

• For the numerical discrete model, *discrete p-fields* correspond to maps from p-simplices to numbers. For instance the discrete electric field is represented by a set of values associated with edges, *i.e.* a column vector $\mathbf{e}$ of size $\sharp\mathcal{E}$, and the discrete magnetic flux density is represented by a set of values associated with facets, *i.e.* a column vector $\mathbf{b}$ of size $\sharp\mathcal{F}$. A practical mesh $\mathcal{T}_h$ is placed in a continuous physical space $\mathbb{R}^3$ such that the tetrahedra are embeddings of a reference tetrahedron. One therefore may define a *projection* Π_h^p from a continuous p-form to a discrete p-field by taking the integral on the corresponding p-simplices. For instance, the continuous electric field $\boldsymbol{\mathcal{E}}$ is projected onto the mesh by taking its line integrals along all of the edges to obtain $\mathbf{e} = \Pi_h^1 \boldsymbol{\mathcal{E}}$.

3.2.3 *Topological operators*

To manipulate quantities one has to define operators that map them to each other. The first class of operators relies only on the topological structure.

• In the continuous model, one defines the *exterior derivative* d, the linear map from $(p-1)$-forms α to p-forms $d\alpha$ that corresponds to the **grad**, **curl**, and div of vector analysis when acting on 0-, 1-, and 2-forms respectively.

The boundary operator ∂ is a map from p-dimensional sets of points to $(p-1)$-dimensional sets of points. If $\partial\Omega$ is regular enough, it is a map from $(p-1)$-forms α to numbers $< \partial\Omega, \alpha >$. The (general differential geometric) Stokes theorem [JÖ1] states that for any p-dimensional submanifold Ω one has $< \Omega, d\alpha > = < \partial\Omega, \alpha >$.

The fact that the boundary of a boundary is an empty set ($\partial\partial\Omega = \emptyset$ for any Ω) leads immediately to the fact that $dd\alpha = 0$ for any p-form α.

The forms with a null exterior derivative are called *closed* and those that can be expressed as the exterior derivative of a form are called *exact*. Exact forms are all closed. Whether the converse is true depends on the topology of Ω. For instance, this is always true if Ω is topologically simple, *i.e.* it can be continuously contracted to a point.

• The *discrete operators* on the mesh are defined as *incidence matrices*, which are representations of the boundary operator. Here the boundary operator associates with a p-simplex the $(p-1)$-simplices that constitute its boundary, *e.g.* it associates with a tetrahedron $\{i, j, k, l\}$ the four triangles that are its faces: $\{i, j, k\} + \{i, j, l\} + \{i, k, l\} + \{j, k, l\}$.

First, G is the node-edge incidence matrix, the rows correspond to edges and the columns to nodes, whose elements are +1, -1 or 0. The $\sharp\mathcal{E} \times \sharp\mathcal{N}$ rectangular array is constructed such that every row corresponds to an edge and every column to a node. The connectivity of the mesh (*i.e.* which nodes are boundaries of which edges) in this array is encoded in the following way: consider a row of the array corresponding to a given edge $\{i, j\}$, then for every element in the row, if the column number corresponds to the node which is the end point $\{j\}$ of the edge, set the element value to +1 or if it is the starting point $\{i\}$, set the value to -1 else if the edge is not connected to the node, set the value to 0.

This incidence matrix is also the discrete equivalent of the gradient and the distinction between the exterior derivative and the boundary operator vanishes at the discrete level in the sense that they are represented by the same matrices.

Consider $\mathbf{v}$, the column array (of size $\sharp\mathcal{N}$) of nodal values of a 0-form (*e.g.* the electric scalar potential) and $\mathbf{e}$, the column array (of size $\sharp\mathcal{E}$) of line integrals of a 1-form (*e.g.* the electric field) along the edges.

The matrix G is now a linear operator from nodal quantities to edge quantities, *e.g.* the fact that electromotive force along an edge is the difference of the potentials on the nodes at the extremities of the edges is expressed by the matrix product $\mathbf{e} = G\mathbf{v}$. The discrete differential operators have then an exact interpretation on integral quantities by means of the Stokes theorem.

The discrete curl C is the $\sharp\mathcal{F} \times \sharp\mathcal{E}$ edge-facet incidence matrix, the rows of which correspond to faces and the columns to edges, and whose elements are +1, -1 or 0 indicating that the given column corresponds to an edge that is on the boundary of the face corresponding to the given row with a similar or opposite orientation. It also encodes the fact that for a discrete 1-form $\mathbf{a}$ and a discrete 2-form $\mathbf{b} = C\mathbf{a}$, the signed sum of values of $\mathbf{a}$ associated with the edges of a facet is equal to the value of $\mathbf{b}$ associated with the facet.

The discrete divergence D is the $\sharp\mathcal{T} \times \sharp\mathcal{F}$ face-tetrahedron incidence matrix, the rows correspond to tetrahedra and the columns to facets, whose elements are +1, -1 or 0 indicating that the given column corresponds to a facet that is on the boundary of the tetrahedron corresponding to the given row with a similar or opposite orientation. It also encodes the fact that for a discrete 2-form $\mathbf{d}$ and a discrete 3-form $\mathbf{c} = D\mathbf{d}$, the signed sum of values of $\mathbf{d}$ associated with the facets of a tetrahedron is equal to the value of $\mathbf{c}$ associated with the tetrahedron.

The fact that the boundary of a boundary is an empty set is now encoded in the matrix products $CG = 0$ and $DC = 0$.

The topological operators now available already allow the fundamental equations to be expressed.

- Using the exterior derivative, the **Maxwell equations** are:

$$\begin{aligned} d\boldsymbol{\mathcal{H}} &= \boldsymbol{\mathcal{J}} + \partial_t \boldsymbol{\mathcal{D}} \\ d\boldsymbol{\mathcal{E}} &= -\partial_t \boldsymbol{\mathcal{B}} \\ d\boldsymbol{\mathcal{D}} &= \rho \\ d\boldsymbol{\mathcal{B}} &= 0 \end{aligned}$$

where ∂_t is the partial derivative with respect to time. A vector potential (1-form) $\boldsymbol{\mathcal{A}}$ and a scalar potential $\mathcal{V}$ can be introduced so that the homogeneous equations $d\boldsymbol{\mathcal{E}} + \partial_t \boldsymbol{\mathcal{B}} = 0$ and $d\boldsymbol{\mathcal{B}} = 0$ are satisfied by $\boldsymbol{\mathcal{E}} = -d\mathcal{V} - \partial_t \boldsymbol{\mathcal{A}}$ and $\boldsymbol{\mathcal{B}} = d\boldsymbol{\mathcal{A}}$. The structure of the equation can be represented on a *Tonti diagram* 3.3 (flattened on the time operator dimension, (See Fig. 2.1 p. 31) for a complete version) where the horizontal lines involve only topological operators. Vertical lines are related to material laws and metric properties, as will be explained in Sec. 3.2.4.

$$\begin{array}{ccccccc}
\mathcal{V} & \xrightarrow{d} & \mathcal{E},\mathcal{A} & \xrightarrow{d} & \mathcal{B} & \xrightarrow{d} & 0 \\
 & & \downarrow{\scriptstyle \sigma,\varepsilon} & & \uparrow{\scriptstyle \mu} & & \\
\rho & \xleftarrow{d} & \mathcal{J},\mathcal{D} & \xleftarrow{d} & \mathcal{H} & \xleftarrow{d} &
\end{array}$$

Fig. 3.3 *Tonti diagram* for the structure of electromagnetism.

• A fundamental issue is that an **exact** formulation of the discrete Maxwell equations for discrete fields (denoted here by the boldface lowercase letters corresponding to the continuous fields) is obtained by means of incidence matrices, namely:

$$\begin{aligned}
C\mathbf{h} &= \mathbf{j} + \partial_t \mathbf{d} \\
C\mathbf{e} &= -\partial_t \mathbf{b} \\
D\mathbf{d} &= \rho \\
D\mathbf{b} &= 0
\end{aligned}$$

Discrete 1-form $\mathbf{a}$ and 0-form $\mathbf{v}$ can be defined such that $\mathbf{e} = -\partial_t \mathbf{a} - G\mathbf{v}$ and $\mathbf{b} = C\mathbf{a}$ satisfy $C\mathbf{e} = -\partial_t \mathbf{b}$ and $D\mathbf{b} = 0$. This can be summarized in the form of a Tonti diagram 3.4 exactly similar to that for the continuous case. A fundamental property of the discrete topological operators is that they

$$\begin{array}{ccccccc}
\mathbf{v} & \xrightarrow{G} & \mathbf{e},\mathbf{a} & \xrightarrow{C} & \mathbf{b} & \xrightarrow{D} & 0 \\
 & & \downarrow{\scriptstyle \sigma,\varepsilon} & & \uparrow{\scriptstyle \mu} & & \\
\boldsymbol{\rho} & \xleftarrow{D} & \mathbf{j},\mathbf{d} & \xleftarrow{C} & \mathbf{h} & \xleftarrow{G} &
\end{array}$$

Fig. 3.4 *Tonti diagram* for the structure of discrete electromagnetism.

commute with the projection operators on the discrete fields as indicated by the commutativity of diagram[4] 3.5.

[4]A *diagram* is a set of points representing spaces connected by arrows, each arrow being associated with an operator such that it acts on an element in the space associated with the initial point of the arrow to give an element of the space associated with the endpoint. If this space is itself the initial point of one or several arrows and so on, the corresponding operators can be composed according to a sequence of arrows. A *diagram* is *commutative* if all the sequences of arrows in the diagram having the same initial and end points give the same result independently of the path followed. The reader should note that (unfortunately for clarity) "Tonti diagrams" are not exactly diagrams according to this definition (See Fig. 2.1 p. 31 for their explanation).

$$\begin{array}{ccccccc}
C^{\infty}(\mathbb{R}^3) & \xrightarrow{\textbf{grad}} & [C^{\infty}(\mathbb{R}^3)]^3 & \xrightarrow{\textbf{curl}} & [C^{\infty}(\mathbb{R}^3)]^3 & \xrightarrow{\text{div}} & C^{\infty}(\mathbb{R}^3) \\
\Pi_h^0 \downarrow & & \Pi_h^1 \downarrow & & \Pi_h^2 \downarrow & & \Pi_h^3 \downarrow \\
\mathbf{w}_h^0 & \xrightarrow{G} & \mathbf{w}_h^1 & \xrightarrow{C} & \mathbf{w}_h^2 & \xrightarrow{D} & \mathbf{w}_h^3
\end{array}$$

Fig. 3.5 Commutative diagram for the discrete topological operators and the projection operators on discrete fields.

3.2.4 *Metric*

In the previous sections, only topological properties have been involved. This section is devoted to the introduction of metric concepts.

• The *Hodge star operator* $*$ is a linear map from p-forms to $(3-p)$-forms. It is necessary to express the usual material *constitutive relations* that relate 1-forms such as electric and magnetic fields to 2-forms such as current density, electric displacement and magnetic flux density. For instance $\boldsymbol{\mathcal{D}} = \varepsilon * \boldsymbol{\mathcal{E}}$, $\boldsymbol{\mathcal{B}} = \mu * \boldsymbol{\mathcal{H}}$, and $\boldsymbol{\mathcal{J}} = \sigma * \boldsymbol{\mathcal{E}}$ where ε is the dielectric permittivity, μ is the magnetic permeability, and σ is the electric conductivity (Do not confuse the Hodge star $*$ with the convolution operator $\star$). In vector analysis, metric and topological aspects are interlaced and usually completely hidden in the use of Cartesian coordinates.
• At the discrete level, it is necessary to consider *dual meshes*. The recipe for constructing a dual mesh from the initial mesh, called therefore the *primal mesh*, is the following:

- Choose one point inside each of the primal tetrahedra. The exact place is not important as it is purely topological for the moment but the barycentre is a good choice. The new points are the nodes of the dual mesh.
- For each facet of the primal mesh, connect by a segment the dual nodes inside the two tetrahedra that have this facet on their boundary. These segments are the edges of the dual mesh.
- For each edge of the primal mesh, consider the facets such that this edge is on the boundary and takes the associated dual edges. They form a loop that delineates a polygonal surface that is a dual facet.
- For each node of the primal mesh, consider the edges such that this node is one of their vertices and takes the associated dual facets. They enclose a polyhedron that is a dual volume.

Note that the dual mesh is no longer simplicial. Nevertheless, all the discrete concepts of discrete p-forms and discrete operators are extended to this structure without any difficulty.

There is a one-to-one correspondence between primal nodes and dual volumes, between primal edges and dual facets, between primal facets and dual edges, and between primal tetrahedra and dual nodes.

The incidence matrices G^*, C^*, D^* of the dual mesh correspond therefore to the transposed versions of the primal ones $G^* = D^T, C^* = C^T, D^* = G^T$.

The key to understanding the discretisation of the Maxwell system is to associate discrete "magnetic" quantities $\mathbf{h}, \mathbf{j}, \mathbf{d}, \rho$ with a mesh and "electric" quantities $\mathbf{e}^*, \mathbf{a}^*, \mathbf{b}^*, \mathbf{v}^*$ with its dual mesh and consequently a new discrete form of the Maxwell equations can be written:

$$\begin{aligned} C\mathbf{h} &= \mathbf{j} + \partial_t \mathbf{d} \\ C^*\mathbf{e}^* &= -\partial_t \mathbf{b}^* \\ D\mathbf{d} &= \boldsymbol{\rho} \\ D^*\mathbf{b}^* &= 0 \end{aligned}$$

Building a discrete Hodge operator H is now a very natural operation since it relates discrete fields having the same number of components. For instance in the case of discrete 1-forms and 2-forms, it has the form of a square $\sharp\mathcal{E} \times \sharp\mathcal{E}$ matrix H_2 that transforms the fluxes through the faces of the primal mesh to the line integrals along the corresponding edges of the dual mesh. Of course, this is regular, with the inverse matrix giving the reverse transformation. In electromagnetism, as indicated above, the Hodge operator is involved through the material laws. It is therefore natural to build matrices that directly take into account both the metric and the material properties, which are naturally interlaced. For instance, conductivity, dielectric, and magnetic properties are respectively given by $H_2(\sigma^{-1})\mathbf{j} = \mathbf{e}^*$, $H_2(\varepsilon^{-1})\mathbf{d} = \mathbf{e}^*$, and $H_1(\mu)\mathbf{h} = \mathbf{b}^*$ and the inverse relations involving inverse matrices are denoted by $\mathbf{j} = H_1^*(\sigma)\mathbf{e}^*$, $\mathbf{d} = H_1^*(\varepsilon)\mathbf{e}^*$, $\mathbf{h} = H_2^*(\mu^{-1})\mathbf{b}^*$.

As an example, the *Finite Difference Time Domain (FDTD) method* based on the Yee algorithm [Yee66] can be introduced. Though this method has been designed, as its name indicates, with finite differences in mind, it perfectly fits into the present framework [Bos01]. Take a mesh K of a domain Ω that is no longer simplicial but rectangular, *i.e.* all the elements are hexahedra with rectangular faces. The centers of the hexahedra are taken as nodes of the dual mesh which is also made of rectangular hexahedra, the dual facets (edges) being orthogonal to the corresponding edges (facets). The omnipresent orthogonality makes the construction of the discrete Hodge operator particularly easy. Consider for instance the magnetic relation in the case of a homogeneous linear medium. The operation to be performed is to relate the magnetomotive force $I_e = \int_e \mathcal{H}$ given by the line integral of the magnetic field $\mathcal{H}$ along a primal edge e to the magnetic flux $\Phi_{f^*} = \int_{f^*} \mathcal{B}$ given by the surface integral of the magnetic flux density $\mathcal{B}$ across the dual facet f^* associated with e. The fundamental assumption is to consider that the mesh is fine enough to suppose that both the magnetic field is uniform on the edge e and that the magnetic flux density is uniform on the facet f^*. This approximation is the only source of error in the present method. The orthogonality of e and f^* assures that components with the same direction are involved, the tangential component to e for $\mathcal{H}$ and the normal component to f^* for $\mathcal{B}$.

Introducing the magnetic relation $\mathcal{B} = \mu * \mathcal{H}$, the following relation concerning the components of the Hodge operator can be stated:

$$I_e = \frac{1}{\mu} \frac{\text{area}(f^*)}{\text{length}(e)} \Phi_{f^*} = \left(H_2^*(\mu^{-1})\right)_{ef^*} \Phi_{f^*} \,.$$

An important property of the "discrete Hodge" matrices presented here is that they are diagonal which is more difficult to obtain on an unstructured mesh.

At this stage, the Maxwell-Ampère and Maxwell-Faraday equations reduce to a system of ordinary differential equations (the divergence conditions are implicitly satisfied):

$$\begin{aligned} C\mathbf{h} &= \mathbf{j} + H_1^*(\varepsilon)\partial_t \mathbf{e}^* \\ C^*\mathbf{e}^* &= -H_1(\mu)\partial_t \mathbf{h} \,. \end{aligned}$$

The FDTD method solves this system via an explicit method called the *leapfrog scheme*. The basic idea is again duality. Take the time t and divide it in regular intervals of duration Δt. Call t^i with an integer index the instants that delimit the intervals. Consider now a dual time subdivision where the instants $t^{i+1/2}$ correspond to the mid time of the previously defined intervals and are given half-integer indices. If the finite differences for the magnetic field are computed with the t^i so that $\partial_t \mathbf{h} \,|_{t=t^{i+1}} = \frac{\mathbf{h}^{i+1} - \mathbf{h}^i}{\Delta t}$ and those for the electric field with the $t^{i+1/2}$ so that $\partial_t \mathbf{e}^* \,|_{t=t^{i+1/2}} = \frac{\mathbf{e}^{*i+1/2} - \mathbf{e}^{*i-1/2}}{\Delta t}$, the following algorithm gives a numerical solution:

- Step 0: Suppose that the current density $\mathbf{j}$ is a known function of time and that the initial values $\mathbf{e}^{*-1/2}$ and $\mathbf{h}^0$ are given. Set $i = 0$.
- Step 1: Compute

$$\mathbf{e}^{*i+1/2} = \mathbf{e}^{*i-1/2} + H_1^*(\varepsilon)^{-1}\, C\, \mathbf{h}^i\, \Delta t$$

- Step 2: Compute

$$\mathbf{h}^{i+1} = \mathbf{h}^i - H_1(\mu)^{-1}\, C^*\, \mathbf{e}^{*i+1/2}\, \Delta t$$

- Step 3: Set $i = i + 1$ and resume at step 1.

This explicit scheme does not require any algebraic system resolution (because the Hodge matrices are diagonal) but is not unconditionally stable. The stability criterion is that the time step Δt must be small enough so that no electromagnetic wave can propagate within a single time step on a distance larger than the smallest side in each element and therefore the finer the mesh, the smaller the maximum time step allowed. This is the *CFL (Courant, Friedrichs and Lewy) condition* [MG80].

One of the strong limitations of this beautiful method is that it is not always possible to find a rectangular mesh that fits correctly the boundaries of the objects involved in the computation. Several attempts were made to palliate this drawback and several methods are now based on this direct construction of Hodge operators with the help of a dual mesh [CW01; Ton01a].

3.2.5 *Differential complexes: from de Rham to Whitney*

Much of the geometrical structure of electromagnetism can be encoded in the *de Rham complex*:

$$\overset{0}{\bigwedge}(\Omega) \xrightarrow{d} \overset{1}{\bigwedge}(\Omega) \xrightarrow{d} \overset{2}{\bigwedge}(\Omega) \xrightarrow{d} \overset{3}{\bigwedge}(\Omega)$$

A *complex* is a sequence of vector spaces (or other convenient structures such as groups for instance) together with operators between two consecutive spaces in the sequence such that the composition of the operators is identically equal to zero. In the case of the de Rham complex, the spaces are the spaces of differential p-forms $\bigwedge^p(\Omega)$ with, for instance, $C_0^\infty(\Omega)$ coefficients. The operators are different avatars of the exterior derivative acting on the various spaces and $dd = 0$ ensures that it is a complex. A complex is *exact* if the image of an operator in a space is equal to the kernel of the next one. In the case of the de Rham complex, this condition corresponds to the topological simplicity of Ω. The complex presented above corresponds of course to a three-dimensional manifold, the only case considered here for practical purposes, and longer sequences correspond to higher dimensions.

Such complexes can also be introduced on discrete structures. Given a simplicial mesh $\mathcal{T}_h$, the sequence of spaces of discrete p-fields together with the incidence matrices form a discrete complex:

$$\mathbf{w}_h^0 \xrightarrow{G} \mathbf{w}_h^1 \xrightarrow{C} \mathbf{w}_h^2 \xrightarrow{D} \mathbf{w}_h^3$$

In the 1950s, Whitney developed a theoretical tool in the form of discrete differential forms [Whi57]. Given a manifold Ω of dimension n and its decomposition in a simplicial mesh $\mathcal{T}_h$, a *Whitney p-form* (also called *Whitney elements* in the context of the finite element method) is a piecewise polynomial form defined as follows: consider a discrete p-field $\in \mathbf{w}_h^p$ and a p-simplex $s_p = \{n_0, \ldots, n_i, \ldots, n_p\}$ of $\mathcal{T}_h$ where the n_i are the vertices of s_p. Consider $\mathcal{T}_{s_p}$, the set of n-simplices s_n of $\mathcal{T}_h$ (*i.e.* of maximal dimension) such that the p-simplex s_p is a subset $s_p \subset s_n$ or, in other words, such that all the vertices of s_p are also vertices of s_n. The restriction to s_n of the Whitney p-form associated with s_p may be expressed as:

$$w_{s_p}|_{s_n} = p! \sum_{i=0}^{p} (-1)^i \lambda_{n_i} d\lambda_{n_0} \wedge \ \ldots \ \wedge d\lambda_{n_{i-1}} \wedge \widehat{d\lambda_{n_i}} \wedge d\lambda_{n_{i+1}} \wedge \ \ldots \ \wedge d\lambda_{n_p}$$

where $\widehat{d\lambda_{n_i}}$ means that this term is omitted and where λ_{n_i} is the barycentric coordinate on s_n associated with vertex n_i. The Whitney p-form associated with s_p is:

$$w_{s_p} = \sum_{s_n \in \mathcal{T}_{s_p}} w_{s_p}|_{s_n}$$

for a point in $\mathcal{T}_{s_p}$ and 0 otherwise. The fundamental property of w_{s_p} is that

$$\int_{s_p} w_{s_p} = 1 \text{ and } \int_{s'_p \neq s_p} w_{s_p} = 0$$

where s'_p is any p-simplex of $\mathcal{T}_h$ distinct from s_p. The space W_h^p of Whitney p-forms on $\mathcal{T}_h$ is the vector space spanned by all the w_{s_p}. The set of coefficients associated with an element of W_h^p are the integrals on the p-simplices and they form a discrete p-field $\mathbf{w}_p$. Building a discrete p-form from a discrete p-field defines the interpolation operator I_h^p while the inverse operation defines the projection operator σ_h^p. Of course, no information is lost in these operations, which are isomorphisms of discrete spaces. The fundamental property of Whitney spaces is that the image of W_h^p by the exterior derivative is in W_h^{p+1}:

$$dW_h^p \subset W_h^{p+1}$$

which can be expressed by saying that the W_h^p form a discrete complex called the *Whitney complex*. Moreover, taking the exterior derivative amounts to applying the incidence matrices on the corresponding discrete field, which is expressed in the commutativity of the lower part of diagram 3.7. Whitney forms have been defined piecewise and nothing guarantees *a priori* their continuity across the boundaries of the n-simplices. Indeed the forms are discontinuous but their tangential traces (see Sec. 3.3.3 for a more detailed discussion of this notion) coincide on the $(n-1)$-simplices that are the boundaries of the n-simplices.

The Whitney forms are now displayed in the particular case $n = 3$ of a Whitney complex on a tetrahedral mesh:

$$W_h^0 \xrightarrow{d} W_h^1 \xrightarrow{d} W_h^2 \xrightarrow{d} W_h^3 .$$

From now on, vector field proxies and **grad**, **curl**, and div operators will be used. W_h^0 is simply the space spanned by the barycentric coordinates λ_i associated with the vertices and it is a continuous scalar field. The degrees of freedom are associated with the nodal values, *i.e.* the values of the scalar field on the vertices.

W_h^1 is spanned by the vector fields

$$\mathbf{w}_e = \lambda_i \,\mathbf{grad}\, \lambda_j - \lambda_j \,\mathbf{grad}\, \lambda_i$$

where $\mathbf{w}_e$ is the expression on the tetrahedron with barycentric coordinates λ_i of the vector field associated with edge e from node i to node j. The vector spaces of W_h^1 are discontinuous across the triangular faces of the mesh but the tangential component is continuous. The degrees of freedom are associated with the line integrals of the vector fields along the edges.

W_h^2 is spanned by the vector fields

$$\mathbf{w}_f = 2(\lambda_i \,\mathbf{grad}\, \lambda_j \times \mathbf{grad}\, \lambda_k + \lambda_k \,\mathbf{grad}\, \lambda_i \times \mathbf{grad}\, \lambda_j + \lambda_j \,\mathbf{grad}\, \lambda_k \times \mathbf{grad}\, \lambda_i)$$

where $\mathbf{w}_f$ is the expression on the tetrahedron with barycentric coordinates λ_i of the vector field associated with triangular facet f having vertices i, j, and k. The vector spaces of W_h^2 are discontinuous across the triangular faces of the mesh but the normal component is continuous. The degrees of freedom are associated with the flux of the vector field across the facets.

W_h^3 is spanned by the scalar fields

$$w_v = 6(\lambda_i(\mathbf{grad}\,\lambda_j \times \mathbf{grad}\,\lambda_k)\cdot\mathbf{grad}\,\lambda_l + \lambda_j(\mathbf{grad}\,\lambda_k \times \mathbf{grad}\,\lambda_l)\cdot\mathbf{grad}\,\lambda_i + \lambda_k(\mathbf{grad}\,\lambda_l \times \mathbf{grad}\,\lambda_i)\cdot\lambda_j + \lambda_l(\mathbf{grad}\,\lambda_i \times \mathbf{grad}\,\lambda_j)\cdot\mathbf{grad}\,\lambda_k)$$

where w_f is the expression on the tetrahedron with vertices i, j, k, and l. This expression is in fact constant on the tetrahedron and W_h^3 is the discontinuous scalar field that is piecewise constant on the tetrahedra. The degrees of freedom are associated with the volume integrals of the scalar field on the tetrahedra.

The elements of W_h^p are not smooth but their continuity properties show that $W_h^0 \subset H^1(\Omega)$, $W_h^1 \subset H(\mathbf{curl},\Omega)$, $W_h^2 \subset H(\mathrm{div},\Omega)$, $W_h^3 \subset L^2(\Omega)$. Therefore, they can be used to discretise an analogue of the de Rham complex involving a sequence of Hilbert spaces:

$$H^1(\Omega) \xrightarrow{\mathbf{grad}} H(\mathbf{curl},\Omega) \xrightarrow{\mathbf{curl}} H(\mathrm{div},\Omega) \xrightarrow{\mathrm{div}} L^2(\Omega)\,.$$

In fact, a discrete form of this complex can be obtained by using the mixed elements introduced in Sec. 3.1:

$$P_1(\mathcal{T}_h) \xrightarrow{\mathbf{grad}} \mathcal{N}_0(\mathcal{T}_h) \xrightarrow{\mathbf{curl}} \mathcal{RT}_0(\mathcal{T}_h) \xrightarrow{\mathrm{div}} P_0(\mathcal{T}_h)$$

but this is actually exactly the same thing: we have the nodal elements $P_1(\mathcal{T}_h) = W_h^0$, the edge elements $\mathcal{N}_0(\mathcal{T}_h) = W_h^1$, the facet elements $\mathcal{RT}_0(\mathcal{T}_h) = W_h^2$, and the volume elements $P_0(\mathcal{T}_h) = W_h^3$. The connection between low order edge and face elements and Whitney forms was first realised by Bossavit (probably after a discussion with Kotiuga [Bos88a]). It provides a new point of view on the mixed formulations and a new framework that yields guidelines for building elements whereas the LBB criterion is just a censor [Bos89]. For a more detailed and deeper discussion of the role of the Whitney complex in computational electrodynamics, we refer the reader to Ref. [Bos88b; Bos93; Bos98; Bos03; GK04].

Given a p-form on Ω, it is now possible to compute its projection on a Whitney p-form defined on the meshing $\mathcal{T}_h$ of Ω by performing the projection Π_h^p on a discrete p-field defined on $\mathcal{T}_h$ and interpolating with I_h^p. By a slight abuse of notation, the same name Π_h^p is used for the projection operators on the Whitney forms and on the discrete fields. The commutative diagram 3.6 expresses the commutativity of the differential and projection operators necessary to make the Π_h^p Fortin operators (see the annex 7.6). Note that this diagram is very similar to diagram 3.5, except that Hilbert spaces are used here instead.

$$\begin{array}{ccccccc} H^1 & \xrightarrow{\mathbf{grad}} & H(\mathbf{curl}) & \xrightarrow{\mathbf{curl}} & H(\mathrm{div}) & \xrightarrow{\mathrm{div}} & L^2 \\ \Pi_h^0 \downarrow & & \Pi_h^1 \downarrow & & \Pi_h^2 \downarrow & & \Pi_h^3 \downarrow \\ W_h^0 & \xrightarrow{\mathbf{grad}} & W_h^1 & \xrightarrow{\mathbf{curl}} & W_h^2 & \xrightarrow{\mathrm{div}} & W_h^3 \end{array}$$

Fig. 3.6 Commutative diagram for the projectors on Whitney forms.

What the differential operators on the Whitney forms do is really similar to the action of the incidence matrices on discrete fields and we have the following identities:

$$G = \sigma_h^1 \circ \mathbf{grad} \circ I_h^0 \quad C = \sigma_h^2 \circ \mathbf{curl} \circ I_h^1 \quad D = \sigma_h^3 \circ \mathrm{div} \circ I_h^2 \,.$$

Everything can be summed up in the large commutative diagram 3.7.

$$\begin{array}{ccccccc} H^1 & \xrightarrow{\mathbf{grad}} & H(\mathbf{curl}) & \xrightarrow{\mathbf{curl}} & H(\mathrm{div}) & \xrightarrow{\mathrm{div}} & L^2 \\ \downarrow \Pi_h^0 & & \downarrow \Pi_h^1 & & \downarrow \Pi_h^2 & & \downarrow \Pi_h^3 \\ W_h^0 & \xrightarrow{\mathbf{grad}} & W_h^1 & \xrightarrow{\mathbf{curl}} & W_h^2 & \xrightarrow{\mathrm{div}} & W_h^3 \\ I_h^0 \uparrow \ \downarrow \sigma_h^0 & & I_h^1 \uparrow \ \downarrow \sigma_h^1 & & I_h^2 \uparrow \ \downarrow \sigma_h^2 & & I_h^3 \uparrow \ \downarrow \sigma_h^3 \\ \mathbf{w}_h^0 & \xrightarrow{G} & \mathbf{w}_h^1 & \xrightarrow{C} & \mathbf{w}_h^2 & \xrightarrow{D} & \mathbf{w}_h^3 \end{array}$$

Fig. 3.7 Complete commutative diagram showing the relations between the infinite dimensional spaces, the interpolated discrete spaces of Whitney forms and the discrete field (sets of numerical coefficients).

The main interest of interpolation is that it makes available the definition of the Hodge operator $*$ on p-forms to define the Hodge operator H_p on discrete fields.

All the tools are now available to construct finite element approximations to electromagnetic problems.

3.3 Some Practical Questions

Everything seems to be pretty abstract up to now but, *in fine*, the finite element process reduces to computing integrals on the elements, which is generally a very simple numerical task, and to filling sparse matrices with these numbers to get the system to be solved. In this section, several tools are presented that are useful in the practical implementation of such a procedure.

3.3.1 *Building the matrices (discrete Hodge operator and material properties)*

Some explicit cases are now given that allow direct computation of the matrices involved in the finite element process and in particular those playing the role of discrete Hodge star operators.

As a first example, consider the weak discrete form of

$$\mathbf{curl}\,\mu^{-1}\,\mathbf{curl}\,\mathbf{E} - \omega^2\varepsilon\mathbf{E} = 0, \tag{3.15}$$

i.e. find $\mathbf{E} \in W_h^1$ such that:

$$\int_\Omega \mu^{-1}\,\mathbf{curl}\,\mathbf{E} \cdot \mathbf{curl}\,\mathbf{E}' d\mathbf{x} - \omega^2 \int_\Omega \varepsilon\mathbf{E} \cdot \mathbf{E}' d\mathbf{x} = 0 \;\; , \forall \mathbf{E}' \in W_h^1. \tag{3.16}$$

Consider first the second term. The linear combination corresponding to $\mathbf{E}$ can be written as $\mathbf{W_1}^T\mathbf{e}$ where $\mathbf{e}$ is a column vector containing numerical coefficients which are the line integrals along the edges and $\mathbf{W_1}$ is another column vector containing the Whitney 1-forms $\mathbf{w}_1$ associated with the corresponding edges. To be precise, given T_e the set of all the tetrahedra sharing the edge e of interest going from vertex i to vertex j, the entry in $\mathbf{W_1}$ corresponding to this edge is $\mathbf{w}_e = \sum_{\mathrm{T} \in \mathrm{T}_e} \lambda_i^{\mathrm{T}}\,\mathbf{grad}\,\lambda_j^{\mathrm{T}} - \lambda_j^{\mathrm{T}}\,\mathbf{grad}\,\lambda_i^{\mathrm{T}}$ where λ_i^{T} is the barycentric coordinate of node i in tetrahedron T. To build the linear system, $\mathbf{E}'$ must scan the basis functions of the discrete space, *i.e.* all the Whitney 1-forms associated with an edge that are in $\mathbf{W_1}$. Therefore, the set of linear relations associated with this second term can be written as $\int_\Omega \varepsilon\mathbf{W_1} \cdot \mathbf{W_1}^T\mathbf{e}\;d\mathbf{x}$. The product $\mathbf{W_1} \cdot \mathbf{W_1}^T$ is in fact a "dyadic dot" product between a column vector and a row vector that produces a square matrix. The dot indicates that the entries in this matrix are computed by performing a dot product between the vector fields that are the elements of $\mathbf{W_1}$. Since $\mathbf{e}$ is a constant vector, it can be extracted from the integral so that $\int_\Omega \varepsilon \mathbf{E} \cdot \mathbf{E}' d\mathbf{x}$ leads to the following contribution in the matrix system:

$$\mathbf{M_1}(\varepsilon)\mathbf{e} \text{ with } \mathbf{M_1}(\varepsilon) = \int_\Omega \varepsilon \mathbf{W_1} \cdot \mathbf{W_1}^T d\mathbf{x}.$$

The matrix $\mathbf{M_1}(\varepsilon)$ is obviously symmetric and the fact that only two edges that share at least a common tetrahedron correspond to shape functions with non-disjoint supports guarantees sparsity.

As for the first term of the weak form, it undergoes a similar treatment but the curl operator must be involved in some way. $\mathbf{curl}\,\mathbf{E}$ is in fact a 2-form that has to be represented by face elements. The commutation properties of diagram 3.7 show that $\mathbf{curl}(\mathbf{W_1}^T\mathbf{e}) = \mathbf{W_2}^T C\mathbf{e}$ using the discrete curl matrix C and a column vector $\mathbf{W_2}$ containing the face shape functions. Similarly the set of curls of the shape functions $\mathbf{curl}\,\mathbf{E}'$ used as weights in the residual can be expressed as $C^T\mathbf{W_2}$. The set of linear relations associated with the first term can be written as $\int_\Omega \mu^{-1} C^T \mathbf{W_2} \cdot \mathbf{W_2}^T C\mathbf{e} d\mathbf{x}$ As $\mathbf{e}$ and C are constant, they can be extracted from the integral so that $\int_\Omega \mu^{-1}\, \mathbf{curl}\,\mathbf{E} \cdot \mathbf{curl}\,\mathbf{E}' d\mathbf{x}$ leads to the following contribution in the matrix system:

$$C^T\mathbf{M_2}(\mu^{-1})C\mathbf{e} \text{ with } \mathbf{M_2}(\mu^{-1}) = \int_\Omega \mu^{-1}\mathbf{W_2} \cdot \mathbf{W_2}^T d\mathbf{x}.$$

Gathering the various terms, Eq. 3.16 can be written as:

$$C^T\mathbf{M_2}(\mu^{-1})C\mathbf{e} - \omega^2\mathbf{M_1}(\varepsilon)\mathbf{e} = 0$$

and this expression has a structure that is quite similar to that of Eq. 3.15.

The $\mathbf{M}_p$ matrices may be called Galerkin-Hodge matrices since they are the Hodge matrices defined in Sec. 3.2.4 built on a tetrahedral mesh via the Galerkin method. It must be emphasized that the computation of these matrices is the only place in the finite element method where the interpolation with Whitney forms is necessary. The metric of the Euclidean space is involved in the dot product that appears in the definition of the matrices. In the language of differential forms, matrix $\mathbf{M}_p(\alpha)$ associated with p-forms and the material characteristic α is given by:

$$\mathbf{M_p}(\alpha) = \int_\Omega \alpha \mathbf{W_p} \wedge *\mathbf{W_p}^T$$

where $\mathbf{W_p}$ is the column vector of Whitney p-forms and the product is a "dyadic exterior" product. This expression shows explicitly the use of the Hodge operator.

The unfortunate point is that the Galerkin-Hodge matrices are not diagonal so that, contrary to the FDTD method, some linear system solution is involved in the numerical modelling. Some attempts have been done to condense the matrix to make it diagonal [BK99] in a way that preserves convergence but these have failed so-far to give a practical algorithm.

3.3.2 *Reference element*

Only the case of tetrahedra will be considered here but it is very easy to transpose it to triangles in the two-dimensional case. For other elements, see for instance [Geu02].

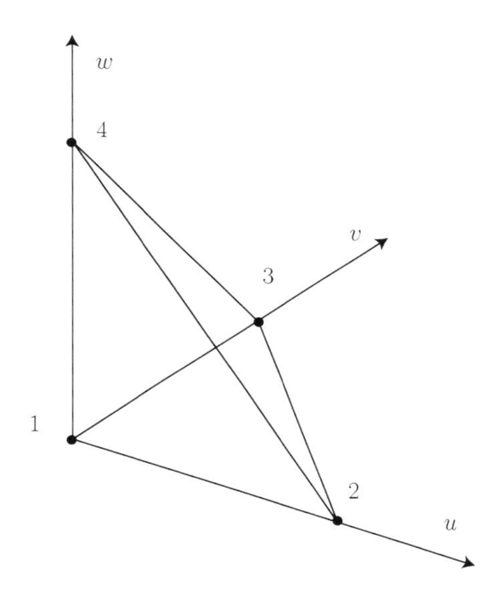

Fig. 3.8 Reference tetrahedron

In order to perform the computations on the various tetrahedra of a simplicial mesh in a systematic way, a *reference element* is introduced:

$$\mathrm{T}_{ref} = \{(u, v, w) \in \mathbb{E}^3 : 0 < u < 1 - v - w, v > 0, w > 0\}\,.$$

Note that u plays no special role in the definition since similar inequalities are automatically satisfied by v and w. The barycentric coordinates are given by:

$$\begin{cases} \lambda_1(u,v,w) = 1 - u - v - w \\ \lambda_2(u,v,w) = u \\ \lambda_3(u,v,w) = v \\ \lambda_4(u,v,w) = w \, . \end{cases} \tag{3.17}$$

A geometrical tetrahedron K in the mesh is given by the affine transformation: $\mathbf{r}(u,v,w) = \sum_{i=1}^{4} \mathbf{r}^i \lambda_i(u,v,w)$ where $\{\mathbf{r}^1, \mathbf{r}^2, \mathbf{r}^3, \mathbf{r}^4\}$ are the position vectors of the vertices of K. Note that the correspondence between T_{ref} and K is not only vertex to vertex but also edge to edge, and face to face. It must be stressed that it is not the physical element that is mapped onto the reference element but on the contrary it is the reference element that is mapped onto the physical one.

In some cases, supplementary control points can be introduced and the transformation involves higher degrees polynomials, *e.g.* quadratic functions that allow the representation of the geometry with curvilinear elements. If the same functions are used for the geometric transformation of the reference element and the basis functions used to approximate a scalar field, the element is called *isoparametric*.

For the extension of Whitney elements to higher degrees, we refer the reader to Ref. [Geu02; Hip01] and chapter 5 of [IPS96] by Peterson and Wilton.

3.3.3 *Change of coordinates*

The problem is now to translate an expression written in terms of Cartesian coordinates $\{x, y, z\}$ to an expression written in terms of coordinates $\{u, v, w\}$ on the reference element. The affine transformations between these coordinates do not preserve norms and orthogonality so that $\{u, v, w\}$ are not orthonormal coordinates. Here it is therefore much simpler to use the formalism of differential geometry.

Considering a map from the coordinate system $\{u, v, w\}$ to the coordinate system $\{x, y, z\}$ given by the functions $x(u, v, w)$, $y(u, v, w)$, and $z(u, v, w)$, the transformation of the differentials is given by:

$$\begin{cases} dx = \frac{\partial x}{\partial u} du + \frac{\partial x}{\partial v} dv + \frac{\partial x}{\partial w} dw \\ dy = \frac{\partial y}{\partial u} du + \frac{\partial y}{\partial v} dv + \frac{\partial y}{\partial w} dw \\ dz = \frac{\partial z}{\partial u} du + \frac{\partial z}{\partial v} dv + \frac{\partial z}{\partial w} dw \, . \end{cases} \tag{3.18}$$

Given a p-form expressed in the $\{x, y, z\}$ coordinate system, it suffices to replace the dx, dy, dz by the corresponding 1-forms involving du, dv, dw in the basis exterior monomials to obtain the expression of the form in the new coordinate system. Note that the form travels naturally counter to the current with respect to the map and this is why this transportation of the forms from x, y, z to u, v, w is called a *pull-back*. This operation can be defined not only between two coordinate systems but also between two different manifolds even if they do not have the same dimensions.

Consider two manifolds (or more simply, two open domains of $\mathbb{R}^m$ and $\mathbb{R}^n$ respectively) M and N and a (regular) map φ from M to N such that (for simplicity) $\varphi(M) = N$. The example above shows that it is very easy to express the differentials of the coordinates on N in terms of the differentials of the coordinates on M and therefore to find the image on M of a 1-form on N given by the dual map φ^*, from N to M, also called, as indicated above, the pull-back. In fact, any covariant object such as a p-form or a metric can be pulled back by translating the differentials on N into differentials on M. Defined in this way, the operation commutes of course with the exterior and tensor products but also with the exterior derivative and the Hodge star (defined with the pulled-back metric).

As for the contravariant objects such as the vector fields, they travel forward just like the geometrical domains. Given a vector $\mathbf{v}$ at a point $\mathbf{p}$ on M, it suffices to choose a curve γ going through the point and such that the vector is the tangent vector to the curve at this point, to take the image of the curve $\varphi(\gamma)$ on N and the vector tangent to this curve at the point $\varphi(\mathbf{p})$ as the image of $\mathbf{v}$. Defined in this way, the map for vectors from M to N, denoted by $\varphi_*(\mathbf{v})$ or $d\varphi(\mathbf{v})$, is called the *differential of* φ or the *push-forward* and it can be extended to any contravariant object.

Another fundamental property of the pull-back is its commutativity with integration in the sense that, for any form α that is integrable on a subset $\varphi(\Omega)$ of N, which is the image of a subset Ω of M, one has:

$$\int_{\varphi(\Omega)} \alpha = \int_{\Omega} \varphi^*(\alpha)\,.$$

In the case of the map $\varphi(\mathrm{T}_{ref}) = \mathrm{T}$ from the reference element to an element of the mesh, the contribution to the finite element system can be written $\int_{\mathrm{T}_{ref}} \varphi^*(\alpha) = \int_{\mathrm{T}} \alpha$ where α is a 3-form (or a 2-form in the two-dimensional case) that can be pulled back to obtain an integral on the reference element. Further details on this process are now given in the rest of this section.

All the information is obviously contained in the Jacobian matrix $\mathbf{J}$ (or maybe we should say matrix field since it depends on the point in space considered) in terms of which 3.18 can be written:

$$\mathbf{J}(u,v,w) = \frac{\partial(x,y,z)}{\partial(u,v,w)} = \begin{pmatrix} \frac{\partial x}{\partial u} & \frac{\partial x}{\partial v} & \frac{\partial x}{\partial w} \\ \frac{\partial y}{\partial u} & \frac{\partial y}{\partial v} & \frac{\partial y}{\partial w} \\ \frac{\partial z}{\partial u} & \frac{\partial z}{\partial v} & \frac{\partial z}{\partial w} \end{pmatrix} \text{ and } \begin{pmatrix} dx \\ dy \\ dz \end{pmatrix} = \mathbf{J} \begin{pmatrix} du \\ dv \\ dw \end{pmatrix} . \quad (3.19)$$

Using matrix notation, the detailed computation of the relation between the coefficients of a 1-form $\mathbf{E}$ in $\{x,y,z\}$ and $\{u,v,w\}$ coordinates is performed as follows:

$$\begin{aligned} \mathbf{E} &= E_x dx + E_y dy + E_z dz = (E_x\, E_y\, E_z) \begin{pmatrix} dx \\ dy \\ dz \end{pmatrix} = (E_x\, E_y\, E_z)\, \mathbf{J} \begin{pmatrix} du \\ dv \\ dw \end{pmatrix} \\ &= E_u du + E_v dv + E_w dw = (E_u\, E_v\, E_w) \begin{pmatrix} du \\ dv \\ dw \end{pmatrix} \end{aligned} \quad (3.20)$$

and the following relation is obtained:

$$(E_x\, E_y\, E_z)\, \mathbf{J} = (E_u\, E_v\, E_w)\,. \quad (3.21)$$

Now the contributions to the finite element system may have the following form:

$$\int_{\mathrm{T}} \mathbf{E} \wedge *\mathbf{E}'$$

where $\mathbf{E}$ and $\mathbf{E}'$ are 1-forms (that can be obtained as gradients of a scalar field, although it really does not matter here). The only remaining question is how to deal with the Hodge operator. A direct attack would be to pull back the metric and use the explicit expression of the operator but it is faster here to take advantage of the simple form of the scalar product in Cartesian coordinates that reduces to the dot product. Again using matrix notation:

$$\begin{aligned} \mathbf{E} \wedge *\mathbf{E}' &= (E_x\, E_y\, E_z)(E'_x\, E'_y\, E'_z)^T dx \wedge dy \wedge dz \\ &= (E_u\, E_v\, E_w)\mathbf{J}^{-1}[(E'_u\, E'_v\, E'_w)\mathbf{J}^{-1}]^T dx \wedge dy \wedge dz \\ &= (E_u\, E_v\, E_w)\mathbf{J}^{-1}\mathbf{J}^{-T}(E'_u\, E'_v\, E'_w)^T \det(\mathbf{J}) du \wedge dv \wedge dw\,. \end{aligned} \quad (3.22)$$

The fact that the transformation of 3-forms only involves the Jacobian, *i.e.* the determinant of the Jacobian matrix, has been used here. Hence, the only difference from the case of Cartesian coordinates is that one of the (column) vectors has to be multiplied (on the left) by a symmetric matrix $\mathbf{T}^{-1}$ before performing the dot product and $\mathbf{T}$ is given by:

$$\mathbf{T} = \frac{\mathbf{J}^T\mathbf{J}}{\det(\mathbf{J})}\,.$$

It is now interesting to look at how a particular 2-form basis monomial transforms, for instance, $dx \wedge dy = [\frac{\partial x}{\partial u}du + \frac{\partial x}{\partial v}dv + \frac{\partial x}{\partial w}dw] \wedge [\frac{\partial y}{\partial u}du + \frac{\partial y}{\partial v}dv + \frac{\partial y}{\partial w}dw] = (\frac{\partial x}{\partial u}\frac{\partial y}{\partial v} - \frac{\partial x}{\partial v}\frac{\partial y}{\partial u})du \wedge dv + (\frac{\partial x}{\partial v}\frac{\partial y}{\partial w} - \frac{\partial x}{\partial w}\frac{\partial y}{\partial v})dv \wedge dw + (\frac{\partial x}{\partial w}\frac{\partial y}{\partial u} - \frac{\partial x}{\partial u}\frac{\partial y}{\partial w})dw \wedge du$. The cofactors of $\mathbf{J}$ are now involved in the transformation. These are the elements of $\mathbf{J}^{-1}\det(\mathbf{J})$.

Given a 2-form:

$$\begin{aligned}\mathbf{D} &= D_x dy \wedge dz + D_y dz \wedge dx + D_z dx \wedge dy \\ &= D_u dv \wedge dw + D_v dw \wedge du + D_w du \wedge dv\end{aligned} \tag{3.23}$$

the following relation is obtained:

$$(D_x\, D_y\, D_z)\,\mathbf{J}^{-1}\det(\mathbf{J}) = (D_u\, D_v\, D_w) \tag{3.24}$$

and the matrix involved in the scalar product is here $\mathbf{T}$.

Everything can now be summarised in the following recipe that takes into account implicitly the Hodge star: consider a 3-form α to be integrated on an element T in order to get $\int_{\mathrm{T}} \alpha$, then:

- If the integrand involves only scalars (0-forms or 3-forms and it does not matter if the 3-forms are expressed as the divergence of a vector field), only $\det(\mathbf{J})$ has to be introduced as a factor.
- If the integrand is the scalar product of two 1-forms (and it does not matter if one or both 1-forms are expressed as the gradient of a scalar field), multiply on the left one of the column vectors of coefficients by the matrix $\mathbf{T}^{-1}$.
- If the integrand is the scalar product of two 2-forms (and it does not matter if one or both 2-forms are expressed as the curl of a vector field), multiply on the left one of the column vectors of coefficients by the matrix $\mathbf{T}$.
- If the integrand is the exterior product of a 1-form and a 2-form (and it does not matter if they are respectively a gradient and a curl), no factor has to be introduced because the expression is actually metric-free.
- The expression obtained for $\varphi^*(\alpha)$ depending on variables u, v and w (coordinates x, y and z have been replaced by the functions $x(u, v, w)$,

$y(u, v, w)$ and $z(u, v, w)$ respectively) is integrated on the reference element to get the desired contribution to the finite element system.

There are several important remarks that must be made here.

Everything has been performed in this section with the transformation between elements of the mesh and the reference element in mind but the treatment was completely general and it works for any non-singular transformation and allows, for instance, the use of any coordinate system.

It can also be interesting to consider a compound transformation, *i.e.* the transformation of a transformation. Consider three systems of coordinates u_i, X_i and x_i (possibly on different manifolds) and the maps $\varphi_{Xu} : u_i \rightarrow X_i$ given by functions $X_i(u_j)$ and $\varphi_{xX} : X_i \rightarrow x_i$ given by functions $x_i(X_j)$. The composition map $\varphi_{xX} \circ \varphi_{Xu} = \varphi_{xu} : u_i \rightarrow x_i$ is given by the functions: $x_i(X_j(u_k))$. If $\mathbf{J}_{xX}$ and $\mathbf{J}_{Xu}$ are the Jacobian matrices of the maps φ_{xX} and φ_{Xu} respectively, the Jacobian matrix $\mathbf{J}_{xu}$ of the composition map φ_{xu} is simply the product of the Jacobian matrices:

$$\mathbf{J}_{xu} = \mathbf{J}_{xX}\mathbf{J}_{Xu} \,.$$

This rule naturally applies for an arbitrary number of maps.

It is also worth noting that the matrix $\mathbf{J}^T\mathbf{J}$ is nothing but the metric tensor whose coefficients are expressed in the local coordinates.

Another interesting interpretation is that the matrix $\mathbf{T}$ and its inverse can be viewed as tensorial characteristics of equivalent materials. In electromagnetism, changing a material can be viewed as changing metric properties and conversely a change of coordinates can be taken into account by introducing a fictitious equivalent material. For a general transformation, the equivalent material is inhomogeneous and anisotropic. It will be seen below that it may be interesting in some cases to introduce non-orthogonal coordinate systems to facilitate the solution of particular problems.

A natural question then arises: div-conforming and curl-conforming elements preserve the continuity of the normal and tangential components respectively of the interpolated vector fields but orthogonality is metric dependent and one may wonder if conformity depends on the metric and the coordinate system. As the conforming elements are Whitney differential forms, the answer is certainly negative but it deserves a little bit more explanation. Consider a surface Σ and a system of coordinates $\{u, v, w\}$ such that, at least locally, the surface is defined by $\Sigma : \{u, v \in \mathbb{R}^2, w = C\}$ where C is a constant. The differential of w vanishes on the surface: $dw|_\Sigma = 0$ and this allows one to write explicitly the restriction (a particular case of pull back) of a form to the surface called the *tangential trace* of the form: given a 1-form $\mathbf{E} = E_u du + E_v dv + E_w dw$, the restriction $\mathbf{E}|_\Sigma = E_u du + E_v dv$ corresponds obviously to the tangential component as defined in vector analysis. As for a 2-form $\mathbf{D} = D_u dv \wedge dw + D_v dw \wedge du + D_w du \wedge dv$, the restriction is $\mathbf{D}|_\Sigma = D_w du \wedge dv$ and corresponds amazingly to the so-called normal component of vector analysis but this notion is really independent of the metric and of the system of coordinates.

Note that the Stokes theorem should be more rigorously written as $\int_\Omega d\alpha = \int_{\partial\Omega} \alpha|_{\partial\Omega}$ where $\alpha|_{\partial\Omega}$ is the restriction (pull-back) of α on $\partial\Omega$ but this is most often omitted as a slight abuse of notation.

3.3.4 *Nédélec edge elements vs. Whitney 1-forms*

It is now a good exercise to prove that Nédélec lowest order elements and Whitney 1-forms are exactly the same thing by showing explicitly their shape functions and demonstrating that they are curl-conforming.

The first step is an explicit computation of the vector shape function. For the Whitney 1-form, consider for instance the edge $1 - 2$ going from vertex 1 to vertex 2 with the numbering of Fig. 3.8. The associated shape function is $\lambda_1 \,\mathbf{grad}\, \lambda_2 - \lambda_2 \,\mathbf{grad}\, \lambda_1$ and the associated degree of freedom is e_1. Similar information for all the edges is given in Table 3.1. For the Nédélec element, $\mathbf{a} + \mathbf{b} \times \mathbf{r}$ is simply introduced (where $\mathbf{a}$ and $\mathbf{b}$ are three-dimensional constant vectors). The explicit computation of all the components gives for the vector shape function:

$$\begin{aligned}
&[e_1 + (-e_1 + e_2 - e_4)v + (-e_1 + e_3 - e_6)w]\mathbf{e}^u + \\
&[e_2 + (+e_1 - e_2 + e_4)u + (-e_2 + e_3 - e_5)w]\mathbf{e}^v + \\
&[e_3 + (+e_1 - e_3 - e_6)u + (+e_2 - e_3 + e_5)v]\mathbf{e}^w \\
&= [a_1 + b_2 w - b_3 v]\mathbf{e}^u + [a_2 + b_3 u - b_1 w]\mathbf{e}^v + [a_3 + b_1 v - b_2 u]\mathbf{e}^w
\end{aligned} \tag{3.25}$$

and the resultant identification between the parameters is given in the third column of Table 3.1.

The second step is to interpret the degrees of freedom. Table 3.2 gives for each edge (in the first column) the equations satisfied by the coordinates on this edge (in the second column), a vector (whose norm is equal to the length of the edge) indicating its direction (in the third column) and the value of the shape function on the edge (in the fourth column) obtained by introducing the second column in 3.25. The projection on the vector of the third column shows that the tangential component of the shape function is constant on the edges and, using the third column of Table 3.1, that the value of its line integral along the edge is exactly e_i on each corresponding edge.

The third step is to prove that the tangential part of the shape function on a face depends only on the degrees of freedom associated with the boundary of this face. Table 3.3 gives for each face (in the first column) the equation satisfied by the coordinates on this face (in the second column), and two linearly independent vectors spanning the tangential directions (in the third column). The value of the shape function on the face is obtained by introducing the second column in 3.25. The projections on the vectors of the third column are given in the fourth column. For the first three lines, it is obvious that the three parameters involved can be computed from the degrees of freedom of the edges that form the boundary of the face. As for the fourth line, the replacement $u = 1-v-w$ is performed in the shape function for the projection on the first vector and $v = 1-u-w$ for the second one. It suffices to remark that $(-a_1+a_2+b_3)-(b_1+b_2+b_3)w = e_4-(e_4+e_5+e_6)w$ and $(-a_2+a_3+b_1)-(b_1+b_2+b_3)u = e_5-(e_4+e_5+e_6)u$ to see that the tangential components depend only on the degrees of freedom of the boundary of the face. Therefore, two tetrahedra sharing a face share also the corresponding edges and degrees of freedom so that the tangential component of the vector shape function on this face is the same for both tetrahedra and is therefore continuous across the face.

It has just been proved that Nédélec edge elements and Whitney 1-forms are the same, that choosing the line integral of the vector field to be interpolated along the edges as degrees of freedom provides curl-conformity.

Table 3.1 Whitney 1-forms and the associated degrees of freedom (DoF).

Edge	Whitney 1-form	DoF
$1-2$	$(1-v-w, u, u)$	$e_1 = a_1$
$1-3$	$(v, 1-u-w, v)$	$e_2 = a_2$
$1-4$	$(w, w, 1-u-v)$	$e_3 = a_3$
$2-3$	$(-v, u, 0)$	$e_4 = b_3 - a_1 + a_2$
$3-4$	$(0, -w, v)$	$e_5 = b_1 - a_2 + a_3$
$4-2$	$(w, 0, -u)$	$e_6 = b_2 - a_3 + a_1$

We leave to the reader the pleasure of a similar proof for the identity of Raviart-Thomas lowest order elements with Whitney 2-forms and their div-conforming property as well as the much simpler similar proofs for triangles

Table 3.2 Tangential components on the edges.

Edge	Equation	Tan. vec.	$\mathbf{a}+\mathbf{b}\times\mathbf{r}\vert_{edge}$
$1-2$	$v=0, w=0$	$(1,0,0)$	$(a_1, a_2+b_3u, a_3-b_2u)$
$1-3$	$u=0, w=0$	$(0,1,0)$	$(a_1-b_3v, a_2, a_3+b_1v)$
$1-4$	$u=0, v=0$	$(0,0,1)$	$(a_1+b_2w, a_2-b_1w, a_3)$
$2-3$	$w=0, v=1-u$	$(-1,1,0)$	$(a_1-b_3(1-u), a_2+b_3u,$ $a_3+b_1(1-u)-b_2u)$
$3-4$	$u=0, w=1-v$	$(0,-1,1)$	$(a_1+b_2(1-v)-b_3v,$ $a_2-b_1(1-v), a_3+b_1v)$
$4-2$	$v=0, u=1-w$	$(1,0,-1)$	$(a_1+b_2w, a_2+b_3(1-w)-b_1w,$ $a_3-b_2(1-w))$

Table 3.3 Tangential components on the faces.

Face	Equation	Tan. vec.	$\mathbf{a}+\mathbf{b}\times\mathbf{r}\vert_{face}$
$1-2-3$	$w=0$	$(1,0,0)$	a_1-b_3u
		$(0,1,0)$	a_2+b_3u
$1-2-4$	$v=0$	$(1,0,0)$	a_1+b_2w
		$(0,0,1)$	a_3-b_2u
$1-3-4$	$u=0$	$(0,1,0)$	a_2-b_1w
		$(0,0,1)$	a_3+b_1v
$2-3-4$	$1-u-$	$(-1,1,0)$	$(-a_1+a_2+b_3)-(b_1+b_2+b_3)w$
	$v-w=0$	$(0,-1,1)$	$(-a_2+a_3+b_1)-(b_1+b_2+b_3)u$

in the two-dimensional case.

3.3.5 *Infinite domains*

A common difficulty in electromagnetic modelling is to deal with unbounded domains where the fields extend to infinity. To mesh an infinite domain with finite elements is of course impractical since it leads to an infinite number of degrees of freedom. In order to discuss the possible solutions it is important to distinguish two cases:

- In the first case, there is no wave propagation in the infinite region. This is the case for static or quasi-static problems because there are no waves at all. This is also the case in open dielectric waveguides because the wave vector is parallel to the open region. The problem is therefore to take into account the vanishing fields at infinity.
- In the second case, waves are propagating inside the open region and

the problem is to find some numerical equivalent to the Sommerfeld or Silver-Müller radiation conditions.

A preliminary idea is to truncate the domain and to apply more or less arbitrary conditions on the artificial boundary such as homogeneous Dirichlet conditions. For the first case, this may not be such a bad idea. The main difficulty is to determine where to put the truncation: the further, the more accurate but also the more expensive. Various kinds of infinite elements have been proposed and the clearest approach is to use a mapping from a bounded domain to the unbounded domain as explained in Sec. 3.3.5.1. Unfortunately, none of these approaches work in the wave propagation case. In the case of the truncation at finite distance, the boundary appears as a mirror and the reflected waves provoke perturbations in the whole domain. In a transient problem, it may be assumed in some cases that the perturbation takes some time to invade the whole domain and that therefore the approximation is correct in some region for some initial time interval but in the harmonic case, it is hopeless. In this case, it seemed natural to look for alternative boundary conditions that minimize the reflection on the boundary, *i.e.* absorbing boundary conditions (ABC) [Taf95]. Unfortunately, such methods are often difficult to design and degrade the sparsity of the system matrix. Nowadays, the most prominent method is the perfectly matched layer (PML) method introduced by Bérenger in 1994 [Ber94]. The initial approach was based on the splitting of the field components into two subcomponents. A more recent version introduces the method as a continuation of the traditional coordinate space on a complex coordinate space [Tei03; TC99; LLS01].

Another technique that will not be discussed here is to couple the finite element method with integral boundary element methods [Poi94].

3.3.5.1 *Transformation method for infinite domains*

To deal with the open problem, a judicious choice of coordinate transformation allows the finite element modelling of the infinite exterior domain [HMH+99]. For simplicity, an example of the method is presented here in the two-dimensional case.

Consider two disks $D(O,A)$ and $D(O,B)$ with centers $O = (0,0)$ and radii A and $B > A$ such that the material characteristics of the problem are homogeneous outside $D(O,A)$. An annulus $An(O,A,B) = D(O,B) \setminus \overline{D(O,A)}$ is defined. Let (x,y) be a point in $\mathbb{R}^2 \setminus \overline{D(O,A)}$ (the infinite outer domain) and (X,Y) be a point in $An(O,A,B)$ and consider the transformation given by:

$$\begin{cases} x = f_1(X, Y) = X[A(B - A)]/[R(B - R)] \\ y = f_2(X, Y) = Y[A(B - A)]/[R(B - R)] \end{cases}$$

where R denotes the Euclidean norm $\sqrt{X^2 + Y^2}$. This transformation may be viewed as a mapping of the finite annulus $An(O, A, B)$ with a non orthogonal coordinate system (X, Y) to the infinite domain with a Cartesian coordinate system (x, y). As explained in Sec. 3.3.3, this can be taken into account by special material properties. Imposing homogeneous Dirichlet conditions on the boundary of $D(O, B)$ leads to a finite element model that solves the unbounded domain with a finite number of elements.

A variant is to map the outer boundary of $An(O, A, B)$ not to infinity but to an arbitrarily large but finite distance C. It suffices to use the transformation:

$$\begin{cases} x = f_1(X, Y) = X[CA(B - A)]/[(C - 1)R(B - R + A(B - A))] \\ y = f_2(X, Y) = Y[CA(B - A)]/[(C - 1)R(B - R + A(B - A))] \,. \end{cases}$$

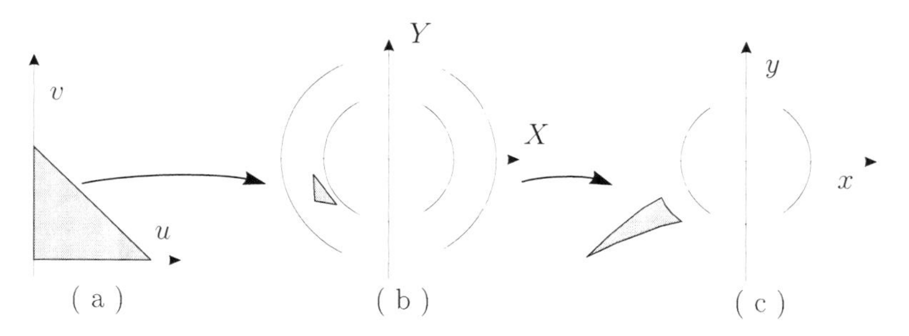

Fig. 3.9 (a) Reference element. (b) Mesh of the annulus equivalent to the unbounded domain with transformed coordinates. (c) Physical unbounded domain in Cartesian coordinates.

Using such coordinate transformations, the finite element discretisation appears as a chained map from the reference space to the transformed space and from the transformed space to the physical space (see Fig. 3.9).

3.3.5.2 *Perfectly Matched Layer (PML)*

As explained in the introduction to this section, the PML amounts to a change of complex coordinates, *i.e.* introducing particular inhomogeneous anisotropic lossy materials. Inside the PML, the spatial coordinates are mapped to a complex domain:

$$X(x) = x_0 + \int_{x_0}^{x} s_x(x')dx'$$

where $s_x(x) = a_x(x) + i\sigma_x(x)/\omega$ and similar expressions hold for y and z [Tei03; TC99]. The complex part transforms waves represented by exponentials with imaginary arguments to evanescent exponentials with a negative real part of the argument. As the permittivity ε and the permeability μ are transformed by the same factors, there is no change in the impedance of the material, hence the matched character of the layer. From the change of coordinates, the equivalent tensors $\underline{\underline{\varepsilon}}$ and $\underline{\underline{\mu}}$ are easily obtained.

3.4 Propagation Modes Problems in Dielectric Waveguides

Everything is now ready to present the study of dielectric waveguides in general with microstructured fibres as a particular case of interest. A waveguide of cross section Ω with translational invariance along the z-axis is considered. In the case of optical fibres, the electromagnetic field in the whole $\mathbb{R}^2$ has to be considered for Ω even if the fibre itself can be described by a finite cross section in the sense that it is given by a dielectric permittivity profile $\varepsilon = \varepsilon_0\varepsilon_r$ (ε_0 being the permittivity of vacuum and ε_r, the relative permittivity of the medium) that is different from ε_0 only on a compact subset of $\mathbb{R}^2$. There is a vast literature about the finite element treatment of such waveguide problems, see Ref. [IPS96; Jin02; GNZ+01; DW94; GNZL02; LSC91; Geu02; Poi94; VD03; VDN03].

In optical fibres, all the materials can usually be considered as non-magnetic, *i.e.* the magnetic permeability $\mu = \mu_r\mu_0$ is equal to that of the vacuum μ_0 but it is not difficult to keep a general permeability in our formal development.

For practical purposes, it is sometimes necessary to consider perfect electric conductors (PEC) that constrain the tangential electric field to be zero on their surfaces Γ_e ($\mathbf{n} \times \mathbf{E} = 0$) and perfect magnetic walls (PMW) that impose the tangential magnetic field to be zero on their surfaces Γ_m ($\mathbf{n} \times \mathbf{H} = 0$). These materials may for instance be useful for the truncation of the problem and also to take symmetries into account. All the materials are supposed to satisfy the classical conditions that provide the coercivity of the operators.

The problem is to find electromagnetic fields that meet the criteria to be guided modes as detailed in Sec. 2.3 p. 1.

Keep in mind that the fields **E** and **H** considered here are complex valued fields depending on two variables (coordinates x and y) but they still have three components (along the three axes).

3.4.1 *Weak and discrete electric field formulation*

An electric formulation is chosen with homogeneous Dirichlet boundary conditions (perfect metallic conditions in this instance). In the case where these conditions are used for the truncation of the domain, this reduces the number of unknowns. Eliminating the magnetic field and denoting by k_0 the wave number $\omega\sqrt{\mu_0\varepsilon_0}$, one is led to the following system:

$$\begin{cases} \mathbf{curl}_\beta(\mu_r^{-1}\,\mathbf{curl}_\beta\,\mathbf{E}) = k_0^2\varepsilon_r\mathbf{E} \\ \mathrm{div}_\beta(\varepsilon_r\mathbf{E}) = 0\,. \end{cases}$$

The solution of the above problem is then given by the weak formulation associated with the following bilinear form in the space $H(\mathbf{curl}_\beta,\Omega) = \{\mathbf{v} \in [L^2(\Omega)]^3, \mathbf{curl}_\beta\,\mathbf{v} \in [L^2(\Omega)]^3\}$:

$$\begin{aligned} \mathcal{R}(\beta;\mathbf{E},\mathbf{E}') = &\int_\Omega \mu_r^{-1}\,\mathbf{curl}_\beta\,\mathbf{E}\cdot\overline{\mathbf{curl}_\beta\,\mathbf{E}'}\,dxdy - k_0^2\int_\Omega \mathbf{E}\cdot\overline{\mathbf{E}'}\,\varepsilon_r dxdy \\ &+ \int_{\Gamma_m} \mu_r^{-1}\,\mathbf{curl}_\beta\,\mathbf{E}\times\overline{\mathbf{E}'}\cdot\mathbf{n}\,dl = 0 \quad, \forall\mathbf{E}' \in H(\mathbf{curl}_\beta,\Omega) \end{aligned}$$

The boundary conditions are $\mathbf{n}\times\mathbf{E} = 0$ on Γ_e (essential boundary conditions) and $\mathbf{n}\times\mu_r^{-1}\,\mathbf{curl}_\beta\,\mathbf{E} = 0$ on Γ_m (natural boundary conditions) so that the boundary integral is equal to zero hence this boundary term is discarded in the further developments.

To make the finite element formulation implicit, the electric field is separated into a transverse component $\mathbf{E}_t$ in a cross-section of the guide (in the xy-plane) and a longitudinal field E_z along the z-axis of invariance so that the total field can be written $\mathbf{E} = \mathbf{E}_t + E_z\mathbf{e}^z$ with $\mathbf{E}_t\cdot\mathbf{e}^z = 0$. The section of the guide is meshed with triangles (see Fig. 3.10) and Whitney finite elements are used, *i.e.* edge elements for the transverse field and nodal elements for the longitudinal field.

$$\begin{cases} \mathbf{E}_t = \sum_{j=1}^{\sharp\mathcal{E}} \mathrm{e}_j^t \, \mathbf{w}_e^j(x,y) \\ \\ E_z = \sum_{j=1}^{\sharp\mathcal{N}} \mathrm{e}_j^z \, w_n^j(x,y) \end{cases}$$

where e_j^t denotes the line integral of the transverse component $\mathbf{E}_t$ on the edges, and e_j^z denotes the line integral of the longitudinal component E_z along one unit of length of the axis of the guide (which is equivalent to the nodal value). Besides, $\mathbf{w}_e^j$ and w_n^j are respectively the basis functions of Whitney 1-forms (associated with edges) and Whitney 0-forms (associated with nodes) on triangles. On the boundary of the PEC, those degrees of freedom corresponding to the tangential electric field are simply taken equal to zero. On the boundary of the PMW, the degrees of freedom are just kept as unknowns.

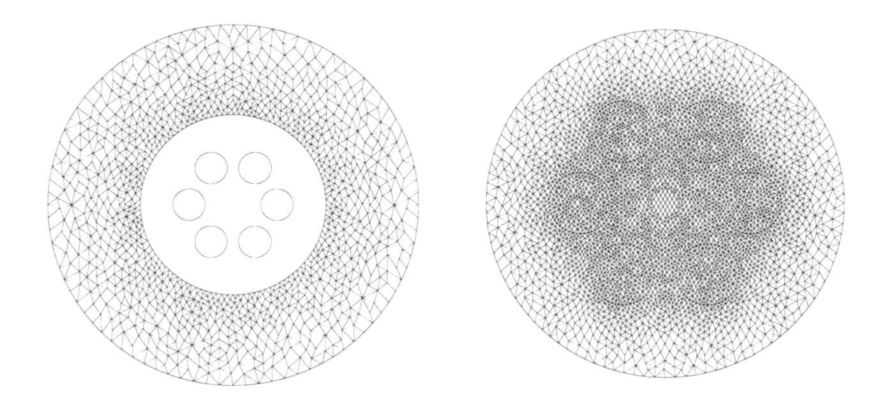

Fig. 3.10 Meshing of the cross section of a six-hole fibre. 10.736 triangles are used to provide an accurate evaluation of the fields associated with the mode. The righthand picture is a magnification of the interior part of the structure in order to make the smaller triangles visible.

The following transverse operators are defined for a scalar function $\varphi(x,y)$ and a transverse field $\mathbf{v} = v_x(x,y)\mathbf{e}^x + v_y(x,y)\mathbf{e}^y$:

$$\mathbf{grad}_t \, \varphi = \frac{\partial \varphi}{\partial x}\mathbf{e}^x + \frac{\partial \varphi}{\partial y}\mathbf{e}^y$$

$$\mathbf{curl}_t\,\mathbf{v} = (\frac{\partial v_x}{\partial y} - \frac{\partial v_y}{\partial x})\mathbf{e}^z$$
$$\mathrm{div}_t\,\mathbf{v} = (\frac{\partial v_x}{\partial x} + \frac{\partial v_y}{\partial y})$$

and are used to separate $\mathbf{curl}_\beta$ and div_β into their transverse and longitudinal components:

$$\begin{aligned}\mathrm{div}_\beta(\mathbf{v} + \varphi\mathbf{e}^z) &= \mathrm{div}_t\,\mathbf{v} + i\,\beta\varphi\\ \mathbf{curl}_\beta\,(\mathbf{v} + \varphi\mathbf{e}^z) &= \mathbf{curl}_t\,\mathbf{v} + (\mathbf{grad}_t\,\varphi - i\,\beta\mathbf{v}) \times \mathbf{e}^z\,.\end{aligned}$$

At this stage, the materials can still be supposed to be anisotropic so that ε and μ are tensorial but they are requested not to mix longitudinal and transverse components. The most general form with this property (called z-anisotropy in this book) is:

$$\underline{\underline{\varepsilon}} = \begin{pmatrix} \varepsilon_{xx} & \varepsilon_{xy} & 0 \\ \varepsilon_{yx} & \varepsilon_{yy} & 0 \\ 0 & 0 & \varepsilon_{zz} \end{pmatrix} = \underline{\underline{\varepsilon_{tt}}}\,\underline{\underline{\varepsilon_{zz}}} = \begin{pmatrix} \varepsilon_{xx} & \varepsilon_{xy} & 0 \\ \varepsilon_{yx} & \varepsilon_{yy} & 0 \\ 0 & 0 & 1 \end{pmatrix} \begin{pmatrix} 1 & 0 & 0 \\ 0 & 1 & 0 \\ 0 & 0 & \varepsilon_{zz} \end{pmatrix} \tag{3.26}$$

so that $\underline{\underline{\varepsilon}}\boldsymbol{E} = \underline{\underline{\varepsilon_{tt}}}\mathbf{E}_t + \underline{\underline{\varepsilon_{zz}}}E_z\mathbf{e}^z$ and a similar form is supposed for the permeability $\underline{\underline{\mu}}$. Note that this property is conserved for the inverse tensors $\underline{\underline{\varepsilon}}^{-1}$ and $\underline{\underline{\mu}}^{-1}$. The discrete weak formulation is given by:

$$\begin{aligned}\mathcal{R}(\beta;\mathbf{E},\mathbf{E}') = \int_\Omega \Big(&(\underline{\underline{\mu_{r\,zz}}}^{-1}\,\mathbf{curl}_t\,\mathbf{E}_t) \cdot \mathbf{curl}_t\,\overline{\mathbf{E}'_t} + \\ &(\underline{\underline{\mu_{r\,tt}}}^{-1}(\mathbf{grad}_t\,E_z \times \mathbf{e}^z)) \cdot (\mathbf{grad}_t\,\overline{E'_z} \times \mathbf{e}^z) - \\ &i\beta(\underline{\underline{\mu_{r\,tt}}}^{-1}(\mathbf{grad}_t\,E_z \times \mathbf{e}^z)) \cdot (\overline{\mathbf{E}'_t} \times \mathbf{e}^z) - \\ &i\beta(\underline{\underline{\mu_{r\,tt}}}^{-1}(\mathbf{E}_t \times \mathbf{e}^z)) \cdot (\mathbf{grad}_t\,\overline{E'_z} \times \mathbf{e}^z) + \\ &\beta^2(\underline{\underline{\mu_{r\,tt}}}^{-1}(\mathbf{E}_t \times \mathbf{e}^z)) \cdot (\overline{\mathbf{E}'_t} \times \mathbf{e}^z)\Big)\,dxdy \\ -k_0^2 \int_\Omega \Big(&(\underline{\underline{\varepsilon_{r\,tt}}}\mathbf{E}_t) \cdot \overline{\mathbf{E}'_t} + (\underline{\underline{\varepsilon_{r\,zz}}}E_z\mathbf{e}^z) \cdot \overline{E'_z}\mathbf{e}^z\Big)\,dxdy = 0\,.\end{aligned}$$

The more general case where transverse and longitudinal components are mixed up by general tensors is detailed in Sec. 3.6. For microstructured optical fibres, the case of scalar ε_r and μ_r $(= 1)$ is the most common so this case will be considered in the sequel in order to avoid heavy notation. In this case, the weak formulation can be written:

$$\begin{aligned}\mathcal{R}(\beta;\mathbf{E},\mathbf{E}') = \quad &\int_\Omega \mu_r^{-1}\Big(\mathbf{curl}_t\,\mathbf{E}_t \cdot \mathbf{curl}_t\,\overline{\mathbf{E}'_t} + \mathbf{grad}_t\,E_z \cdot \mathbf{grad}_t\,\overline{E'_z} \\ &-i\beta\mathbf{E}_t \cdot \mathbf{grad}_t\,\overline{E'_z} + i\beta\,\mathbf{grad}_t\,E_z \cdot \overline{\mathbf{E}'_t} + \beta^2\,\mathbf{E}_t \cdot \overline{\mathbf{E}'_t}\Big)\,dxdy\end{aligned}$$

$$-k_0^2 \int_\Omega \varepsilon_r \left(\mathbf{E}_t \cdot \overline{\mathbf{E}_t'} + E_z \, \overline{E_z'} \right) dxdy = 0$$

where the approximations and weight factors are taken in the appropriate discrete spaces: edge element space for transverse parts and node element space for longitudinal parts.

Even if the uniqueness of the problem in the infinite dimensional vector space $H(\mathbf{curl}_\beta, \Omega)$ requires the addition of a penalty term $s \int_\Omega \mathrm{div}_\beta \, \mathbf{E} \, \overline{\mathrm{div}_\beta \, \mathbf{E}'} \, \varepsilon_r \, dxdy$ in the weak formulation (See lemma 2.1 p. 67) which acts in fact as a constraint that imposes the nullity of $\mathrm{div}_\beta(\varepsilon_r \mathbf{E})$, this term is not necessary in the present finite element formulation since the solution of the weak discrete formulation satisfies a weak divergence condition. To show this, one has to take

$$\mathbf{E}' = \mathbf{grad}_\beta \, \phi = \mathbf{grad}_t \, \phi + i\beta\phi\mathbf{e}^z$$

in the residual, where ϕ is a Whitney 0-form (this is allowed since $\mathbf{grad}_t \, \phi$ is a Whitney 1-form). One has $\mathbf{curl}_\beta \, \mathbf{grad}_\beta \, \phi = 0$ and it follows that for all $k_0 \neq 0$:

$$\int_\Omega \mathbf{E} \cdot \overline{\mathbf{grad}_\beta \, \phi} \, \varepsilon_r \, dxdy = 0 \ ,$$

which is a weak form of $\mathrm{div}_\beta(\varepsilon_r \mathbf{E}) = 0$.

It is now straightforward to build the matrix system and the value of β is fixed to obtained a generalized eigenvalue problem in k_0^2. The following matrices are defined:

$$\mathrm{A}_{ij}^{tt} = \int_\Omega \mu_r^{-1} \, \mathbf{curl}_t \, \mathbf{w}_e^i \cdot \mathbf{curl}_t \, \mathbf{w}_e^j \, dxdy$$

$$\mathrm{A}_{ij}^{zz} = \int_\Omega \mu_r^{-1} \, \mathbf{grad}_t \, w_n^i \cdot \mathbf{grad}_t \, w_n^j \, dxdy$$

$$\mathrm{A}_{ij}^{tz} = \mathrm{A}_{ji}^{zt} = \int_\Omega \mu_r^{-1} \mathbf{w}_e^i \cdot \mathbf{grad}_t \, w_n^j \, dxdy$$

$$\mathrm{B}_{ij}^{tt} = \int_\Omega \mu_r^{-1} \mathbf{w}_e^i \cdot \mathbf{w}_e^j \, dxdy$$

$$\mathrm{C}_{ij}^{tt} = \int_\Omega \varepsilon_r \mathbf{w}_e^i \cdot \mathbf{w}_e^j \, dxdy \qquad \mathrm{C}_{ij}^{zz} = \int_\Omega \varepsilon_r w_n^i \cdot w_n^j \, dxdy$$

The corresponding Hermitian matrix system can be written in the form

$$\begin{pmatrix} \mathrm{A}^{tt} + \beta^2 \mathrm{B}^{tt} & i\beta \mathrm{A}^{tz} \\ -i\beta \mathrm{A}^{zt} & \mathrm{A}^{zz} \end{pmatrix} \begin{pmatrix} \mathbf{e}_t \\ \mathbf{e}_z \end{pmatrix} = k_0^2 \begin{pmatrix} \mathrm{C}^{tt} & 0 \\ 0 & \mathrm{C}^{zz} \end{pmatrix} \begin{pmatrix} \mathbf{e}_t \\ \mathbf{e}_z \end{pmatrix} \tag{3.27}$$

where $\mathbf{e}_t$ and $\mathbf{e}_z$ are the column arrays of transversal and longitudinal degrees of freedom respectively.

Such problems involving large sparse Hermitian matrices can be solved for instance using a Lànczos algorithm[TI95], which permits the computation of their largest eigenvalues. Since only the smallest eigenvalues are relevant here, the inverse of the matrices must be used in the iterations. Of course, the inverse is never computed explicitly but the matrix-vector products are replaced by system solutions via a GMRES method [TI95]. It is therefore obvious that the numerical efficiency of the process relies strongly on Krylov subspace techniques and the Arnoldi iteration algorithm [TI95]. In practice, there exist excellent free numerical libraries such as Sparskit [Saa99] for GMRES and Arpack [LSY98] for the eigenvalue computation with variants of the Lànczos method. Solving this eigenvalue problem for a sampling of β gives the shape of the dispersion relations $\beta(k_0)$. All the numerical examples presented in this chapter have been performed using the GetDP finite element freeware [DGHL98].

3.4.2 *Numerical comparisons*

In this section, a finite element computation is compared with that of a multipole method on the same structure (a six hole photonic crystal fibre (PCF) introduced by White [NGZ+02]).

As the geometric domain is unbounded, one of the pitfalls in this problem is the presence of a continuous spectrum due to the lack of compacity of the resolvent of the associated operator (see Sec. 2.5). It has been proved (see Sec. 2.5) that the eigenvalues k_0^2 in the discrete spectrum for a given β satisfy the following criterion:

$$k_0^2 \in \left] \frac{\beta^2}{\sup\limits_{\mathbf{r}\in\mathbb{R}^2} \varepsilon_r(\mathbf{r})}, \beta^2 \right] . \tag{3.28}$$

By means of 3.28, it is known that every eigenvalue greater than β^2 belongs to the essential spectrum, which gives us a numerical criterion to eliminate those modes that do not correspond to guided waves. But the situation is a little more subtle here since leaky modes are considered with a complex propagation constant such that the real part may not respect 3.28 and the imaginary part is very small. Therefore this does not strictly respect the definition of a guided mode given initially. Taking Dirichlet boundary conditions at a finite distance from the fibre via an artificial metallic jacket with PEC walls [GNZL02] allows us to consider an operator with a compact resolvent (thus artificially eliminating the continuous spectrum) but this would add the modes of a non-physical metallic guide. In practice, these modes correspond to frequencies different enough to not perturb the leaky mode that is projected on the real axis.

On the other hand, with the multipole method, an infinite silica matrix is considered and therefore one deals with complex frequencies [KWR+02]. In both cases, the PCF consists of six air-holes ($n = 1$) of radius $2.5\mu m$ distributed in a hexagonal arrangement with pitch $\Lambda = 6.75\mu m$ in a matrix of silica ($n = 1.45$). The finite element method (Fig. 3.11, left) is used (with the mesh presented in Fig. 3.10) to retrieve the real part of the leaky mode's frequency calculated with the multipole method for the six hole honeycomb structure (Fig. 3.11, right). More precisely, the wavelength $\lambda = 1.45\mu m$ is set up in the multipole method and one obtains $n_{eff} = \beta/k_0 = \beta\lambda/2\pi = 1.445395345 + i\ 3.15\ 10^{-8}$ for the degenerate fundamental mode of the above C_{6v} MOF structure [Isa75a]. Then, for a given propagating constant $\beta_{FEM} = 1.445395k_0 = 6.263232\mu m^{-1}$, a leaky mode is caught (Fig. 3.11, left) whose eigenvalue $\omega_{FEM} = 2\pi c/\lambda_{FEM}$ corresponds to a wavelength $\lambda_{FEM} = 1.449996\mu m$ that matches the initial value. Fig. 3.12 illustrates the impact of a perturbation on the geometry of the PCF on this leaky mode.

The main drawback of the finite element method used here is the impossibility of computing the imaginary part of the propagation constant for the leaky modes. In order to perform such a computation, two modifications would be necessary. The first one would be to consider an unbounded domain, *e.g.* by means of a transformation method. The operator associated with the problem then has a continuous spectrum but this is not enough to find complex eigenvalues. Indeed the matrices of the finite element formulation are still Hermitian. The point here is that the eigenvectors that would be associated with the complex eigenvalues do not have a finite energy (*i.e.* are not in $H(\mathbf{curl}_\beta, \mathbb{R}^2)$). It is therefore necessary to work with an extension of the operator to a larger functional space that includes such functions. In this space, the operator is no longer self-adjoint and the finite element matrices which allow the computation of the complex eigenvalues associated with the leaky modes are no longer

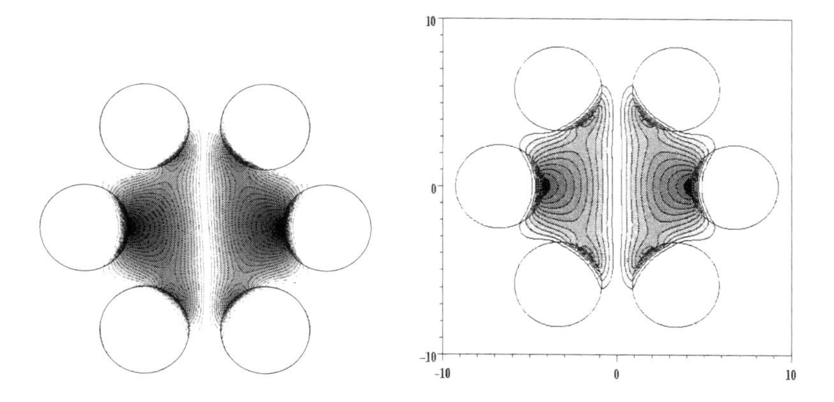

Fig. 3.11 Longitudinal component E_z (absolute value) of the electric field $\mathbf{E}$ for the degenerate fundamental mode (E_z null along the vertical y-axis) in a 6-air-hole PCF (radii $= 2.5\mu m$) of pitch $\Lambda = 6.75\mu m$ for a normalized propagation constant $\beta\Lambda = 42.276817$ (left: finite element modelling, right: multipole method with $M = 5$).

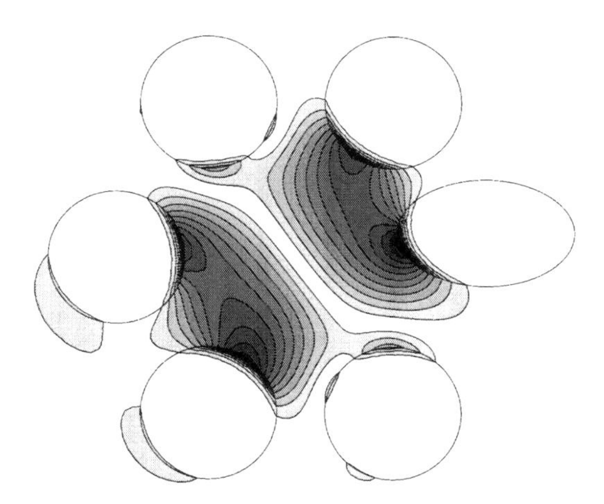

Fig. 3.12 Eigenfield of Fig. 3.11 when breaking the hexagonal symmetry of the structure (the rightmost inclusion is now a shifted ellipse) obtained with the finite element method.

Hermitian. The numerical determination of such eigenvalues may be a delicate task. The Arnoldi algorithm and its various improvements [TI95; LSY98] or the techniques presented in Chap. 5.11 may be used.

3.4.3 *Variants*

3.4.3.1 *Looking for β with k_0 given*

As the square of the wave number k_0^2 appears alone in the problem when the propagation constant β and its square β^2 are involved, it seems natural to fix the value of β to obtain a classical generalized eigenvalue problem for k_0. Nevertheless, it would often be more natural to do the opposite since the pulsation is usually given *a priori* and due to the chromatic dispersion, ε does depend itself on k_0. A first approach is to transform the problem to a generalised eigensystem for β with a technique similar to that used to transform high order differential equations to first order systems (this is in fact the Fourier transform of this technique). Given a quadratic eigenvalue problem of the form $P\mathbf{u} = \beta^2 Q\mathbf{u} + \beta R\mathbf{u}$ where P, Q, and R are $n \times n$ matrices and β and $\mathbf{u}$ are the eigenvalue and eigenvector respectively, it suffices to take $\mathbf{w} = \beta\mathbf{u}$ and to write the equivalent classical generalized eigenvalue problem involving $2n \times 2n$ matrices:

$$\begin{pmatrix} I & 0 \\ 0 & P \end{pmatrix} \begin{pmatrix} \mathbf{w} \\ \mathbf{u} \end{pmatrix} = \beta \begin{pmatrix} Q & R \\ 0 & I \end{pmatrix} \begin{pmatrix} \mathbf{w} \\ \mathbf{u} \end{pmatrix} . \tag{3.29}$$

But in the present case there is a nice trick that involves only a slight modification of the system 3.27: it consists in dividing the longitudinal degrees of freedom by $i\beta$, which amounts to multiplying righthand columns of matrices by $i\beta$, and in multiplying the longitudinal equations by $-i\beta$, which amounts to multiplying lower rows of the matrices by $-i\beta$. Explicitly, the new column vector of unknowns is:

$$\begin{pmatrix} \mathbf{f}_t \\ \mathbf{f}_z \end{pmatrix} = \begin{pmatrix} \mathbf{e}_t \\ (i\beta)^{-1}\mathbf{e}_z \end{pmatrix} . \tag{3.30}$$

With such a modification, the system 3.27 becomes:

$$\begin{pmatrix} \mathrm{A}^{tt} + \beta^2\mathrm{B}^{tt} & \beta^2\mathrm{A}^{tz} \\ \beta^2\mathrm{A}^{zt} & \beta^2\mathrm{A}^{zz} \end{pmatrix} \begin{pmatrix} \mathbf{f}_t \\ \mathbf{f}_z \end{pmatrix} = k_0^2 \begin{pmatrix} \mathrm{C}^{tt} & 0 \\ 0 & \beta^2\mathrm{C}^{zz} \end{pmatrix} \begin{pmatrix} \mathbf{f}_t \\ \mathbf{f}_z \end{pmatrix} \tag{3.31}$$

which is a real symmetric system that does not have terms linear in β and therefore involves only β^2.

3.4.3.2 *Discrete magnetic field formulation*

The magnetic field instead of the electric field can be chosen as a variable. The equation to solve is obtained by eliminating the electric field:

$$\begin{cases} \mathbf{curl}_\beta(\varepsilon_r^{-1}\,\mathbf{curl}_\beta\,\mathbf{H}) = \mu_r k_0^2 \mathbf{H} \\ \mathrm{div}_\beta(\mu_r \mathbf{H}) = 0\,. \end{cases}$$

The corresponding weak formulation is then given by:

$$\begin{aligned} \mathcal{R}(\beta;\mathbf{H},\mathbf{H}') &= \int_\Omega \frac{1}{\varepsilon_r}\,\mathbf{curl}_\beta\,\mathbf{H}\cdot\overline{\mathbf{curl}_\beta\,\mathbf{H}'}\,dxdy \\ &-k_0^2\int_\Omega \mu_r\mathbf{H}\cdot\overline{\mathbf{H}'}\,dxdy \quad ,\forall\mathbf{H}'\in H(\mathbf{curl}_\beta,\Omega) \end{aligned}$$

and its decomposition in transverse and longitudinal components is [GNZ+01; Mey99]:

$$\begin{aligned} \mathcal{R}(\beta;\mathbf{H},\mathbf{H}') = \int_\Omega \varepsilon_r^{-1}\Big(&\mathbf{curl}_t\,\mathbf{H}_t\cdot\mathbf{curl}_t\,\overline{\mathbf{H}'_t} + \mathbf{grad}_t\,H_z\cdot\mathbf{grad}_t\,\overline{H'_z} \\ &-i\,\beta\mathbf{H}_t\cdot\mathbf{grad}_t\,\overline{H'_z} + i\,\beta\,\mathbf{grad}_t\,H_z\cdot\overline{\mathbf{H}'_t} + \beta^2\,\mathbf{H}_t\cdot\overline{\mathbf{H}'_t}\Big)dxdy \\ &-k_0^2\int_\Omega \mu_r\Big(\mathbf{H}_t\cdot\overline{\mathbf{H}'_t} + H_z\,\overline{H'_z}\Big)dxdy \end{aligned}$$

The numerical treatment is point-to-point as in the numerical treatment of the electric field model. It is worth noting the duality between the electric and magnetic formulations: the electric field exchanges its role with the magnetic field, the electric permittivity with the magnetic permeability, the PEC boundary conditions with the PMW boundary conditions. Nevertheless, this mathematically perfect symmetry is lost at the physical level since at optical frequencies, the permeabilities μ_r are most often equal to 1 and the properties of the guide rely rather on dielectric and conducting properties.

3.4.3.3 *Eliminating one component with the divergence*

In the concern to reduce the number of degrees of freedom to obtain faster solutions and lower memory requirements, it has been proposed that one should use the nullity of the divergence to eliminate the longitudinal component directly from the problem (see [DW94] and chapter 1 of [IPS96] by Davies). The transverse part of $\mathbf{curl}_\beta(\varepsilon_r^{-1}\,\mathbf{curl}_\beta\,\mathbf{H}) = \mu_r k_0^2\mathbf{H}$ can be written:

$$\mathbf{curl}_z(\varepsilon_r^{-1}\,\mathbf{curl}_t\,\mathbf{H}_t) + i\beta\,\mathbf{grad}_t\,H_z + (\beta^2\varepsilon_r^{-1} - k_0^2\mu_r)\mathbf{H}_t = 0 \tag{3.32}$$

with $\mathbf{curl}_z(\varphi\mathbf{e}^z) = \mathbf{grad}_t\,\varphi \times \mathbf{e}^z$. The expression of the divergence:

$$\mathrm{div}_\beta(\mu_r \boldsymbol{H}) = \mathrm{div}_t(\mu_r\mathbf{H}_t) + i\beta\mu_r H_z = 0$$

permits us to write Eq. 3.32 in a way that depends only on the transverse component:

$$\mathbf{curl}_z(\varepsilon_r^{-1}\,\mathbf{curl}_t\,\mathbf{H}_t) + i\beta\,\mathbf{grad}_t(\mu_r^{-1}\,\mathrm{div}_t(\mu_r\mathbf{H}_t)) + (\beta^2\varepsilon_r^{-1} - k_0^2\mu_r)\mathbf{H}_t = 0\,. \tag{3.33}$$

The weak formulation of this equation is:

$$\int_\Omega \Big(\varepsilon_r^{-1}\,\mathbf{curl}_t\,\mathbf{H}_t \cdot \mathbf{curl}_t\,\overline{\mathbf{H}}_t' + (\beta^2\varepsilon_r^{-1} - k_0^2\mu_r)\mathbf{H}_t \cdot \overline{\mathbf{H}}_t' \\ - \mu_r^{-1}\,\mathrm{div}_t(\mu_r\mathbf{H}_t)\overline{\mathrm{div}_t(\varepsilon_r^{-1}\mathbf{H}_t')}\Big)\,dxdy \\ + \int_\Gamma \varepsilon_r^{-1}\Big(\mathbf{curl}\,\mathbf{H}_t \times \overline{\mathbf{H}_t'} + (\mu_r^{-1}\,\mathrm{div}_t\,bf H_t)\overline{\mathbf{H}_t'}\Big) \cdot \mathbf{n}\,dl = 0\,.$$

where Γ is the boundary of Ω together with the material interfaces, *i.e.* anywhere there is any discontinuity in ε_r or μ_r. In fact, the line integral runs twice (with opposite orientation) over interior lines of discontinuity so that the integral is evaluated for the values of the integrand on both sides which amounts to integrating the jump of the integrand along its lines of discontinuity.

The fundamental point is that between these discontinuities $\mathbf{H}_t$ and $\mathbf{H}_t'$ must be continuous since they are involved in both curls and divergences. Therefore it is no longer possible to use edge elements and hence it is necessary to approximate Cartesian components of the transverse field with nodal elements. Anyway, since the divergence condition is inserted exactly in the formulation, this formulation produces no spurious modes. Apparently this formulation is far less expensive since only two components instead of three have to be evaluated. Nevertheless, the situation is not so simple and several factors have to be taken into account: there are more edges than nodes in the mesh but the connectivity of the edges is lower (for a given edge, the average number of edges that share a tetrahedron or a triangle with this edge is lower than the similar quantity for nodes) and the number of non-zero coefficients in the matrix is smaller. In the end, the algorithmic efficiency depends on the number of algebraic operations required to perform an iteration (which is related to the cost of the product of the matrix with a vector) and the number of iterations that are necessary to reach the requested accuracy (which is related to the condition number of the matrix) and it is not obvious that the transverse nodal method is much better than the edge element approach. Edge elements have been proved to provide a robust algorithm and it may not be a good idea to try to gain memory space and computation time at the price of a less general and more cumbersome method.

Another possibility for the elimination of the longitudinal component is to perform it at the matrix level [DW94] but this destroys the sparsity of the resulting system and it is much better to use an efficient iterative solver on the non-reduced system.

3.4.3.4 E_z, H_z *formulation*

A classical approach (this is the one used in the setting-up of the multipole method) is to eliminate the transverse components and to keep E_z and H_z as unknowns (See Sec. 2.3.1 p. 57). Unfortunately, this method is heavily hounded by spurious modes [DW94].

3.5 Periodic Waveguides

3.5.1 *Bloch modes*

Though real structures are finite and one is often interested in the study of defects, the determination of modes in ideal periodic structures is of foremost importance. The Floquet-Bloch theory reduces the problem to the study of a single cell [LHHD95] as recalled in Sec. 2.6 p. 72 in this book. The purpose of this section is to show how to combine this feature with finite element modelling in order to obtain numerical solutions for propagating modes in periodic structures. We consider a structure still invariant along the z-axis but now also periodic in the xy-plane. Given two linearly independent vectors $\mathbf{a}$ and $\mathbf{b}$ in the xy-plane, the set of points $n\mathbf{a} + m\mathbf{b}$ is called the *lattice*. The *primitive cell* Y is a subset of $\mathbb{R}^2$ such that for any point $\mathbf{r}'$ of $\mathbb{R}^2$ there exist unique $\mathbf{r} = x\mathbf{e}_x + y\mathbf{e}_y \in Y$ and $n, m \in \mathbb{Z}$ such that $\mathbf{r}' = \mathbf{r} + n\mathbf{a} + m\mathbf{b}$. A function $U(\mathbf{r})$ is Y*-periodic* if $U(\mathbf{r} + n\mathbf{a} + m\mathbf{b}) = U(\mathbf{r})$ for any $n, m \in \mathbb{Z}$. The waveguide is Y-periodic if $\varepsilon_r(x, y)$ and $\mu_r(x, y)$ are Y-periodic functions. Possible PEC's and PMW's have boundaries that form a Y-periodic pattern.

The problem reduces to looking for *Bloch wave* solutions $\mathbf{U_k}$ that have

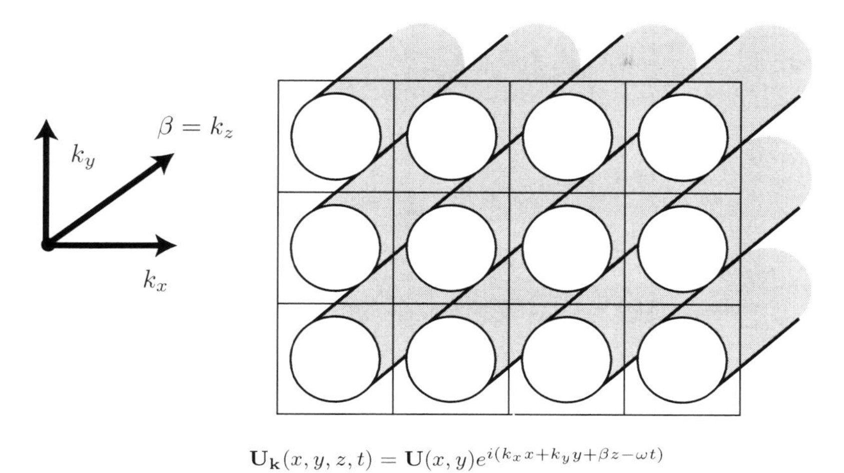

Fig. 3.13 A system with a continuous translational invariance along the z-axis together with a two-dimensional periodicity in the xy-plane and the general form of propagating modes $\mathbf{U_k}(x, y, z, t)$.

the form (Bloch theorem, see Sec. 2.6 p. 78):

$$\mathbf{U_k}(\mathbf{r}) = e^{i\mathbf{k}\cdot\mathbf{r}}\mathbf{U}(\mathbf{r}) = e^{i(k_x x + k_y y)}\mathbf{U}(x, y) \ , \ \forall \ (x, y) \text{ in } \mathbb{R}^2 \tag{3.34}$$

where $\mathbf{U}(x, y)$ is a Y-periodic function and $\mathbf{k} = k_x \mathbf{e}^x + k_y \mathbf{e}^y \in Y^* \subset \mathbb{R}^2$ is a parameter (the *Bloch vector* or quasi-momentum in solid state physics). $Y^* \subset \mathbb{R}^2$ is the *dual cell (first Brillouin zone), i.e.* the primitive cell of the *reciprocal lattice* determined by the two vectors $\mathbf{a}^*$ and $\mathbf{b}^*$ such that $\mathbf{a}^* \cdot \mathbf{a} = 2\pi$, $\mathbf{a}^* \cdot \mathbf{b} = 0$, $\mathbf{b}^* \cdot \mathbf{a} = 0$, $\mathbf{b}^* \cdot \mathbf{b} = 2\pi$ (it is worth noting that this dot product is in fact a duality product: $\mathbf{k} \cdot \mathbf{r} =< \mathbf{k}, \mathbf{r} >$). Such solutions $\mathbf{U_k}$ are said to be $(\mathbf{k}, Y)$*-periodic* in the sequel (though they are not periodic but almost-periodic).

To specify the class of solutions of our spectral problem, one introduces the Hilbert space

$$\left[L^2_\sharp(\mathbf{k}, Y)\right]^3 = \left\{ \mathbf{U_k}|_Y \in [L^2(Y)]^3 \ , \ \mathbf{U_k} \ \text{ is } (\mathbf{k}, Y)\text{-periodic} \right\} \tag{3.35}$$

of $(\mathbf{k}, Y)$-periodic square integrable functions with values in $\mathbb{C}^3$.

The pair $(\mathbf{E_k}, \mathbf{H_k})$ associated with the Bloch vector $\mathbf{k}$ is called an *electromagnetic propagating Bloch mode* if $\mathbf{E_k}$ and $\mathbf{H_k}$ are $(\mathbf{k}, Y)$-periodic fields satisfying the spectral problem:

$$\begin{cases} \mathbf{curl}_\beta \, \mathbf{H_k} = -i\omega\varepsilon_0\varepsilon_r(x,y)\mathbf{E_k} \\ \mathbf{curl}_\beta \, \mathbf{E_k} = i\omega\mu_0\mu_r(x,y)\mathbf{H_k} \end{cases} \tag{3.36}$$

with

$$\begin{cases} (\beta, \omega, \mathbf{k}) \in \mathbb{R}_+ \times \mathbb{R}_+ \times Y^* \\ (\mathbf{E_k}, \mathbf{H_k}) \neq (\mathbf{0}, \mathbf{0}) \\ \mathbf{E_k}, \mathbf{H_k} \in \left[L^2_\sharp(\mathbf{k}, Y)\right]^3 . \end{cases} \tag{3.37}$$

Looking for solutions that are Bloch functions in $\left[L^2_\sharp(\mathbf{k}, Y)\right]^3$ ensures the well-posedness of this spectral problem, as a replacement for the Sommerfeld radiation condition (or other decaying conditions for the far field) which is usually imposed in the presence of compact obstacles in the medium. The finite element formulation is completely identical to the non-periodic one. The only difference is that the study is now reduced to the primitive cell Y which is meshed and in which the integrations are performed. Some technique must be found to ensure that the solution is a $(\mathbf{k}, Y)$-periodic Bloch mode. This can be imposed by using special boundary conditions as explained in the next section.

3.5.2 *The Bloch conditions*

In order to find Bloch modes with the finite element method, some changes have to be made with respect to classical boundary value problems which will be named *Bloch conditions* [LHHD95; NGGZ]. For the sake of simplicity, one considers first a square cell $Y =]0,1[\times]0,1[$ as an example. To avoid tedious notation, the case of a scalar field $U_\mathbf{k}(x,y)$ (time and z dependence are irrelevant here and it is no particular problem to extend this method to vector quantities and edge elements) is considered on the square cell Y with Bloch conditions relating the lefthand and the righthand sides (Fig. 3.14). The set of nodes is separated into three subsets: the nodes on the left side, *i.e.* with $x = 0$, corresponding to the column array of unknowns $\mathbf{u_l}$, the nodes on the right side, *i.e.* with $x = 1$, corresponding to the column array of unknowns $\mathbf{u_r}$, and the internal nodes, *i.e.* with $x \in]0,1[$, corresponding to the column array of unknowns $\mathbf{u}$. One has the following structure for the matrix problem (corresponding in fact to natural boundary conditions, *i.e.* Neumann homogeneous boundary conditions, as the degrees of freedom on

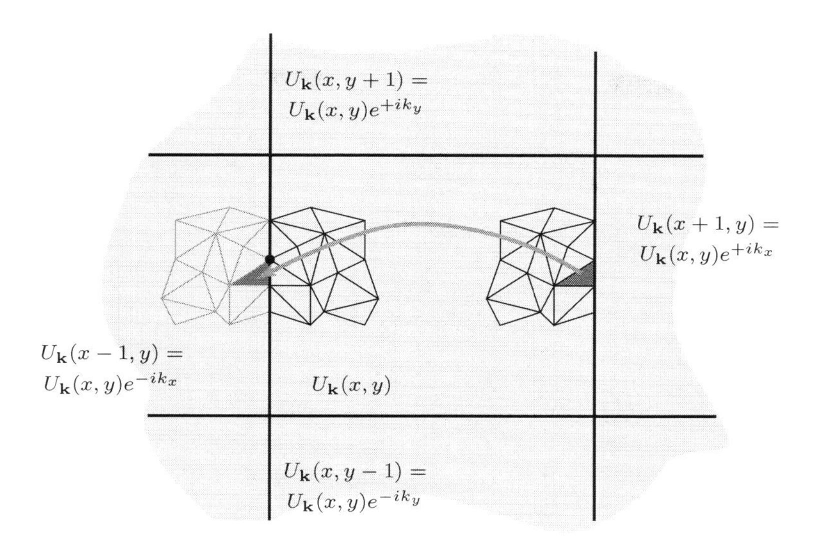

Fig. 3.14 Bloch theorem and virtual periodic mesh.

the boundaries have to be kept as unknowns in the problem):

$$\mathbf{A} \begin{pmatrix} \mathbf{u} \\ \mathbf{u_l} \\ \mathbf{u_r} \end{pmatrix} = \mathbf{b} \tag{3.38}$$

where $\mathbf{A}$ is the (square Hermitian) matrix of the system and $\mathbf{b}$ is the right hand side. The solution to be approximated by the numerical method is a Bloch function $U_{\mathbf{k}}(x, y) = U(x, y)e^{i(k_x x + k_y y)}$ with U being Y-periodic and in particular $U(x+1, y) = U(x, y)$. Therefore, the relation between the lefthand and the righthand sides is:

$$U_{\mathbf{k}}(1, y) = U(1, y)e^{i(k_x + k_y y)} = U_{\mathbf{k}}(0, y)e^{ik_x} \Rightarrow \mathbf{u_r} = \mathbf{u_l}e^{ik_x} \, . \tag{3.39}$$

The set of unknowns can thus be expressed as a function of the reduced set $\mathbf{u}$ and $\mathbf{u_l}$ via:

$$\begin{pmatrix} \mathbf{u} \\ \mathbf{u_l} \\ \mathbf{u_r} \end{pmatrix} = \mathbf{P} \begin{pmatrix} \mathbf{u} \\ \mathbf{u_l} \end{pmatrix} \quad \text{with } \mathbf{P} = \begin{pmatrix} \mathbf{1} & \mathbf{0} \\ \mathbf{0} & \mathbf{1} \\ \mathbf{0} & \mathbf{1}e^{ik_x} \end{pmatrix} \tag{3.40}$$

where $\mathbf{1}$ and $\mathbf{0}$ are identity and null matrices respectively with suitable dimensions. The finite element equations related to the eliminated nodes have now to be taken into account. Due to the periodicity of the structure, the elements on the left of the right side correspond to elements on the left of the left side (Fig. 3.14). Therefore their contributions (i.e. the equations corresponding to $\mathbf{u_r}$) must be added to the equations corresponding to $\mathbf{u_l}$ with the right phase factor, i.e. e^{-ik_x}, which amounts to multiplying the system matrix by $\mathbf{P}^*$ (the Hermitian of $\mathbf{P}$). Finally, the linear system to be solved is:

$$\mathbf{P}^*\mathbf{A}\mathbf{P}\begin{pmatrix}\mathbf{u}\\ \mathbf{u_l}\end{pmatrix} = \mathbf{P}^*\mathbf{b} \tag{3.41}$$

where it is worth noting that the system matrix is still Hermitian, which is important for numerical computation. Now a generalized eigenvalue problem (with natural boundary conditions) $\mathbf{Au} = \lambda\mathbf{Bu}$ is transformed to a Bloch mode problem according to $\mathbf{P}^*\mathbf{APu}' = \lambda\mathbf{P}^*\mathbf{BPu}'$ which is still a large sparse Hermitian generalized eigenvalue problem.

3.5.3 *A numerical example*

As an illustration, the Bloch finite element method will be used to reproduce the results presented in [MM94], where they were obtained using a plane wave method.

The basic cell is a rhombus made of two equilateral triangles: the lattice vectors are $\mathbf{a} = \Lambda\mathbf{e}_x$ and $\mathbf{b} = \frac{\Lambda}{2}\mathbf{e}_x + \frac{\Lambda\sqrt{3}}{2}\mathbf{e}_y$ where Λ is the nearest neighbour distance, *i.e.* the length of the sides of the cells. This cell contains a circular air inclusion (radius $R = 0.48\Lambda$, so that the filling fraction $f = 0.8358$, and $\varepsilon_r = 1.0$) surrounded by solid dielectric material ($\varepsilon_r = 13.0$). The vectors of the reciprocal lattice are $\mathbf{a}^* = \frac{2\pi}{\Lambda}\mathbf{e}^x - \frac{2\pi\sqrt{3}}{3\Lambda}\mathbf{e}^y$ and $\mathbf{b}^* = \frac{4\pi\sqrt{3}}{3\Lambda}\mathbf{e}^y$ and the first Brillouin zone is hexagonal. The irreducible part can be represented by the triangle with vertices $\Gamma = (0,0)$, $M = (0, \frac{2\pi}{\sqrt{3}\Lambda})$, and $K = (\frac{2\pi}{3\Lambda}, \frac{2\pi}{\sqrt{3}\Lambda})$. The basic cell is meshed with 4628 triangles. All these data are summarised in Fig. 3.15. Note that the circular inclusion is too large to fit as a single piece inside the basic cell hence the splitting into four parts in the corners.

The Bloch boundary conditions connect the degrees of freedom on opposite sides of the rhombus: the degrees of freedom on the lower lefthand side are equal to the corresponding ones on the upper righthand side multiplied by a phase factor equal to $e^{i(-k_x\frac{\Lambda}{2} - k_y\frac{\sqrt{3}\Lambda}{2})}$ and the degrees of freedom on the lower righthand side are equal to the corresponding ones on the upper lefthand side multiplied by a phase factor equal to $e^{i(+k_x\frac{\Lambda}{2} - k_y\frac{\sqrt{3}\Lambda}{2})}$.

The dispersion curves shown on Figs. 3.16 and 3.17 correspond to pulsations ω (only the ω such that $\omega < \frac{2\pi c}{\Lambda}$ are represented here) of the prop-

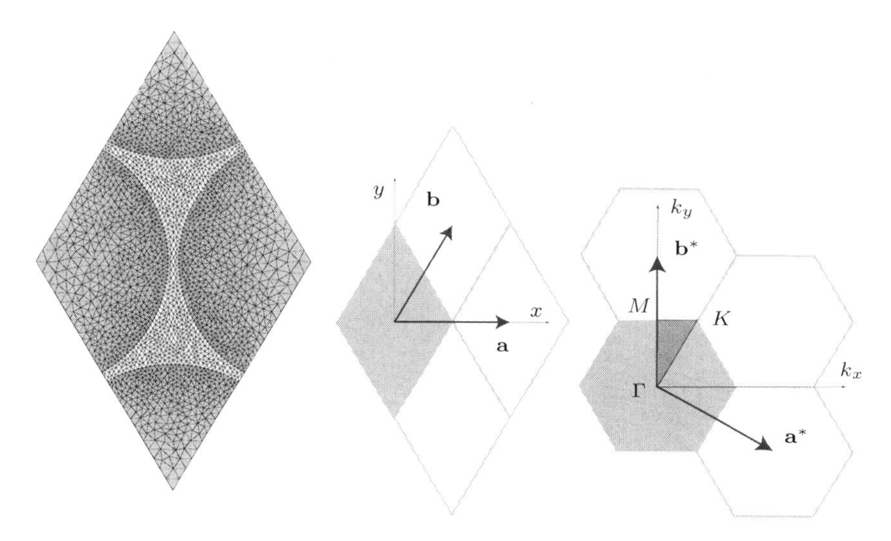

Fig. 3.15 Two-dimensional periodic structure (the basic cell is rhombic with a side length Λ) with a circular air inclusion (radius $R = 0.48\Lambda$, $\varepsilon_r = 1.0$) surrounded by solid dielectric material ($\varepsilon_r = 13.0$): meshing of a basic rhombic cell with 4628 triangles (left), representation of some lattice cells with the lattice vectors $\mathbf{a} = \Lambda\mathbf{e}_x$ and $\mathbf{b} = \frac{\Lambda}{2}\mathbf{e}_x + \frac{\Lambda\sqrt{3}}{2}\mathbf{e}_y$ (center), representation of some cells of the reciprocal lattice with the lattice vectors $\mathbf{a}^* = \frac{2\pi}{\Lambda}\mathbf{e}^x - \frac{2\pi\sqrt{3}}{3\Lambda}\mathbf{e}^y$ and $\mathbf{b}^* = \frac{4\pi\sqrt{3}}{3\Lambda}\mathbf{e}^y$ and the irreducible part of the first Brillouin zone represented by the triangle with vertices $\Gamma = (0,0)$, $M = (0, \frac{2\pi}{\sqrt{3}\Lambda})$, and $K = (\frac{2\pi}{3\Lambda}, \frac{2\pi}{\sqrt{3}\Lambda})$ (right).

agation modes associated with a given value of the propagation constant β ($\beta\Lambda = 0.0, 2.0, 4.0, 2\pi$).

The value 1.0 is given to Λ for the numerical computations. The boundary of the irreducible Brillouin zone is sampled with 120 points (40 on each side of the triangle). The computation of the eigenvalues associated with a particular Bloch vector takes a few minutes on a typical 2.6GHz desktop micro-computer. The results are in good agreement with those of [MM94].

3.5.4 *Direct determination of the periodic part*

Another possible approach is to solve an equation with periodic boundary conditions that gives directly the periodic vector field $\mathbf{U}(\mathbf{r})$ involved in the Bloch mode $\mathbf{U}(\mathbf{r})e^{i\mathbf{k}\cdot\mathbf{r}}$.

The equation $\mathbf{curl}_\beta(\mu_r^{-1}\,\mathbf{curl}_\beta\,\mathbf{E_k}) = k_0^2\varepsilon_r\mathbf{E_k}$ with $\mathbf{E_k}(\mathbf{r}) = \mathbf{E}(\mathbf{r})e^{i\mathbf{k}\cdot\mathbf{r}}$

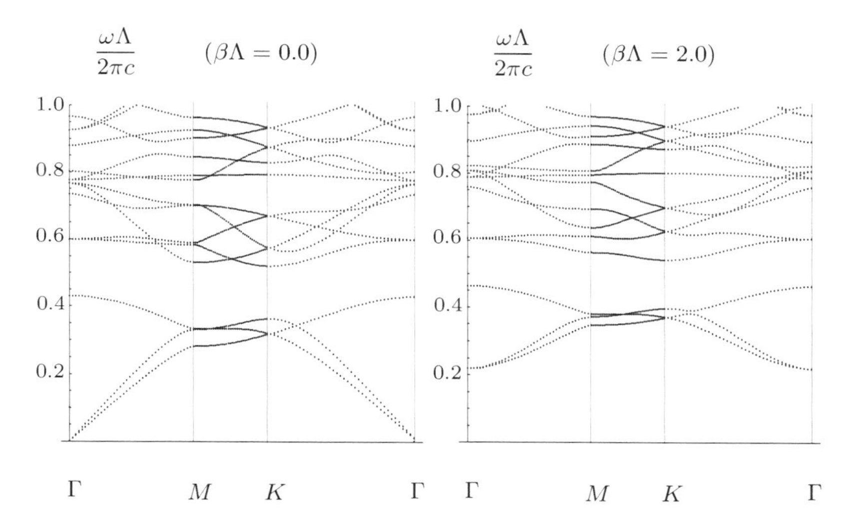

Fig. 3.16 Dispersion curves corresponding to Bloch waves in conical mounting in the lattice of Fig. (3.15) for $\beta\Lambda = 0.0$ (left) and $\beta\Lambda = 2.0$ (right).

gives for **E**:

$$\mathbf{curl}_{\beta,\mathbf{k}}(\mu_r^{-1}\,\mathbf{curl}_{\beta,\mathbf{k}}\,\mathbf{E}) = k_0^2\varepsilon_r\mathbf{E}$$

where $\mathbf{curl}_{\beta,\mathbf{k}}\,\mathbf{U} = \mathbf{curl}_\beta(\mathbf{U}(\mathbf{r})e^{i\mathbf{k}\cdot\mathbf{r}})e^{-i\mathbf{k}\cdot\mathbf{r}}$. The following transverse operators are defined for a scalar function $\varphi(x,y)$ and a transverse field $\mathbf{v} = v_x(x,y)\mathbf{e}^x + v_y(x,y)\mathbf{e}^y$:

$$\begin{aligned}
\mathbf{grad}_{t,\mathbf{k}}\,\varphi &= \mathbf{grad}_t\,\varphi + i\mathbf{k}\varphi \\
\mathbf{curl}_{t,\mathbf{k}}\,\mathbf{v} &= \mathbf{curl}_t\,\mathbf{v} + i\mathbf{k}\times\mathbf{v} \\
\mathrm{div}_{t,\mathbf{k}}\,\mathbf{v} &= \mathrm{div}_t\,\mathbf{v} + i\mathbf{k}\cdot\mathbf{v}
\end{aligned}$$

and one has:

$$\begin{aligned}
\mathbf{curl}_{\beta,\mathbf{k}}\,(\mathbf{v} + \varphi\mathbf{e}^z) &= \mathbf{curl}_{t,\mathbf{k}}\,\mathbf{v} + (\mathbf{grad}_{t,\mathbf{k}}\,\varphi - i\,\beta\mathbf{v})\times\mathbf{e}^z \\
&= \mathbf{curl}_t\,\mathbf{v} + i\mathbf{k}\times\mathbf{v} + (\mathbf{grad}_t\,\varphi + i\mathbf{k}\varphi - i\,\beta\mathbf{v})\times\mathbf{e}^z\,.
\end{aligned}$$

The weak formulation is now:

$$\begin{aligned}
\mathcal{R}(\beta;\mathbf{E},\mathbf{E}') &= \int_\Omega \mu_r^{-1}\,\mathbf{curl}_{\beta,\mathbf{k}}\,\mathbf{E}\cdot\overline{\mathbf{curl}_{\beta,\mathbf{k}}\,\mathbf{E}'}\,dxdy \\
-k_0^2 &\int_\Omega \mathbf{E}\cdot\overline{\mathbf{E}'}\,\varepsilon_r dxdy = 0 \quad ,\forall\,\mathbf{E}'\in H(\mathbf{curl}_{\beta,\mathbf{k}},\Omega)
\end{aligned}$$

with periodic boundary conditions on **E** [BGN] (which can be seen as a particular case of the Bloch boundary conditions with $k_x = k_y = 0$).

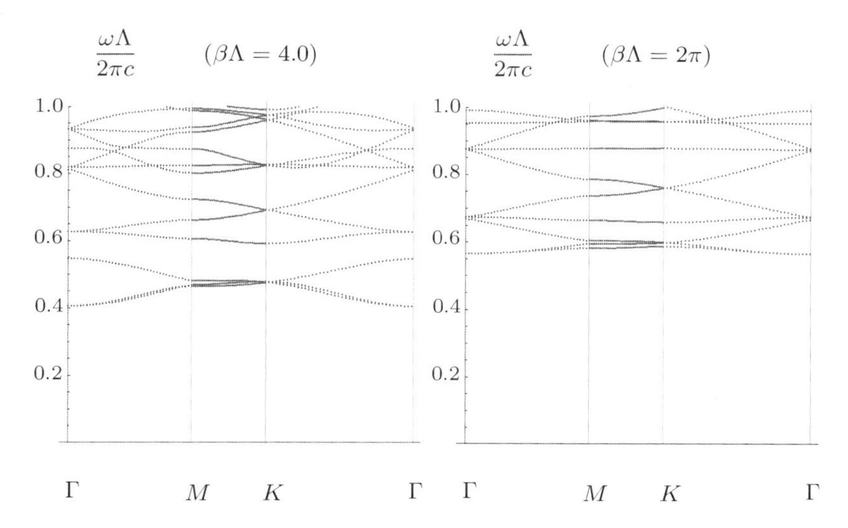

Fig. 3.17 Dispersion curves corresponding to Bloch waves in conical mounting in the lattice of Fig. (3.15) for $\beta\Lambda = 4.0$ (left) and $\beta\Lambda = 2\pi$ (right).

3.6 Twisted Fibres

During the production process of microstructured fibres, a matrix reproducing the desired cross section is stretched and a thin fibre is obtained that reproduces the initial pattern at a smaller scale (See Fig. 3.18). Unfortunately, some unwanted transformations may also occur and, in practice, an uncontrolled twist often appears. This phenomenon is particularly annoying in microstructured fibre since it destroys the translational invariance and experimental tests have shown its negative impact on the performance of the fibre. This section presents a possible approach to the modelling of twisted fibres that calls for the flexibility of the finite element method [NZG].

Fig. 3.18 Schematic of twisted fibres.

The central feature of the model is a non-orthogonal change of coordinates that restores the invariance of the fibre along a co-ordinate. But the question whether the very concept of a propagation mode is still valid in such a structure or not must be addressed. Ideal waveguides are translationally invariant structures along an axis (say z), for the material properties (here $\varepsilon(x, y)$) as well as for the geometry (the surfaces of discontinuity of the material properties in fact). In the case of the ideal waveguide, the operators arising in the differential equations involve only coefficients that are functions of x and y but are independent of z. The Fourier transform of the equation with respect to z is therefore immediately apparent and it amounts simply to replacing the derivative with respect to z by a multiplication by $i\beta$. The propagation modes are solutions of the form: $\boldsymbol{\mathcal{E}}(x, y, z, t) = \Re e\{\mathbf{E}(x, y)e^{-i(\omega t - \beta z)}\}$ which are therefore merely solutions to the problem where the two transverse co-ordinates x and y are kept in the position space while the time t and the longitudinal co-ordinate z are transformed to the pulsation ω and momentum β respectively by a Fourier transform.

If a similar situation arises for another structure in a particular co-ordinate system, propagation modes can be associated with such a structure. This is the case for axi-symmetrical structures in cylindrical coordinates: modes can be found in a structure looping along azimuthal co-ordinate θ.

An *helicoidal co-ordinate system* u, v, w is introduced that is related to Cartesian co-ordinates by:

$$\begin{cases} x = u\cos(\alpha w) + v\sin(\alpha w) \\ y = -u\sin(\alpha w) + v\cos(\alpha w) \\ z = w \end{cases} \tag{3.42}$$

where x, y, z are the Cartesian co-ordinates and α is a parameter characterizing the twist. A twisted electromagnetic problem can be defined as a problem such that the geometry (the boundary conditions and the material properties $\varepsilon, \mu, \ldots$) depends only on the co-ordinates u and v for some given value of α.

Helicoidal co-ordinates are not orthonormal and there are no *a priori* guaranties that twisted structures will lead to a nice propagation mode problem. It has been seen in Sec. 3.3.3, that changing the coordinate system amounts to replacing the various materials (that are often piecewise isotropic and homogeneous) by equivalent materials that are anisotropic and inhomogeneous. The guidelines of Sec. 3.3.3 are followed to obtain the replacements corresponding to helicoidal co-ordinates. The first step is to compute the Jacobian matrix associated with Eq. 3.42:

$$\mathbf{J} = \frac{\partial(x,y,z)}{\partial(u,v,w)} = \begin{pmatrix} \cos(\alpha w) & \sin(\alpha w) & \alpha v\cos(\alpha w) - \alpha u \sin(\alpha w) \\ \sin(\alpha w) & \cos(\alpha w) & -\alpha u \cos(\alpha w) - \alpha v \sin(\alpha w) \\ 0 & 0 & 1 \end{pmatrix}. \tag{3.43}$$

The fact that $\det(\mathbf{J}) = 1$ expresses the conservation of volume. The fundamental property of the helicoidal co-ordinate system is that despite the fact that the Jacobian matrix depends on the co-ordinate w, the corresponding $\mathbf{T}$ matrix depends only on the transverse co-ordinates u and v:

$$\mathbf{T} = \frac{\mathbf{J}^T\mathbf{J}}{det(\mathbf{J})} = \begin{pmatrix} 1 & 0 & \alpha v \\ 0 & 1 & -\alpha u \\ \alpha v & -\alpha u & 1+\alpha^2(u^2+v^2) \end{pmatrix} \tag{3.44}$$

$$\mathbf{T}^{-1} = \begin{pmatrix} 1+\alpha v^2 & -\alpha^2 uv & -\alpha v \\ -\alpha^2 uv & 1+\alpha u^2 & \alpha u \\ -\alpha v & \alpha u & 1 \end{pmatrix}. \tag{3.45}$$

The twisted problem is set up by replacing the actual material characteristic by new tensorial ones given by: $\underline{\underline{\varepsilon'}} = \varepsilon \mathbf{T}^{-1}$ and $\underline{\underline{\mu'}} = \mu \mathbf{T}^{-1}$. Note once again that the permittivity and permeability undergo the same transformation so that the impedances of the media remain unchanged.

The finite element method is based on the annulation of the following residuals:

$$\int_\Omega (\underline{\underline{\mu'}})^{-1} \,\mathbf{curl}_\beta\, \mathbf{E} \cdot \mathbf{curl}_\beta\, \overline{\mathbf{E'}} - \omega^2 \underline{\underline{\varepsilon'}} \mathbf{E} \cdot \overline{\mathbf{E'}}\, d\Omega = 0\,. \tag{3.46}$$

By means of the independence of this problem on w, the pairs (ω, β) that give a solution to this two-dimensional problem correspond to a propagation mode in the twisted guide. What is important here is to note that the transverse and longitudinal components are rotated by the action of the tensors so that all the terms have to be kept in the scalar products. Separating into transverse and longitudinal components, the following expression is obtained:

$$\begin{aligned}
\mathcal{R}(\beta;\mathbf{E},\mathbf{E}') = \int_\Omega \Big((\underline{\underline{\mu_r}}^{-1}\,\mathbf{curl}_t\,\mathbf{E}_t)\cdot\mathbf{curl}_t\,\overline{\mathbf{E}'_t} + \\
(\underline{\underline{\mu_r}}^{-1}\,\mathbf{curl}_t\,\mathbf{E}_t)\cdot(\mathbf{grad}_t\,\overline{E'_z}\times\mathbf{e}^z) - i\beta(\underline{\underline{\mu_r}}^{-1}\,\mathbf{curl}_t\,\mathbf{E}_t)\cdot(\overline{\mathbf{E}'_t}\times\mathbf{e}^z) + \\
(\underline{\underline{\mu_r}}^{-1}(\mathbf{grad}_t\,E_z\times\mathbf{e}^z))\cdot\mathbf{curl}_t\,\overline{\mathbf{E}'_t} + \\
(\underline{\underline{\mu_r}}^{-1}(\mathbf{grad}_t\,E_z\times\mathbf{e}^z))\cdot(\mathbf{grad}_t\,\overline{E'_z}\times\mathbf{e}^z) - \\
i\beta(\underline{\underline{\mu_r}}^{-1}(\mathbf{grad}_t\,E_z\times\mathbf{e}^z))\cdot(\overline{\mathbf{E}'_t}\times\mathbf{e}^z) - i\beta(\underline{\underline{\mu_r}}^{-1}(\mathbf{E}_t\times\mathbf{e}^z))\cdot\mathbf{curl}_t\,\overline{\mathbf{E}'_t} - \\
i\beta(\underline{\underline{\mu_r}}^{-1}(\mathbf{E}_t\times\mathbf{e}^z))\cdot(\mathbf{grad}_t\,\overline{E'_z}\times\mathbf{e}^z) + \beta^2(\underline{\underline{\mu_r}}^{-1}(\mathbf{E}_t\times\mathbf{e}^z))\cdot(\overline{\mathbf{E}'_t}\times\mathbf{e}^z)\Big)\,dxdy \\
-k_0^2\int_\Omega \Big((\underline{\underline{\varepsilon_r}}\mathbf{E}_t)\cdot\overline{\mathbf{E}'_t} + (\underline{\underline{\varepsilon_r}}\mathbf{E}_t)\cdot\overline{E'_z}\mathbf{e}^z + \\
(\underline{\underline{\varepsilon_r}}E_z\mathbf{e}^z)\cdot\overline{\mathbf{E}'_t} + (\underline{\underline{\varepsilon_r}}E_z\mathbf{e}^z)\cdot\overline{E'_z}\mathbf{e}^z\Big)\,dxdy = 0\,.
\end{aligned}$$

3.7 Conclusion

In this chapter, several variants of the finite element method have been presented and we strongly advocated the use of the Whitney elements. Indeed this approach avoids spurious modes and leads to a robust algorithm. Moreover, the various boundary conditions can be included in a very natural way and the approximations naturally respect the discontinuities of the fields themselves due to material discontinuities. Changing the system of coordinates is easy, which proved to be useful in the treatment of infinite domains, PML and twisted fibres.

Challenging problems for the finite element method would certainly be the determination of losses in the case of leaky modes and also the study of non-linear fibres and non-linear photonic crystals.

Chapter 4

The Multipole Method

4.1 Introduction

We will now present the *Multipole Method* which has been used to study microstructured optical fibers (MOFs) since 2001. The multipole method presented here [WKM+02; KWR+02] is a generalization to mode searching in conical mounting of a previous multipole method already developed a few years earlier [FTM94]. In fact, these multipole methods are natural extension of the usual method used to find the modes in step-index optical fibres (we will clarify this in the following sections). This method has five advantages which have proved to be useful in the investigation of MOF properties.

- The longitudinal or axial propagation constant of the mode β may be complex, and this is crucial since the imaginary part of the propagation constant β is not null due to the leaky nature of the MOF modes (See the section 1.5 on leaky modes in the introduction p. 19).
- The angular frequency ω, related to the free space wavenumber by $\omega = k_0 c$, is an input parameter and β is given by the calculations. The method is hence well suited for computations involving material dispersion.
- As the MOF represents a new type of waveguide, MOF research needs methods that can deal with a wide variety of structures, allowing systematic studies of MOF properties to be performed. It appears that this is partially the case for the Multipole Method: it can be used on a wide range (several orders of magnitude) of wavelengths relative to MOF dimensions.
- It can deal with the two main types of MOF: solid core MOF and air core photonic crystal fibres.

- The last advantage comes from McIsaac's theoretical work [Isa75a; Isa75b] on the symmetry properties of waveguide modes according to waveguide symmetries. These powerful theoretical results are well suited to MOF due to the usual structure of their inclusions (generally a subset of a triangular lattice, see Fig. 4.1). These symmetry properties permit us to reduce the number of numerical computations, which is useful for the systematic study of MOF, and they will be described more precisely in the following sections of this chapter.

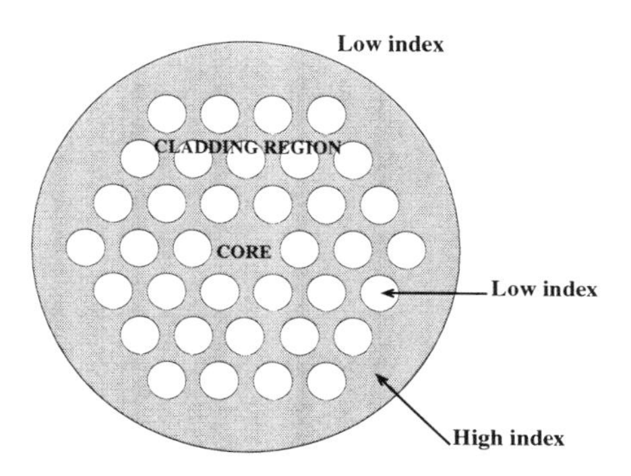

Fig. 4.1 Cross section of a typical microstructured optical fiber. The inclusions have a lower refractive index than the background medium, or matrix. The inclusions are arranged following a subset of a triangular lattice. In this example, there are three rings of inclusions. The solid core consists of one missing inclusion at the center of the structure. We call the region containing the inclusions the *cladding region* .

4.2 The Multipole Formulation

In order to make the Multipole Method intuitive and to avoid an overload of notation and calculus, which can be quite tedious, we start with a *simplified approach* [Kuh03]. First of all, we have to define the geometry of the microstructured optical fiber to be studied and we have to fix the choice of the propagating electromagnetic fields.

4.2.1 *The geometry of the modelled microstructured optical fiber*

In this chapter we will limit our study to MOF models involving non-overlapping *circular inclusions.* This restriction to circular inclusions is not fundamental for the formulation but it allows more straightforward computations. On the contrary, the fact that the inclusions are *non-overlapping* is a fundamental hypothesis of our method (the justification comes from the necessity to fulfill the hypotheses of Graf's theorems). The other fundamental hypothesis is the longitudinal invariance of the MOF model. The geometry we are dealing with is described in Fig. 4.2, which represents a transverse xy cross section of the fibre (the z axis being along the fiber axis). This shows a silica matrix of refractive index n_e, perforated with a finite number N_i of inclusions indexed by j and of diameter d_j, whose centers are specified by $\mathbf{c}_j$. The refractive index of inclusion j is n_j. Outside this hole region, the MOF is enclosed in a *jacket* (radius $r > R_0$, region *(d)* in Fig. 4.2), the index n_0 of which may be complex. One possibility is to take a jacket with refractive index equal to unity, simulating a MOF in air or vacuum.

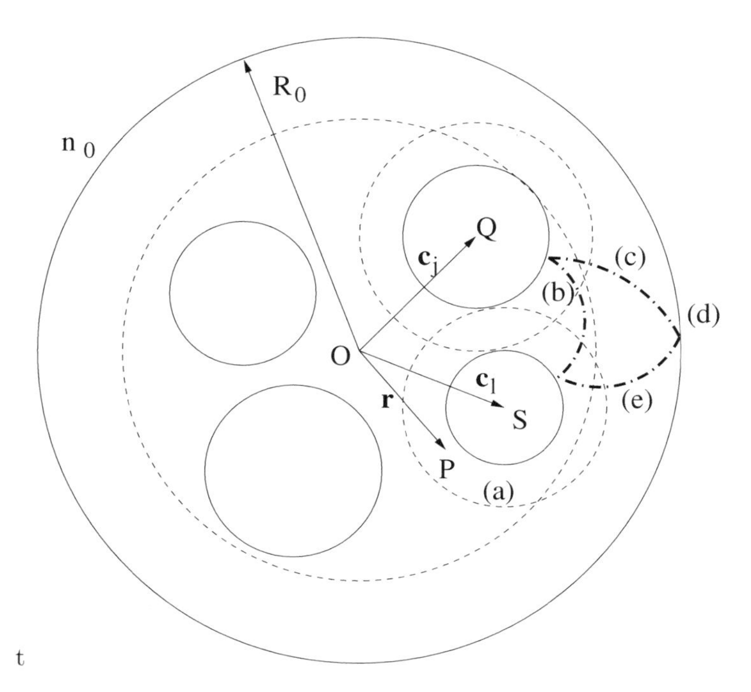

Fig. 4.2 Geometry of the MOFs considered, together with the contributions to the fields just outside a generic hole i. Regions of convergence of multipole expansions are indicated by dashed lines. Note that **QP** is $\mathbf{r}_j$ in (4.22), while **SP** is $\mathbf{r}_l$ and $\vec{OP}$ is $\mathbf{r}$. Solid lines indicate physical boundaries, dashed lines indicate regions of convergence (see section 4.2.4 for a complete description).

4.2.2 *The choice of the propagating electromagnetic fields*

We characterize in the complex representation the electric and magnetic fields $\boldsymbol{\mathcal{E}}$ and $\boldsymbol{\mathcal{H}}$ in the MOF by specifying the components $\mathcal{E}_z$ and $\mathcal{H}_z$ along the fibre axis, with transverse fields following from Maxwell's equations [SL83]. In fact, it is convenient to work with *scaled magnetic fields*: $\boldsymbol{\mathcal{K}} = Z\boldsymbol{\mathcal{H}}$, where $Z = (\mu_0/\varepsilon_0)^{1/2}$ denotes the *impedance of free space.* Each mode is characterized by its *propagation constant* β, and the transverse dependence of the fields is such that

$$\boldsymbol{\mathcal{E}}(r,\theta,z,t) = \mathbf{E}(r,\theta)e^{i(\beta z-\omega t)} , \tag{4.1}$$

$$\boldsymbol{\mathcal{K}}(r,\theta,z,t) = \mathbf{K}(r,\theta)e^{i(\beta z-\omega t)} , \tag{4.2}$$

with ω denoting the angular frequency, related to the free space wavenumber by $\omega = kc$. Note that β is complex for leaky modes, the imaginary part of β accounting for attenuation along the z axis. Here we will use the modes'*effective index effective index*, which is related to β by $n_{\text{eff}} = \beta/k$.

Each of the fields ($V = E_z$ or $V = K_z$) satisfies the *Helmholtz equation*

$$(\triangle + (k_\perp^{\mathrm{M}})^2)V = 0 \tag{4.3}$$

in the matrix, where $k_\perp^{\mathrm{M}} = \sqrt{k^2 n_{\mathrm{M}}^2 - \beta^2}$, and

$$(\triangle + (k_\perp^{i})^2)V = 0 \tag{4.4}$$

in inclusion i, where $k_\perp^i = \sqrt{k^2 n_i^2 - \beta^2}$. Care is required when computing the complex square roots [Nev80].

4.2.3 *A simplified approach of the Multipole Method*

The multipole method simply results from considering the balance of incoming and outgoing fields. Its aim is to solve the problem of scattering from a system consisting of multiple inclusions. In this section we go through each step of the multipole method in a very simplified manner, with simplified notations, to extract the physics behind the Multipole Method.

4.2.3.1 *Fourier-Bessel series*

We consider a single inclusion in the matrix (see the hashed region in Fig. 4.3), with its center at the origin of the coordinate system O. In cylindrical coordinates a field $V(r,\theta)$ is 2π periodic along the angular coordinate ($V(r,\theta+2\pi) = V(r,\theta)$). In any homogeneous annulus around the inclusion (the grey region delimited by the two dashed circles in Fig. 4.3), for fixed r, $V(r,\theta)$ is a regular and 2π-periodic function of θ, so that we can expand $V(r,\theta)$ in a Fourier series:

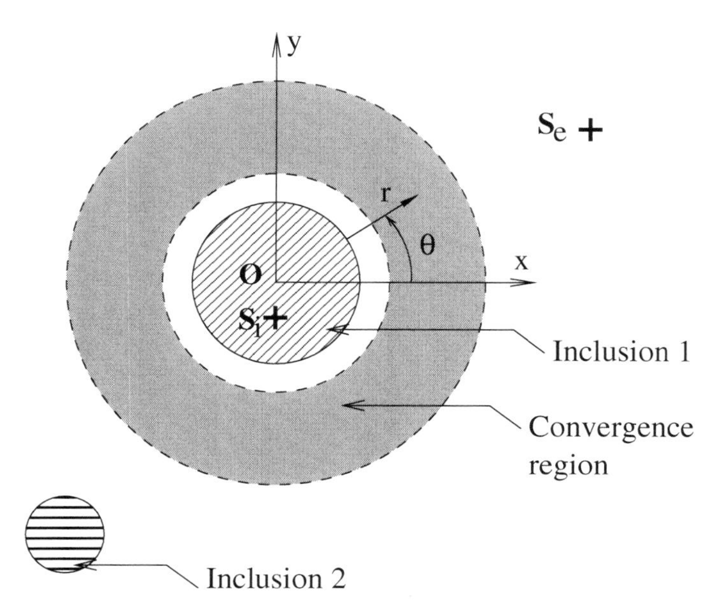

Fig. 4.3 Scheme for the simplified approach of the multipole method. The first inclusion is the hashed disk in the matrix, with center at the origin O. The dashed circles represent the borders of an homogenous annulus (grey region) around the inclusions, this region is also called the convergence region. S_i is a source localized inside the inclusion and S_e represents a source outside the convergence region. The second inclusion is the small hashed disk in the lower part of the schematic diagram.

$$V(r,\theta) = \sum_{n\in\mathbb{Z}} f_n(r)\exp(in\theta). \tag{4.5}$$

Note that since $V(r,\theta)$ is regular in the annulus, the Fourier coefficients $f_n(r)$ are regular functions of r. Using the Fourier expansion in the Helmholtz equation Eq. 4.3, the following identity is obtained:

$$\sum_{n\in\mathbb{Z}}\left[\frac{d^2 f_n(r)}{dr^2} + \frac{1}{r}\frac{df_n(r)}{dr} + \left((k_\perp^{\mathrm{M}})^2 - \frac{n^2}{r^2}\right) f_n(r)\right]\exp(in\theta) = 0\ . \tag{4.6}$$

By making use of the uniqueness of the Fourier expansion, we are led to an equation valid for all n

$$\frac{d^2 f_n(r)}{dr^2} + \frac{1}{r}\frac{df_n(r)}{dr} + \left((k_\perp^{\mathrm{M}})^2 - \frac{n^2}{r^2}\right) f_n(r) = 0 \; . \tag{4.7}$$

With a linear change of variables $u = k_\perp^{\mathrm{M}} r$ this equation becomes

$$\frac{d^2 f_n(u)}{du^2} + \frac{1}{u}\frac{df_n(u)}{du} + \left(1 - \frac{n^2}{u^2}\right) f_n(u) = 0 \; . \tag{4.8}$$

Equation 4.8 is the *Bessel differential equation* of order n [AS65]. The functions $f_n(u)$ are therefore linear combinations of Bessel functions of the first and second kind of order n ($J_n(u)$ and $Y_n(u)$ respectively), or, equivalently, of Bessel and Hankel functions of the first kind of order n, the latter being defined by $H_n^{(1)}(u) = J_n(u) + iY_n(u)$:

$$f_n(u) = A_n J_n(u) + B_n H_n^{(1)}(u) \; . \tag{4.9}$$

Replacing $f_n(r)$ in the Fourier expansion Eq. 4.5, we have

$$V(r,\theta) = \sum_{n\in\mathbb{Z}} \left(A_n J_n(k_\perp^{\mathrm{M}} r) + B_n H_n^{(1)}(k_\perp^{\mathrm{M}} r)\right) \exp(in\theta) \; . \tag{4.10}$$

The expansion of the field V in Eq. 4.10 is called a *Fourier-Bessel series*. Any function which is regular and satisfies the Helmholtz equation in an annulus can be expressed as a Fourier-Bessel series.

4.2.3.2 *Physical interpretation of Fourier-Bessel series (no inclusion)*

The Fourier-Bessel series can be split into two very different parts: the Bessel functions of the first kind are regular everywhere, whereas the Hankel functions have a singularity at 0 where they diverge. Furthermore, Hankel functions of the first kind satisfy the outgoing wave equation, whereas Bessel functions of the first kind do not.

To understand the meaning of the two parts of the Fourier-Bessel function, we consider the same annulus as above, but without the inclusion. The whole space is now homogeneous. If a source is placed inside the inner circle of the annulus (S_i in Fig. 4.3), the field it radiates has a singularity inside the inner circle of the annulus, and it satisfies the outgoing wave condition. In the annulus, it consequently cannot be represented by a Bessel series, but only by a superposition of Hankel functions. Conversely, a source placed beyond the outer ring of the annulus (S_e in Fig. 4.3) radiates a field which is regular in the annulus and in the region delimited by the inner circle of the annulus. Its field expansion in the annulus cannot therefore contain Hankel functions, but only Bessel functions.

Eq. 4.10 can be written as

$$V(r, \theta) = \mathcal{R}(r, \theta) + \mathcal{O}(r, \theta) \tag{4.11}$$

with

$$\mathcal{R}(r, \theta) = \sum_{n \in \mathbb{Z}} A_n J_n(k_\perp^{\mathrm{M}} r) \exp(in\theta) \tag{4.12}$$

$$\mathcal{O}(r, \theta) = \sum_{n \in \mathbb{Z}} B_n H_n^{(1)}(k_\perp^{\mathrm{M}} r) \exp(in\theta) \ . \tag{4.13}$$

$\mathcal{R}$ is the *regular* part of V. It describes fields radiated from sources situated beyond the outer circle of the annulus. $\mathcal{O}$ is the *singular* part of V. It describes fields radiated from sources situated inside the inner circle of the annulus. Note that if a source is placed inside the annulus, the field it radiates has a singularity in the annulus. A field radiated by a source inside the annulus cannot therefore be described by a Fourier-Bessel series in that annulus.

4.2.3.3 *Change of basis*

In the local coordinate system with origin in S_e, the field radiated by S_e is an outgoing, singular field. In an annulus surrounding S_e, the radiated field is described by a series of Hankel functions $\mathcal{O}_s(r_s, \theta_s)$, with (r_s, θ_s) being the local coordinates associated with S_e. In the coordinate system with center O, the same field is regular and incident into the annulus surrounding O defined by the two dashed circles in Fig. 4.3 : the nature of the field depends on the system of coordinates. We can construct a linear operator associating the outgoing field in one coordinate system with the resulting incoming field in another coordinate system. We define the operator $\mathcal{H}$ by

$$\mathcal{R} = \mathcal{H}\mathcal{O}_s \ , \tag{4.14}$$

where $\mathcal{R}$ is the regular field in an annulus around O (Eq. 4.13). In practice, operator $\mathcal{H}$ will be represented by a matrix linking the Fourier-Bessel coefficients A_n of $\mathcal{R}$ to the Fourier-Bessel coefficients B_n of $\mathcal{O}_e$. The coefficients of the matrix are well known, and given by Graf's theorem, as will be shown later in the mathematical derivation of the Multipole Method (see section 4.2.4).

4.2.3.4 *Fourier-Bessel series and one inclusion: scattering operator*

We now put the inclusion back in the annulus, and consider fields originating outside the annulus, *e.g.* from S_e. In the annulus, the field radiated from S_e is regular and follows from Eq. 4.14. The field reaching the inclusion will be scattered. The scattered field radiates away from the inclusion: there are now sources inside the region delimited by the inner circle of the annulus. The scattered field is hence described in the annulus by an outgoing field $\mathcal{O}$ while the incoming field is associated with $\mathcal{R}$. Since we only consider linear scattering, $\mathcal{R}$ and $\mathcal{O}$ are linked by a *linear scattering operator*, $\mathcal{S}$, defined by

$$\mathcal{O} = \mathcal{S}\mathcal{R} . \tag{4.15}$$

Once $\mathcal{H}$ and $\mathcal{S}$ are known, we can compute the scattered field using Eq. 4.14 and 4.15. In practice, the scattering operator is represented by a matrix linking the Fourier-Bessel coefficients A_n of $\mathcal{R}$ and B_n of $\mathcal{O}$. For simple geometries of inclusions (*e.g.* circular inclusions), the coefficients of the matrix can be expressed in exact analytic form. For inclusions with arbitrary geometry, the matrix can be computed numerically[MV72; NP03].

4.2.3.5 *Fourier-Bessel series and two inclusions: the Multipole Method*

We now consider two inclusions (1 and 2), and a source S_e exterior to both inclusions. $\mathcal{R}_1$, the incoming regular field for inclusion 1 now results from the superposition of the field $\mathcal{O}_s$ radiated from S_e and the scattered field $\mathcal{O}_2$ from inclusion 2. Using the change of basis operators $\mathcal{H}_{s,1}$ and $\mathcal{H}_{2,1}$ defined as in Eq. 4.14, we have

$$\mathcal{R}_1 = \mathcal{H}_{2,1}\mathcal{O}_2 + \mathcal{H}_{s,1}\mathcal{O}_s . \tag{4.16}$$

Similarly, $\mathcal{R}_2$, the regular incoming field for inclusion 2 is given by

$$\mathcal{R}_2 = \mathcal{H}_{1,2}\mathcal{O}_1 + \mathcal{H}_{s,2}\mathcal{O}_s . \tag{4.17}$$

The two equations above simply make it explicit that the incoming field on one inclusion results from the superposition of the field radiated by the other inclusion and the source. Using the scattering operators $\mathcal{S}_1$ and $\mathcal{S}_2$ for inclusions 1 and 2 respectively, we have

$$\begin{cases} \mathcal{O}_1 = \mathcal{S}_1(\mathcal{H}_{2,1}\mathcal{O}_2 + \mathcal{H}_{s,1}\mathcal{O}_s) \\ \mathcal{O}_2 = \mathcal{S}_2(\mathcal{H}_{1,2}\mathcal{O}_1 + \mathcal{H}_{s,2}\mathcal{O}_s) \ . \end{cases} \tag{4.18}$$

This linear system of equations links the two unknown scattered fields $\mathcal{O}_1$ and $\mathcal{O}_2$ to the known source field $\mathcal{O}_s$ through change of basis and scattering operators. Once the scattering and change of basis operators are computed, one can deduce the fields scattered from the system constituted of both cylinders through solving Eq. 4.18. It is straightforward to generalize the technique used here to more than two cylinders.

In practice, all operators are represented by matrices and the fields $\mathcal{O}$ and $\mathcal{R}$ by the vectors consisting of the Fourier-Bessel coefficients of the fields. The matrices are readily computed, so that given the Fourier-Bessel expansion of a source field $\mathcal{O}_s$, the Fourier-Bessel coefficients describing $\mathcal{O}_1$ and $\mathcal{O}_2$ follow from solving the matrix equations equivalent to Eq. 4.18. Once $\mathcal{O}_1$ and $\mathcal{O}_2$ are known, the regular part of the field around inclusions 1 and 2 can be deduced *e.g.* from the scattering matrices through Eq. 4.15. The fields are then known in any homogeneous annulus surrounding the inclusions. In fact it appears that the superposition of outgoing fields $\mathcal{O}_s$, $\mathcal{O}_1$ and $\mathcal{O}_2$ describes the field accurately everywhere.

Using change of basis operators, we have converted the computation of the field scattered from a complex system consisting of several inclusions to the computation of scattering operators of single inclusions. Guided modes of a structure consisting of several inclusions correspond to non-zero fields around the inclusion in the absence of any exterior sources. To find modes, one therefore has to find inclusion-parameters for which Eq. 4.18 has solutions with non-zero $\mathcal{O}_1$ and $\mathcal{O}_2$ in the absence of the $\mathcal{O}_s$ term. This definition of a mode for a guiding structure is essentially the same as that which can be used to compute the modes of an ordinary step index optical fiber [Mar91].

In the next few subsections we describe the multipole method more rigorously, explicitly defining all required fields, operators and vectors, and detailing the domains of validity of the expansions. Furthermore, we adapt the method to the case in which the matrix containing the inclusions is surrounded by a jacket.

4.2.4 *Rigourous formulation of the field identities*

In the vicinity of the lth cylindrical inclusion (see Fig. 4.2), we represent the fields in the matrix in local coordinates $\mathbf{r}_l = (r_l, \theta_l) = \mathbf{r} - \mathbf{c}_l$ and express the fields in Fourier-Bessel series. With $J_m(z)$ and $H_m^{(1)}(z)$ being the usual Bessel function of order m and the Hankel function of the first kind of order m respectively, we have for the electric field

$$E_z = \sum_{m \in \mathbb{Z}} \left[A_m^{\mathrm{E}l} J_m(k_\perp^{\mathrm{M}} r_l) + B_m^{\mathrm{E}l} H_m^{(1)}(k_\perp^{\mathrm{M}} r_l) \right] e^{im\theta_l} \tag{4.19}$$

and similarly for the z-component of the scaled magnetic field K_z, but with coefficients $A_m^{\mathrm{K}l}$ and $B_m^{\mathrm{K}l}$. In 4.19 the J_m terms represent the regular incident part[1] $\mathcal{R}^{\mathrm{E}l}$ of the field E_z for cylinder l since it is finite everywhere, including inside the inclusion, while the $H_m^{(1)}$ terms represent the outgoing wave part[2] $\mathcal{O}^{\mathrm{E}l}$ of the field, associated with a source inside the cylinder. We thus have $E_z = \mathcal{R}^{\mathrm{E}l} + \mathcal{O}^{\mathrm{E}l}$.

Local expansion 4.19 is valid only in an annulus extending from the surface of the cylinder to the nearest cylinder or source (region (a) in Fig. 4.2). The same expression may be used around the jacket boundary which we designate by the superscript 0 (region d in Fig. 4.2).

Another description of the fields is originally due to Wijngaard [Wij73]. He reasoned that a field in a region can be written as a superposition of outgoing waves from all source bodies in the region. If the waves originate outside the region, their expansion is in terms of J-type waves, which are source free. Of course this physical argument can be supplemented by rigorous mathematical arguments [LMBM94; Wij73; CC94], as discussed in Appendix B.1. For MOFs, the *Wijngaard expansion* takes the form

$$E_z = \sum_{l=1}^{\mathrm{N}_i} \sum_{m \in \mathbb{Z}} B_m^{\mathrm{E}l} H_m^{(1)}(k_\perp^{\mathrm{M}} |\mathbf{r}_l|) e^{im \arg(\mathbf{r} - \mathbf{c}_l)} + \sum_{m \in \mathbb{Z}} A_m^{\mathrm{E}0} J_m(k_\perp^{\mathrm{M}} r) e^{im\theta} , \tag{4.20}$$

and is valid everywhere in the matrix. Each term of the first m series is an outgoing wave field with a source at cylinder l, while the final term,

[1] The Bessel functions J_m are continuous and finite in a bounded domain, so the field they describe must therefore have its origin in sources outside that domain.

[2] Hankel functions $H_m^{(1)}$ satisfy the outgoing wave condition and diverge at 0; their contribution to the field in an annulus surrounding an inclusion is therefore associated with fields originating in sources in or on the inclusion, and radiating away from it.

indexed by 0, is the regular field originating at the jacket boundary.

Equating 4.19 and 4.20, thus enforcing consistency, yields, in the vicinity of cylinder l,

$$\sum_{m\in\mathbb{Z}} A_m^{\mathrm{E}l} J_m(k_\perp^{\mathrm{M}} r_l) e^{im\theta_l} = \sum_{\substack{j=1\\ j\neq l}}^{N} \sum_{m\in\mathbb{Z}} B_m^{\mathrm{E}j} H_m^{(1)}(k_\perp^{\mathrm{M}} r_j) e^{im\theta_j} + \sum_{m\in\mathbb{Z}} A_m^{\mathrm{E}0} J_m(k_\perp^{\mathrm{M}} r) e^{im\theta}, \tag{4.21}$$

since the $H_m^{(1)}(k_\perp^{\mathrm{M}} r_l)$ terms are common to both expansions. Note that the sum on the left hand side of Eq. 4.21 is associated with the regular incident field for inclusion l, while the double sum on the right hand side is associated with the outgoing field originating from all other inclusions ($j \neq l$), and the last sum represents the field coming from the jacket. Hence Eq. 4.21 simply results from considering in detail the origin of the field incident on inclusion l.

Evaluating Eq. 4.21 is not straightforward since different terms refer to different origins. We therefore use the Graf's addition theorem [AS65] which lets us transform the cylindrical waves between different origins. A full discussion is given in Appendix B.2, where we show that it may be viewed as a change of basis transformation. For example the contribution to the local regular field in the vicinity of cylinder l due to cylinder j (line b, Fig. 4.2) is

$$\sum_{n\in\mathbb{Z}} A_n^{\mathrm{E}lj} J_n(k_\perp^{\mathrm{M}} r_l) e^{in\arg(\mathbf{r}_l)} = \sum_{m\in\mathbb{Z}} B_m^{\mathrm{E}j} H_m^{(1)}(k_\perp^{\mathrm{M}} r_j) e^{im\arg(\mathbf{r}_j)}, \tag{4.22}$$

where

$$A_n^{\mathrm{E}lj} = \sum_{m\in\mathbb{Z}} \mathcal{H}_{nm}^{lj} B_m^{\mathrm{E}j}, \tag{4.23}$$

$$\mathcal{H}_{nm}^{lj} = H_{n-m}^{(1)}(k_\perp^{\mathrm{M}} c_{lj}) e^{-i(n-m)\arg(\mathbf{c}_{lj})}, \tag{4.24}$$

and $\mathbf{c}_{lj} = \mathbf{c}_j - \mathbf{c}_l$, as shown in Appendix B.2.1. The physics behind Eq. 4.22 is quite intuitive, and corresponds to the *Change of Basis* paragraph in Sec. 4.2.3: the right hand term is associated with an outgoing wave originating from sources inside inclusion j. In any annulus not intersecting or including inclusion j, and in particular in an annulus centered on inclusion l, this field is regular and satisfies the Helmholtz equation. It can hence be expressed in terms of a series of Bessel functions, which is exactly what Eq. 4.22 does.

At this point we introduce the notation $\mathbf{A}^{\mathrm{E}lj} = [A_n^{\mathrm{E}lj}]$, which lets us generate vectors of mathematical objects. A similar notation is used for matrices, *i.e.*, $\boldsymbol{\mathcal{H}}^{lj} = [\mathcal{H}_{nm}^{lj}]$. In matrix form, then, we represent the basis change 4.23 as

$$\mathbf{A}^{\mathrm{E}lj} = \boldsymbol{\mathcal{H}}^{lj}\mathbf{B}^{\mathrm{E}j}. \tag{4.25}$$

Similarly, the contribution to the regular incident field at cylinder l due to the jacket (line e, Fig. 4.2) is

$$\sum_{n\in\mathbb{Z}} A_n^{\mathrm{E}l0} J_n(k_\perp^{\mathrm{M}} r_l) e^{in\arg(\mathbf{r}_l)} = \sum_{m\in\mathbb{Z}} A_m^{\mathrm{E}0} J_m(k_\perp^{\mathrm{M}} r) e^{im\theta} , \tag{4.26}$$

where the change of basis (derived in Appendix B.2.2) is

$$\mathbf{A}^{\mathrm{E}l0} = \boldsymbol{\mathcal{J}}^{l0}\mathbf{A}^{\mathrm{E}0} , \tag{4.27}$$

with

$$\boldsymbol{\mathcal{J}}^{l0} = \left[\mathcal{J}_{nm}^{l0}\right] = \left[(-1)^{(n-m)} J_{n-m}(k_\perp^{\mathrm{M}} c_l) e^{i(m-n)\arg(\mathbf{c}_l)}\right]. \tag{4.28}$$

Accumulating these contributions for all cylinders and the jacket we have, in the annulus (a) around cylinder l (see Fig. 4.2)

$$\mathbf{A}^{\mathrm{E}l} = \sum_{\substack{j=1\\ j\neq l}}^{\mathrm{N}_i} \mathbf{A}^{\mathrm{E}lj} + \mathbf{A}^{\mathrm{E}l0} = \sum_{\substack{j=1\\ j\neq l}}^{\mathrm{N}_i} \boldsymbol{\mathcal{H}}^{lj}\mathbf{B}^{\mathrm{E}j} + \boldsymbol{\mathcal{J}}^{l0}\mathbf{A}^{\mathrm{E}0}, \tag{4.29}$$

a result that holds for both the E_z and K_z fields.

In a similar way, the outgoing field in the vicinity of the jacket boundary due to cylinder j (line c, Fig. 4.2) is

$$\sum_{n\in\mathbb{Z}} B_n^{\mathrm{E}0j} H_n^{(1)}(k_\perp^{\mathrm{M}} r) e^{in\theta} = \sum_{m\in\mathbb{Z}} B_m^{\mathrm{E}j} H_m^{(1)}(k_\perp^{\mathrm{M}} r_j) e^{im\arg(\mathbf{r}_j)} , \tag{4.30}$$

with the change of basis represented by

$$\mathbf{B}^{\mathrm{E}0j} = \boldsymbol{\mathcal{J}}^{0j}\mathbf{B}^{\mathrm{E}j} , \tag{4.31}$$

where

$$\boldsymbol{\mathcal{J}}^{0j} = \left[\mathcal{J}_{nm}^{0j}\right] = \left[J_{n-m}(k_\perp^{\mathrm{M}} c_j) e^{-i(n-m)\arg(\mathbf{c}_j)}\right] , \tag{4.32}$$

as shown in Appendix B.2.3. By adding the contributions for all cylinder sources we reexpress the first term on the right-hand side of the Wijngaard expansion 4.20 in a form valid just inside the jacket (region d)

$$\sum_{l=1}^{\mathrm{N}_i} \mathcal{O}^{\mathrm{E}l} = \sum_{n\in\mathbb{Z}} B_n^{\mathrm{E}0} H_n^{(1)}(k_\perp^{\mathrm{M}} r) e^{in\theta} = \mathcal{O}^{\mathrm{E}0} , \tag{4.33}$$

where

$$\mathbf{B}^{\mathrm{E}0} = \sum_{l=1}^{\mathrm{N}_i} \mathbf{B}^{\mathrm{E}0l} = \sum_{l=1}^{\mathrm{N}_i} \mathcal{J}^{0l} \mathbf{B}^{\mathrm{E}l}, \tag{4.34}$$

a result that also holds for both E_z and K_z.

4.2.5 *Boundary conditions and field coupling*

While the field identities of the previous section apply individually to each field component, cross coupling between them occurs at boundaries. In what follows, it is most convenient to formulate the boundary conditions in terms of cylindrical reflection coefficients as derived in Appendix B.3. For circular inclusions, for the reflected fields outside each cylinder we have

$$\begin{aligned} B_n^{\mathrm{E}l} &= R_n^{\mathrm{EE}l} A_n^{\mathrm{E}l} + R_n^{\mathrm{EK}l} A_n^{\mathrm{K}l} , \\ B_n^{\mathrm{K}l} &= R_n^{\mathrm{KE}l} A_n^{\mathrm{E}l} + R_n^{\mathrm{KK}l} A_n^{\mathrm{K}l} , \end{aligned} \tag{4.35}$$

where the expression for the reflection coefficients are given in Eqs B.29 in Appendix B.3. The reflection matrices are derived for each inclusion treated in isolation, and are thus known in closed form for circular inclusions, in which case they are diagonal. For non-circular inclusions, they could be replaced by either analytic expressions for other special cases, or numerical estimates from a differential or integral equation treatment [FTM94; MV72]. In these cases they generally also have off-diagonal elements.

Equations (4.35) can be written as

$$\begin{bmatrix} \mathbf{B}^{\mathrm{E}l} \\ \mathbf{B}^{\mathrm{K}l} \end{bmatrix} = \begin{bmatrix} \mathbf{R}^{\mathrm{EE}l} & \mathbf{R}^{\mathrm{EK}l} \\ \mathbf{R}^{\mathrm{KE}l} & \mathbf{R}^{\mathrm{KK}l} \end{bmatrix} \begin{bmatrix} \mathbf{A}^{\mathrm{E}l} \\ \mathbf{A}^{\mathrm{K}l} \end{bmatrix}, \tag{4.36}$$

or

$$\tilde{\mathbf{B}}^l = \tilde{\mathbf{R}}^l \tilde{\mathbf{A}}^l , \tag{4.37}$$

with $\mathbf{R}^{\mathrm{EE},l} = \mathrm{diag}(R_n^{\mathrm{EE}l})$ and similar definitions for the other reflection matrices. We also need an interior form at the jacket boundary (point d in

Fig. 4.2),

$$\tilde{\mathbf{A}}^0 = \tilde{\mathbf{R}}^0\tilde{\mathbf{B}}^0 \tag{4.38}$$

where $\tilde{\mathbf{A}}^0$, $\tilde{\mathbf{B}}^0$ and $\tilde{\mathbf{R}}^0$ are defined as in Eqs. 4.36–4.37, and the coefficients of $\tilde{\mathbf{R}}^0$ are given by Eqs. B.25. In this form the outgoing field ($\tilde{\mathbf{B}}^0$) generated by all inclusions (line c) is reflected by the jacket to generate the regular field ($\tilde{\mathbf{A}}^0$), which feeds into the incident field for inclusion l (line e in Fig. 4.2). It is straightforward to adapt $\tilde{\mathbf{R}}^0$ to cases where multiple films surround the hole region.

4.2.6 *Derivation of the Rayleigh Identity*

With the structure of the field coupling derived in Section 4.2.5, we now form field identities applying to the vector components $\tilde{\mathbf{A}}^l$ and $\tilde{\mathbf{B}}^l$. From Eq. 4.29 we have

$$\tilde{\mathbf{A}}^l = \sum_{\substack{j=1\\ j\neq l}}^{\mathrm{N}_i} \tilde{\mathcal{H}}^{lj}\tilde{\mathbf{B}}^j + \tilde{\mathcal{J}}^{l0}\tilde{\mathbf{A}}^0\,, \tag{4.39}$$

where $\tilde{\mathcal{H}}^{lj} = \mathrm{diag}(\mathcal{H}^{lj}, \mathcal{H}^{lj})$, and $\tilde{\mathcal{J}}^{l0} = \mathrm{diag}(\mathcal{J}^{l0}, \mathcal{J}^{l0})$. Equation 4.39 is the representation of the regular incident field at cylinder l in terms of outgoing components $\tilde{\mathbf{B}}^j$ from all other cylinders and an incident field contribution $\tilde{\mathbf{A}}^0$ from the jacket.

Combining Eq. 4.39 for all cylinders $l = 1\ldots\mathrm{N}_i$ and introducing $\boldsymbol{\mathcal{A}} = \left[\tilde{\mathbf{A}}^l\right]$ and $\boldsymbol{\mathcal{B}} = \left[\tilde{\mathbf{B}}^l\right]$, we derive

$$\boldsymbol{\mathcal{A}} = \tilde{\mathcal{H}}\boldsymbol{\mathcal{B}} + \tilde{\mathcal{J}}^{\mathrm{B}0}\tilde{\mathbf{A}}^0\,, \tag{4.40}$$

where $\tilde{\mathcal{H}} = \left[\tilde{\mathcal{H}}^{lj}\right]$ for $l, j = 1...\mathrm{N}_i$ with $\tilde{\mathcal{H}}^{ll} \equiv \mathbf{0}$ and

$$\tilde{\mathcal{J}}^{\mathrm{B}0} = \left[\tilde{\mathcal{J}}^{l0}\right] = \left[(\tilde{\mathcal{J}}^{10})^T, (\tilde{\mathcal{J}}^{20})^T, \cdots, (\tilde{\mathcal{J}}^{\mathrm{N}_i0})^T\right]^T, \tag{4.41}$$

where the T indicates the transpose. Similarly, the vector outgoing field in the vicinity of the jacket due to all the cylinders is given by

$$\tilde{\mathbf{B}}^0 = \sum_{j=1}^{\mathrm{N}_i} \tilde{\mathcal{J}}^{0l}\tilde{\mathbf{B}}^l = \tilde{\mathcal{J}}^{0\mathrm{B}}\boldsymbol{\mathcal{B}} \tag{4.42}$$

from Eq.4.34. Here

$$\tilde{\mathcal{J}}^{0\mathrm{B}} = \left[\tilde{\mathcal{J}}^{0l}\right] = \left[\tilde{\mathcal{J}}^{01}, \tilde{\mathcal{J}}^{02}, \cdots \tilde{\mathcal{J}}^{0\mathrm{N}_i}\right] . \tag{4.43}$$

Combining Eqs. 4.37, Eqs. 4.38, 4.40 and 4.42 and eliminating $\tilde{\mathbf{A}}^0$ and $\tilde{\mathbf{B}}^0$ we form a homogeneous system of equations (which represents the *Rayleigh identity*, which we will also call the *field identity*) in the source coefficients

$$\left[\mathbf{I} - \mathcal{R}\left(\tilde{\mathcal{H}} + \tilde{\mathcal{J}}^{\mathrm{B}0}\tilde{\mathbf{R}}^0\tilde{\mathcal{J}}^{0\mathrm{B}}\right)\right]\mathcal{B} \equiv \mathcal{M}\mathcal{B} = \mathbf{0}, \tag{4.44}$$

where the right-hand side indicates the absence of external sources, and

$$\mathcal{R} = \mathrm{diag}\left[\tilde{\mathbf{R}}^1, \tilde{\mathbf{R}}^2, \cdots, \tilde{\mathbf{R}}^{\mathrm{N}_i}\right]. \tag{4.45}$$

Non-trivial solutions to the homogeneous system (4.44) correspond to non-zero fields propagating in the z-direction. The solutions represent a non-zero field existing without any exterior source of energy, in other words propagating (possibly leaky) fibre modes. We will see in the next Chapter how this equation can be solved to obtain the modes of a MOF structure.

4.3 Symmetry Properties of MOF

4.3.1 *Symmetry properties of modes*

In the previous section, we use a generalization of the study carried out by Lord Rayleigh [Ray92] more than a century ago (a more extensive comparison is presented at the beginning of the next Chapter p. 205). In the following we will use the formalization of a work initiated in the same period, *i.e.* in 1894, by Pierre Curie [Cur94] concerning symmetry properties in electromagnetics.

This formalization was realized by P. McIsaac in two seminal articles [Isa75a; Isa75b]. All the theoretical results presented here can be found in much more detail in these two articles.

Using group representation theory, he classified the electromagnetic modes of waveguide structures, according to the symmetry properties of the configuration. Due to the symmetry properties exhibited by the structure of most MOF, his results are very useful for classifying, before any computation, the modes existing in the studied MOF.

A structure which posseses only rotationnal symmetry with no reflection symmetry, and for which $2\pi/n$ (n being an integer) is the smallest angle associated with a symmetry operation is said to possess[3] the symmetry group C_n of order n. Fig. 4.4 shows the cross sections of several structures with C_n symmetry. If a structure has an n-fold rotation symmetry together

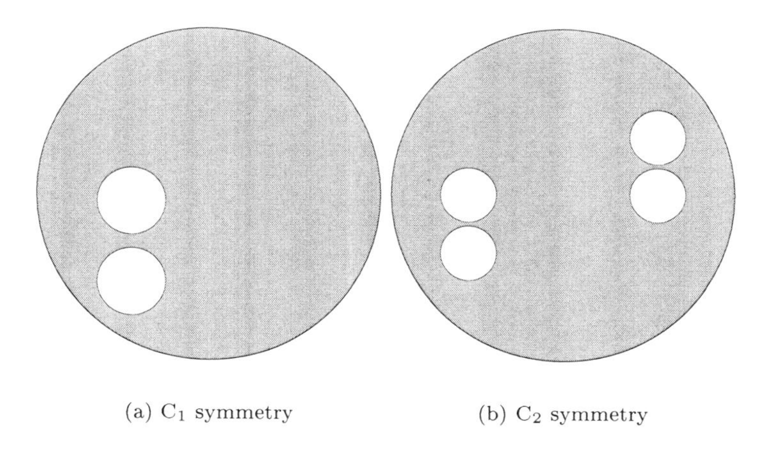

(a) C_1 symmetry (b) C_2 symmetry

Fig. 4.4 Two examples of a structure with a C_n symmetry. The structure is invariant through a rotation of $2\pi/n$ and there is no reflection symmetry.

with at least one plane of relection symmetry, then there are exactly n planes of reflection symmetry. All these planes intersect along the axis of the rotational symmetry and the angular space between them is equal to π/n. The symmetry group for this kind of structure is called C_{nv} (see Fig. 4.5).

McIsaac assigned the modes of a waveguide to classes depending on the azimuthal symmetry of the modal fields patterns, and he called these classes *mode classes.* Each mode class contains an infinite number of modes. For a waveguide with C_n symmetry there are n different mode classes, and for a waveguide with $C_n v$ symmetry there are $n+1$ distinct mode classes if n is odd and $n+2$ classes if n is even. In each class, all the modes are either non-degenerate or two-fold degenerate (see table 4.1).

The special case of $C_{\infty v}$ which corresponds to the limiting case $n \to \infty$ is also treated in McIsaac's study [Isa75a; Isa75b].

In his nomenclature, McIsaac labels the two mode classes containing nondegenerate modes as the first and the second mode classes (p=1,2). The two remaining mode classes containing nondegenerate modes existing

[3] A structure with the C_1 symmetry has actually no symmetry.

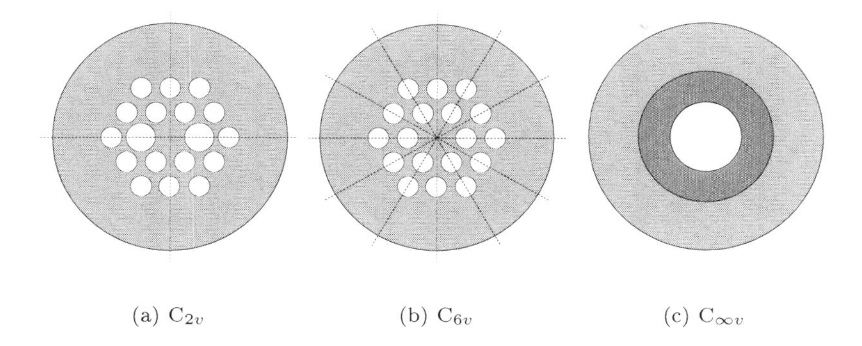

(a) C_{2v} (b) C_{6v} (c) $C_{\infty v}$

Fig. 4.5 Three examples of a structure with a C_{nv} symmetry. The structure is invariant through a rotation of $2\pi/n$ and there is a least one reflection symmetry. The dashed lines represent the symmetry planes. The $C_{\infty v}$ symmetry is the limit case $n \to \infty$.

Table 4.1 Table of mode classes and mode degeneracies for uniform waveguides with $C_n v$ symmetry.

n	Number of non-degenerate mode classes	Number of pairs of two-fold degenerate mode classes	Total number of mode classes
odd	2	(n-1)/2	n+1
even	4	(n-2)/2	n+2

if n is even are placed at the end of the list (p=n+1,n+2). In each class of symmetry, modes have different symmetry properties which can be used to simplify the computation of the modal fields and propagation constants. Indeed, McIsaac showed that for each symmetry class these computations need to take into account only a limited angular sector, the minimum sector, with vertex at the center of the waveguide and appropriate boundary conditions along the edges of the sector.

It must be noted that only the nondegenerate modes satisfy all the symmetries of the waveguide, and on the contrary the degenerate modes will possess only a subgroup of the symmetries. We now focus on C_{nv} waveguides (the C_n case being also treated in McIsaac's articles) since this point group is the most frequently encountered group in MOF studies. Both the angle of the minimum sector of each mode class and the boundary conditions are prescribed by the theory. Two kinds of boundary conditions are defined: the short-circuit boundary condition, *i.e.* the tangential electric field has to vanish at the boundary, and the open-circuit boundary condition, *i.e.* the tangential magnetic field has to vanish at the boundary.

The boundary lines of the minimum sector must coincide with two of the planes of reflection symmetry of the waveguide. In Fig. 4.6, we give the minimun sectors for waveguides with C_{6v} symmetry. For this point group, there are 8 mode classes. For non-degenerate mode classes $p = 1, 2, 7, 8$ the angle of the minimum sector is equal to $\pi/6$, and for the four degenerate mode classes $p = 3, 4$ and $p = 5, 6$ the angle is equal to $\pi/2$. In Fig. 4.7,

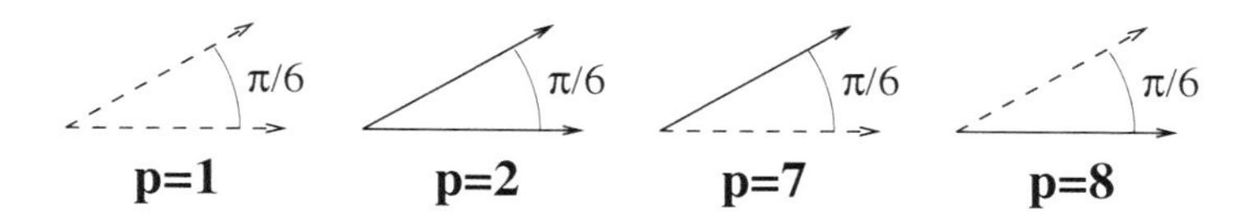

(a) Mode classes $p = 1, 2, 7$ and 8 are non-degenerate.

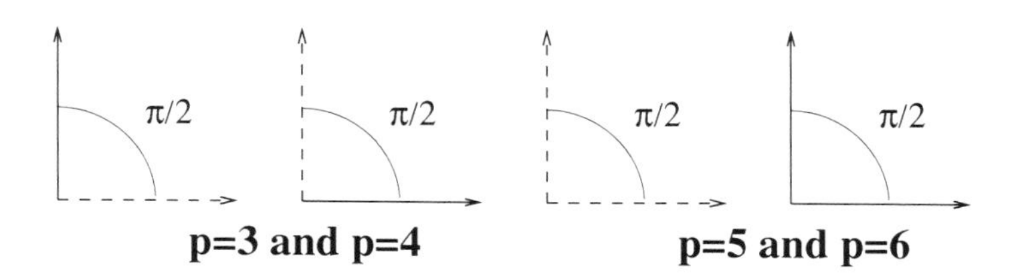

(b) Mode classes $p = 3, 4$ and $p = 5, 6$ are two-fold degenerate.

Fig. 4.6 Minimum sectors for waveguides with C_{6v} symmetry. Solid lines indicate zero tangential electric field, and dashed lines indicate zero tangential magnetic field.

we give the minimum sectors for waveguides with C_{4v} symmetry. For this point group, there are 6 mode classes. For non-degenerate mode classes $p = 1, 2, 5, 6$ the angle of the minimun sector is equal to $\pi/4$, and for the two degenerate mode classes $p = 3, 4$ the angle is equal to $\pi/2$.

In Fig. 4.8, we give the minimun sectors for waveguides with C_{2v} symmetry. For this point group, there are 4 mode classes, all of them being non-degenerate. The angle of the minimun sector is equal to $\pi/2$ for the 4 mode classes.

The application of McIsaac's theoretical results for C_{6v} MOFs has already been used to show that the fundamental mode of these fibres is degenerate [SWdS$^+$01]. Consequently, these perfect fibers are not birefringent. Clear illustrations of this property will be given in section 4.6.1.

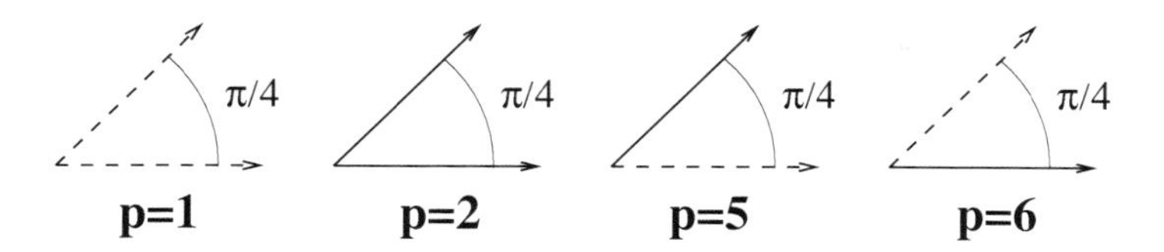

(a) Mode classes $p = 1, 2, 5$ and 6 are non-degenerate.

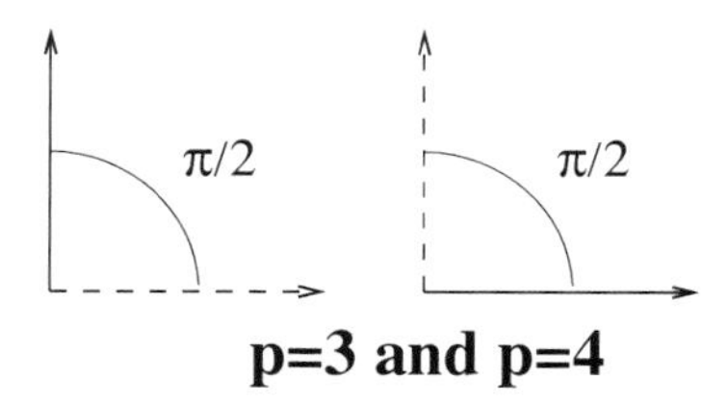

(b) Mode classes $p = 3, 4$ are two-fold degenerate.

Fig. 4.7 Minimum sectors for waveguides with C_{4v} symmetry. Solid lines indicate zero tangential electric field, and dashed lines indicate zero tangential magnetic field.

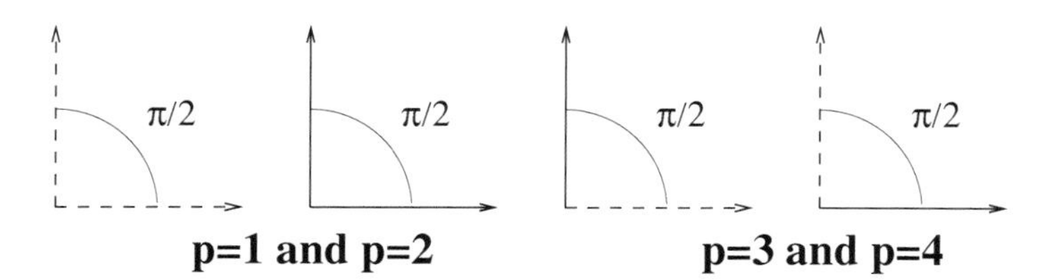

Fig. 4.8 Minimum sectors for waveguides with C_{2v} symmetry. Mode classes $p = 1, 2, 3$ and 4 are non-degenerate. Solid lines indicate zero tangential electric field, and dashed lines indicate zero tangential magnetic field.

4.4 Implementation

In this section, we discuss the implementation of the Multipole Method to find the modes and dispersion characteristics of a given fibre structure. The homogeneous equation 4.44 corresponds to a non-trivial field vector $\boldsymbol{\mathcal{B}}$ only if the determinant of the matrix $\boldsymbol{\mathcal{M}}$ is effectively zero. Once the structure and wavelength are given, the matrix $\boldsymbol{\mathcal{M}}$ depends only on β, or, equivalently, its effective index n_{eff}. The search for modes therefore becomes a matter of finding zeros of the complex function $\det(\mathcal{M})$ of the complex variable n_{eff}. To investigate this numerically, field expansions such as Eq. 4.19 must be truncated, say with m running from $-M$ to M. M will be called the *truncation order parameter*.

We know from the previous section on symmetry properties, that the modes of fibres with a C_{nv} symmetry are either nondegenerate or doubly degenerate. Since $\det(\mathcal{M})$ is the product of the eigenvalues of $\mathcal{M}$, we must look for minima in which one or two of the eigenvalues have magnitudes that are substantially smaller than the others. However, a minimum of the determinant may also correspond to an artefact resulting from many eigenvalues being small simultaneously (which we call a false minimum). To distinguish these from genuine solutions, we consider the singular values [PBTV86] of $\mathcal{M}$, which, for our case, correspond to the magnitudes of the eigenvalues. False minima can be distinguished by a singular value decomposition at the putative minimum.

The null vectors corresponding to small singular values are approximate solutions to the field identity 4.44. For non-degenerate modes, the null vector is unique to within an arbitrary multiplicative constant. For a two-fold degenerate mode, we let the basis states be prescribed by symmetry properties (see Section 4.3), though any linear combination of these is equally justified.

4.4.1 *Finding modes*

For this task we need an algorithm aimed at finding all the zeros of the determinant of $\mathcal{M}$ in a region of the complex n_{eff} plane. The algorithm should be economical in function calls as each evaluation of the determinant is computationally very expensive for large structures. As it can be seen in Fig. 4.9, the zeros are very sharp, so that a very accurate first estimate of the zero is necessary for most simple root-finding routines. More specific algorithms for finding zeros of analytic or meromorphic functions [KB00; BCM83] have good convergence for simple structures (with six cylinders for example) but fail for more complex structures, even with good initial guesses. The present approach to root finding seems computationally efficient. We first compute a map of the modulus of the determinant over the region of interest, and then use the local minima of this map as initial points for a mixed zooming and modified Broyden [Bro65] algorithm (an iterative minimization algorithm, guaranteed to converge for parabolic minima).

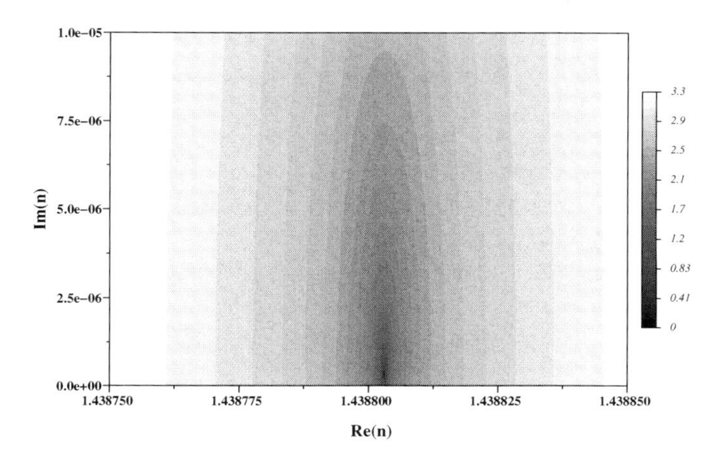

Fig. 4.9 Map of the logarithm of the magnitude of the determinant of $\mathcal{M}$ versus the real and the imaginary parts of the complex refractive index n. The minimum (in the black region) is associated with the fundamental mode of the test MOF described in section 4.5.1.

The initial scanning region has to be chosen in accordance with the physical problem studied (see section 1.5.5 page 25): if the fibre is air cored and air guided modes are sought, we choose $\Re e(n_{\text{eff}}) < 1$, whereas if the fibre has a solid core we usually choose to search for modes in a region where $\Re e(n_{\text{eff}})$ is between the optical indices of the inclusions and the matrix. In the latter case hundreds of modes may exist for small $\Re e(n_{\text{eff}})$ which are of little interest because of their high losses. We therefore often concentrate on a smaller n_{eff} scanning region near the highest index of the structure. A scanning region for $\Im m(n_{\text{eff}})$ giving good results in almost any case is $10^{-12} < \Im m(n_{\text{eff}}) < 10^{-3}$.

4.4.2 *Dispersion characteristics*

The above process of finding modes is carried out for a specific wavelength. We could reiterate the search for modes for many different wavelengths to obtain dispersion characteristics, but this process would be quite laborious. We have found two alternative methods to be of value. One computes and identifies the modes for three or four different wavelengths, then uses a spline interpolation to estimate the n_{eff} values for other wavelengths and refines the estimate with the mixed zooming and Broyden algorithm. Each newly determined point of the $n_{\rm eff}(\lambda)$ curve can be used to enhance the spline estimate. The second (and often more efficient) method is to compute the modes for only one wavelength λ_0, then slightly perturb the wavelength to get $n_{\rm eff}(\lambda_0 + \delta_\lambda)$ using $n_{\rm eff}(\lambda_0)$ as a first guess, and then use a first order estimate of $n_{\rm eff}$ at the next wavelength. One can then compute $n_{\rm eff}(\lambda + m\delta_\lambda)$ using a first order estimate computed from the two preceding values. For both methods, the wavelength step has to be chosen to be very small: For small steps more points are necessary to compute the dispersion characteristics in a given wavelength range, but for large steps the first order guess becomes inaccurate and the convergence of the zooming and the Broyden algorithm becomes unacceptably slow. Having small steps and therefore numerous numerical values on the dispersion curve is also of benefit when evaluating second order derivatives, as is necessary when computing the group velocity dispersion.

Material dispersion can be included easily by changing the optical indices according to the current wavelength at each step using, for example, Sellmeier approximation [Agr01] for silica.

The method described here can be adapted to study the change of $n_{\rm eff}$ of a mode for any continuously varying parameter, for example cylinder radius, cylinder spacing or optical index. One problem that can occur when following the evolution of a mode with a continously varying parameter is mode crossing. This results in wrong data, but can easily be detected in most cases through a discontinuity of derivatives, and can also easily be eliminated by restarting the algorithm with the correct mode on the other side of the crossing. The correct choice for δ_λ strongly depends on wavelength and structure, so that no general advice can be given.

4.4.3 *Using the symmetries within the Multipole Method*

The incorporation of field symmetry in the multipole formulation has two benefits. Firstly, it enables definitive statements to be made about the degeneracies of modes, even in the presence of the accidental degeneracies which arise when normally distinct modal trajectories cross each other. Secondly, it greatly reduces computational burdens, enabling accurate answers for quite large MOF structures to be obtained rapidly on PC's.

In applying the multipole formulation to large C_{6v} MOFs, it is highly advantageous to exploit the symmetry properties in Fig. 4.5(b) to reduce the size of the matrix $\boldsymbol{\mathcal{M}}$. This can be achieved since only multipole coefficients for inclusions lying in the minimum sector indicated in Fig. 4.6 need be specified; those for holes outside the minimum sector can be obtained by multiplying by the appropriate geometric phase factor (related to $\exp(2\pi i/6)$). The resulting reduction in the order of the matrix $\boldsymbol{\mathcal{M}}$ depends on the type of the mode, being maximal for the non-degenerate modes in Fig. 4.6(a), and still being around 3.5 for degenerate modes (see Fig. 4.6(b)), leading to considerable reductions in processing time.

4.4.3.1 *Another way to obtain $\Im m(\beta)$*

One of the challenges of MOF consists in finding leaky modes with losses as small as possible. We are thus faced with a practical problem. The real part of β ($\Re e(\beta)$) is of the same magnitude as k_0 whereas its imaginary part ($\Im m(\beta)$) can be extraordinary smaller, say $10^{-15}\, k_0$! In such circumstances, how can we obtain β'' with a sufficiently good precision? For this purpose, the losses cannot be determined directly via the poles of the scattering matrix. The fact that we are unable to do this is, of course, linked with the computer precision. We are therefore "doomed" to adopt a new strategy by making use of an energy conservation criterion. More precisely, let us consider the closed surface depicted in Fig. 4.10. If we assume that we will only be dealing with lossless materials, we have the following identity;

$$\int_{\Sigma} \Re e\{\underset{\sim}{\boldsymbol{S}}\} \cdot \mathbf{n}_{|\Sigma}\, ds = 0 \tag{4.46}$$

where $\boldsymbol{\mathcal{S}}$ is the complex Poynting vector, namely $\boldsymbol{\mathcal{S}} = \frac{1}{2}\boldsymbol{\mathcal{E}} \times \overline{\boldsymbol{\mathcal{H}}}$. As a result, $\boldsymbol{\mathcal{S}}$ can be expressed as follows:

$$\underset{\sim}{\boldsymbol{S}}(x, y, z) = \boldsymbol{S}(x, y) e^{-2\Im m(\beta) z} \tag{4.47}$$

with $\boldsymbol{S} = \frac{1}{2}\boldsymbol{E} \times \overline{\boldsymbol{H}}$. The equation 4.46 can be recast as per:

$$\int_{\Sigma_1} \Re e\{\boldsymbol{S}\} \cdot (-\boldsymbol{e}^z) \rho\, d\rho d\theta + e^{-2\Im m(\beta) z_0} \int_{\Sigma_2} \Re e\{\boldsymbol{S}\} \cdot (\boldsymbol{e}^z) \rho\, d\rho d\theta$$
$$+ \int_{\Sigma_R} \Re e\{\boldsymbol{S}\} e^{-2\Im m(\beta) z} \cdot (\boldsymbol{e}^r) R\, d\theta dz = 0\,. \tag{4.48}$$

If we let $S_z = \boldsymbol{S} \cdot \boldsymbol{e}^z$ and $S_r = \boldsymbol{S} \cdot \boldsymbol{e}^r$, we are led to:

$$(e^{-2\Im m(\beta) z_0} - 1) \int_{(\theta,\rho)\in[0,2\pi[\times[0,R]} \Re e\{S_z\} \rho \, d\rho d\theta + \left(\int_0^{z_0} e^{-2\Im m(\beta) z} \, dz \right) R \int_0^{2\pi} \Re e\{S_r\} \, d\theta = 0 \, . \tag{4.49}$$

Now $\int_0^{z_0} e^{-2\Im m(\beta) z} \, dz = -\frac{1}{2\Im m(\beta)} (e^{-2\Im m(\beta) z_0} - 1)$, therefore $\Im m(\beta)$ can be expressed as follows:

$$\Im m(\beta) = \frac{R \int_0^{2\pi} \Re e\{S_r\} \, d\theta}{\int_{(\theta,\rho)\in[0,2\pi[\times[0,R]} \Re e\{S_z\} \rho \, d\rho d\theta} \tag{4.50}$$

Via the previous identity, we then deduce the imaginary part of n_{eff} by

$$\Im m(n_{\text{eff}}) = \frac{\Im m(\beta)}{k_0} \, . \tag{4.51}$$

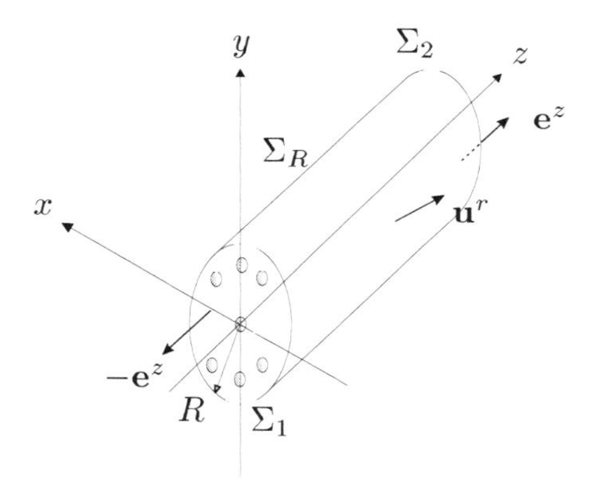

Fig. 4.10 The surface Σ is defined as the boundary of a "slice" of a circular cylinder of radius R and length z. Σ is also defined in the following way: $\Sigma = \Sigma_1 \cup \Sigma_2 \cup \Sigma_R$, where $\Sigma_1 = \{(r, \theta, z) \in \mathbb{R}^3, r \leq R, \theta \in [0, 2\pi[, z = 0\}$, $\Sigma_2 = \{(r, \theta, z) \in \mathbb{R}^3, r \leq R, \theta \in [0, 2\pi[, z = z_0\}$ and $\Sigma_R = \{(r, \theta, z) \in \mathbb{R}^3, r = R, \theta \in [0, 2\pi[, z \in [0, z_0]\}$

4.4.4 *Software and computational demands*

We have developed a FORTRAN 90 code to exploit the above considerations. For symmetric structures the suggested optimizations are used and the software can therefore deal, even on PC's, with large structures (modes for fibres with 180 holes have so far been computed on current personal computers). Once the structure has been defined, the software is able to find automatically all the modes within a given region of the complex plane for n_{eff} and can optionally track a mode as a function of wavelength to obtain dispersion characteristics. Material dispersion can be included, if desired.

Computational demands are relatively modest: the complete set of modes with the truncation order parameter $M = 5$ in the region of interest $1.4 < \Re e(n_{\text{eff}}) < 1.45$ for the structure (with a silica matrix) used in Fig. 4.14 can be computed on a current personal computer in less than one minute using less than 2Mb of memory. Of course the load rises for larger structures, but the complete set of modes for a structure having three layers of holes in a hexagonal arrangement takes less than a quarter of an hour (and about 15Mb memory) on a workstation. Dispersion curves can be computationally more expensive: the loss curves in Chapter 7 took from about half an hour (for $d/\Lambda = 0.075$ where we used $M = 5$) to several hours (for $d/\Lambda = 0.7$ where $M = 7$ was needed for accuracy).

4.5 Validation of the Multipole Method

Even though the Multipole Method is mathematically rigorous, the computed results must be validated. Indeed, the computations are numerical, not analytic and the implementation itself could be erroneous. This validation will be realized in two steps: first the numerical convergence and the self-consistency of the method will be checked, and finally we will compare results obtained from the Multipole Method with those compused using other numerical methods.

4.5.1 *Convergence and self-consistency*

The main approximation used in numerical computations via the Multipole Method is the truncation of the Fourier-Bessel series used to described the electromagnetic fields. In order to check the convergence of the numerical results, the effective indices of the computed modes must converge as M, the truncation order parameter (see section 4.4), is increased. Our test configuration will be the following: The MOF has a C_{6v} symmetry, the distance between the centres of a pair of adjacent holes Λ, called the pitch, is equal to $6.75\,\mu m$, the hole diameter d is fixed at $5.0\,\mu m$, and $\lambda = 1.55\,\mu m$. The silica matrix is considered as infinite and its relative permittivity is taken from the Sellmeier expansion at the corresponding wavelength [Agr01] ($\varepsilon_r = 2.0852042$). As can be seen in the first three columns of table 4.2, the convergence of n_{eff} with respect to M is indeed verified. The data are plotted in figure 4.11. For $M < 17$, the determinant is always below 1.0e-10, and for $M \geq 17$ the determinant is greater than 2.0e-9. This means that a value of M that is too high can lead to a bad determination of the effective index due to cumulative numerical errors in matrix computations. An other example of the convergence of the Multipole Method can be found in the following publication [KWR+02].

Table 4.2 Convergence of n_{eff} with the truncation order parameter M. Results are for the p=3 fundamental mode of the test MOF described in section 4.5.1. $W^E_{l,m}$ is also defined in this section: it measures the accuracy of the equality between the local and the global expansions of the z-component of the electric field.

M	$\Re e(n_{\text{eff}})$	$\Im m(n_{\text{eff}})$	$W^E_{1,1}$	$W^E_{2,1}$
3	1.43880301866122	3.31002759474698e-07	0.1329E-01	0.4563E-01
4	1.43877685791377	3.85061499591580e-08	0.5049E-02	0.8909E-02
5	1.43877422802377	4.27582918316796e-08	0.4549E-02	0.7941E-02
6	1.43877411171115	4.32504231569041e-08	0.1890E-02	0.3811E-02
7	1.43877411102638	4.32512291538762e-08	0.3527E-03	0.7641E-03
8	1.43877410938293	4.32574507538176e-08	0.2330E-03	0.4270E-03
9	1.43877410904527	4.32582456673292e-08	0.1574E-03	0.3146E-03
10	1.43877410903106	4.32581736812954e-08	0.4535E-04	0.9578E-04
11	1.43877410902931	4.32582277732338e-08	0.1086E-04	0.2190E-04
12	1.43877410902817	4.32582649286026e-08	0.1132E-04	0.2290E-04
13	1.43877410902807	4.32582290535797e-08	0.4683E-05	0.9681E-05
14	1.43877410902806	4.32582111271408e-08	0.8615E-06	0.1803E-05
15	1.43877410902810	4.32582350418105e-08	0.7263E-06	0.1504E-05
16	1.43877410902803	4.32581931220263e-08	0.4118E-06	0.8553E-06
17	1.43877410902643	4.32595358437832e-08	0.1005E-06	0.1869E-06

In the multipole formulation we have two different expansions for the field around each inclusion, the local one (Eq. 4.19) and the global one also called the Wijngaard expansion (Eq. 4.20). These two representations coincide only for untruncated Fourier-Bessel series so their numerical difference can be used as an accurate indicator of truncation errors and of the precision of the mode location. We introduce for the z-component of the electric field, the quantity $W^E_{l,k}$ defined by:

$$W^E_{l,k} = \frac{\int_{C_{l,k}} |E_z^{\text{local}}(\theta_{l,k}) - E_z^{\text{Wijngaard}}(\theta_{l,k})| d\theta_{l,k}}{\int_{C_{l,k}} |E_z^{\text{Wijngaard}}(\theta_{l,k})| d\theta_{l,k}}. \tag{4.52}$$

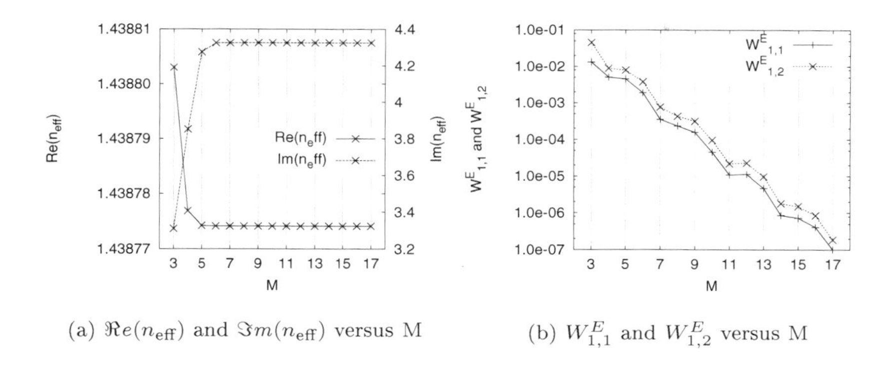

(a) $\Re e(n_{\text{eff}})$ and $\Im m(n_{\text{eff}})$ versus M (b) $W_{1,1}^E$ and $W_{1,2}^E$ versus M

Fig. 4.11 Convergence of n_{eff} and of W^E with respect to M. The data are the same as in table 4.2.

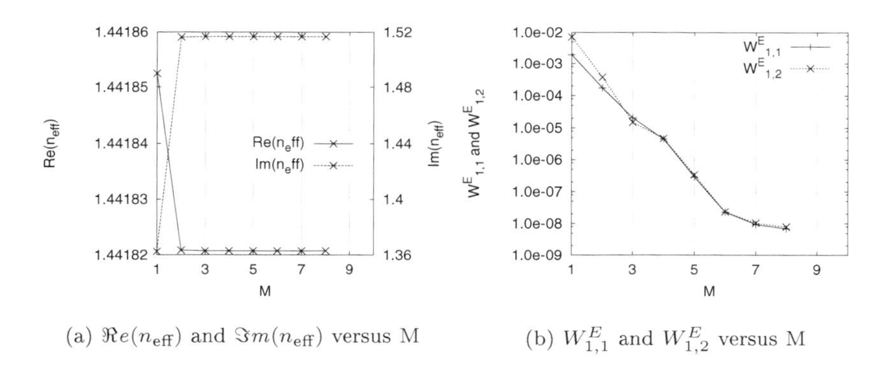

(a) $\Re e(n_{\text{eff}})$ and $\Im m(n_{\text{eff}})$ versus M (b) $W_{1,1}^E$ and $W_{1,2}^E$ versus M

Fig. 4.12 Convergence of n_{eff} and of W^E with M, the MOF is the same as in Fig. 4.11 except that the hole diameters are equal to 1.5 μm.

The doublet l, k stands for the lth cylinder of the kth elementary sector of the MOF structure. A similar quantity can be defined for the scaled magnetic field. As can be seen in table 4.2 or in figures 4.11(b) and 4.12(b) $W_{l,k}^E$ decreases while n_{eff} stabilises, as expected.

From numerical simulations, we have found that the truncation order parameter M should be made larger by a factor of around 1.5 times the largest argument of the Bessel functions on the boundary of the inclusions.

This empirical rule is clearly illustrated in Fig. 4.11 and Fig. 4.12: in the first case where the hole diameter is equal to 5 μm, M=8 is required to get the correct n_{eff} with a 10^{-9} accuracy and in the second case where the hole diameter is equal to 1.5 μm, M=3 is enough.

4.5.2 *Comparison with other methods*

The results obtained by applying the Multipole Method to MOF were compared to one of the first results obtained through a finite element method [BMPR00]. The results are in very good agreement and the existing differences are mainly due to the limitations caused by the numerical implementation (the mesh does not follow the symmetry properties of the structure) of the finite element method [KRM03]. Another numerical implementation of the same multipole method was successfully compared to a beam propagation method [WMdS+01] and to a plane-wave method [SWdS+01] (the two implementations of the Multipole Method giving the same results). For the last case, it was shown that if enough plane-waves (more than 2^{16}) are considered then the two methods give the same results concerning the fundamental mode degeneracy[4] for the simple MOF structure described in the beginning of the previous section.

4.6 First Numerical Examples

We now illustrate the numerical implementation of the Multipole Method through several basic numerical examples. These examples will exhibit the most frequently encountered symmetries in MOF: C_{2v}, C_{4v} and C_{6v}. We will not follow an arithmetical order but start with the most pedagogical example: the C_{6v} six hole MOF.

4.6.1 *A detailed C_{6v} example: the six hole MOF*

The cross-section of this MOF can be seen in Fig. 4.15. The geometrical parameters of the fibre are the same as the ones of the six hole MOF used in section 4.5.1 ($\Lambda = 6.75\ \mu m$, $d = 5.0\ \mu m$), and the wavelength is also the same: $\lambda = 1.55\,\mu m$, but the relative permittivity ε_r is set to 2.1025. M is equal to 8 in all the numerical simulations of this section. In table 4.3, we give the effective indices n_{eff} of the main modes of this C_{6v} MOF. It is

[4]The plane-wave method is not able to take into account the mode losses, so the imaginary part of the mode effective index cannot be computed.

easy to see that the modes labelled C3/4-a of mode classes 3 and 4 have both the highest real part of n_{eff} and the lowest imaginary part, *i.e.* the lowest losses. These modes correspond to the fundamental mode of the MOF. We already know from the results described in section 4.3.1 that the modes of mode classes 3 and 4 are two-fold degenerate in a C_{6v} structure: this is exactly what is found for the C_{6v} example. In Fig. 4.14, we give the moduli of the electromagnetic field longitudinal components E_z and H_z for the two mode classes 3 and 4. As expected, the fields of these two degenerate modes do not have all the symmetry properties of the MOF structure but only a subset. It can be easily checked that these fields fulfil the symmetry properties described in Fig. 4.6(b). The longitudinal component of the Poynting vector is shown in Fig. 4.13, and it is the same for the two degenerate modes whilst being well localized in the core of the MOF.

In order to complete the description of the degenerate fundamental mode, the transverse electromagnetic vector fields for the two mode classes are given in Fig. 4.15. Since in the Multipole Method the electromagnetic fields are complex (see Eqs. (4.2)), we must consider both the real part and the imaginary part of the fields $\mathbf{E}(r, \theta)$ and $\mathbf{K}(r, \theta)$ because of the factor $e^{i(\beta z - \omega t)}$ in the full expression of the propagating electromagnetic fields. This explains why they are both drawn in the vector field figures. Since the vector fields built from the real and the imaginary parts of $\mathbf{E}(r, \theta)$ (or $\mathbf{K}(r, \theta)$) are parallel in the MOF core, we conclude that the physical field, *i.e.* the real part of $\boldsymbol{\mathcal{E}}(r, \theta, z, t)$, is linearly polarized in the MOF core for the degenerate fundamental mode.

We now describe more briefly the main higher order modes of this six hole MOF. First of all we must recall that there is no order relation in $\mathcal{C}$, and consequently it is not possible *a priori* to classify the leaky higher order modes found in this structure with the definition used when the effective indices are purely real. If the imaginary part varied monotonicaly with respect to the real part, it would have been possible to keep a simple ordering for the modes. But as can be seen in Fig. 4.16, this is no longer the case after the sequence C3/4-a (fundamental mode), C2-a, C5/6-a, and C1-a. In the following pages we give for the higher order modes (C2-a, C5/6-a, and C1-a) the moduli of the electromagnetic fields and Poynting vector longitudinal components followed by the transverse electromagnetic vector fields. It has not yet been proven that this order (C2-a, C5/6-a, and

C1-a) is conserved for all C_{6v} MOF[5].

Table 4.3 Mode class, degeneracy, and effective index for the main modes of the C_{6v} six hole MOF of section 4.5.1. The results are computed at $\lambda = 1.55\,\mu m$.

Mode class	Degeneracy	$\Re e(n_{\text{eff}})$	$\Im m(n_{\text{eff}})$	Label in Fig. 4.16
1	1	0.14307554E+01	0.19457921E-05	C1-a
		0.14061823E+01	0.88085482E-03	C1-b
2	1	0.14310182E+01	0.70048952E-06	C2-a
		0.14041826E+01	0.29291417E-03	C2-b
3, 4	2	0.14387741E+01	0.43257457E-07	C3/4-a
		0.14211904E+01	0.21562081E-04	C3/4-b
		0.14176892E+01	0.46170513E-04	C3/4-c
5, 6	2	0.14308483E+01	0.13214492E-05	C5/6-a
		0.14138932E+01	0.17521791E-03	C5/6-b
		0.14025688E+01	0.90274454E-04	C5/6-c
		0.14052517E+01	0.51396894E-03	C5/6-d
7	1	0.14203103E+01	0.11466473E-04	C7-a
8	1	0.14217343E+01	0.29251040E-04	C8-a

Another useful issue is the possible correspondence between C_{6v} MOF and conventional $C_{\infty v}$ step-index optical fiber modes. Due to their different symmetry properties, it is not possible to follow completely for C_{6v} MOF modes the well known classification of guided modes in conventional step-index optical fiber: $HE_{\nu,m}$, $EH_{\nu,m}$, TE_m, TM_m. On one hand, in a $C_{\infty v}$ waveguide, the HE/EH classification uses modes such that the components E_z and H_z are given by a single Fourier-Bessel component when the center of the fiber is the origin of the coordinates. On the other hand, in a C_{6v} waveguide, several Fourier-Bessel components are required to described a mode (see Eq. 4.31 in section 4.2.4 and section B.2.2 in Appendix B). Consequently, if both E_z and H_z of a MOF mode can be approximated by a single Fourier-Bessel component then the correspondence with HE/EH modes is straightforward. Most of the time, however, this is not the case. Nevertheless for a few modes such correspondence can bo done, the results are given in the following table.

A more general analysis can be done by comparing the general forms of the HE and EH longitudinal components as presented in table 12-3 of chapter 12 in reference [SL83] with the expansions of the longitudinal components in terms of Fourier series given in table IV of reference [Isa75a]. Equating these expansions show that MOF modes of a given symmetry class

[5]mode crossing could appear between these MOF modes similarly as for conventional step-index optical fibre higher order modes (see Fig.12-4 of chapter 12 in reference [SL83]).

Table 4.4 Correspondence between the main four C_{6v} MOF modes for the six hole MOF example as defined in the text and conventional step-index fiber modes.

MOF		Conventional fibre
Symmetry class	Mode label	Mode
3/4	C3/4-a	$HE_{1,1}$
2	C2-a	TE_1
5/6	C5/6-a	$HE_{2,1}$
1	C1-a	TM_1

may correspond to the superposition of specific HE/EH modes. Conversely, some EH/HE modes can only be of specific symmetry classes. Nevertheless as already stated no general one to one relationship can be deduced from these results.

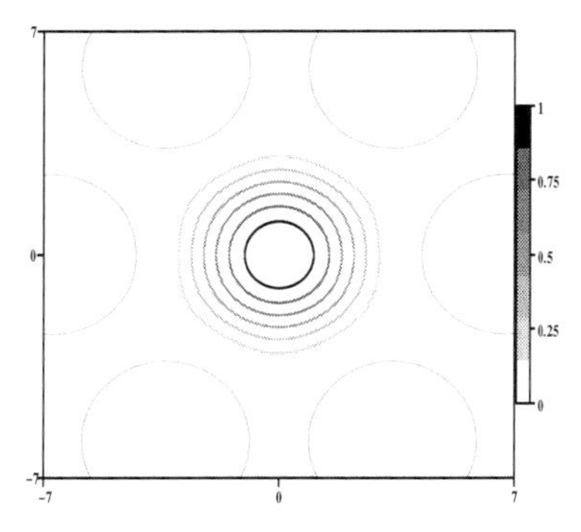

Fig. 4.13 Modulus of the Poynting vector for the C_{6v} six hole MOF example: it is the same for the two symmetry classes p=3 and p=4. The modulus is normalized to unity.

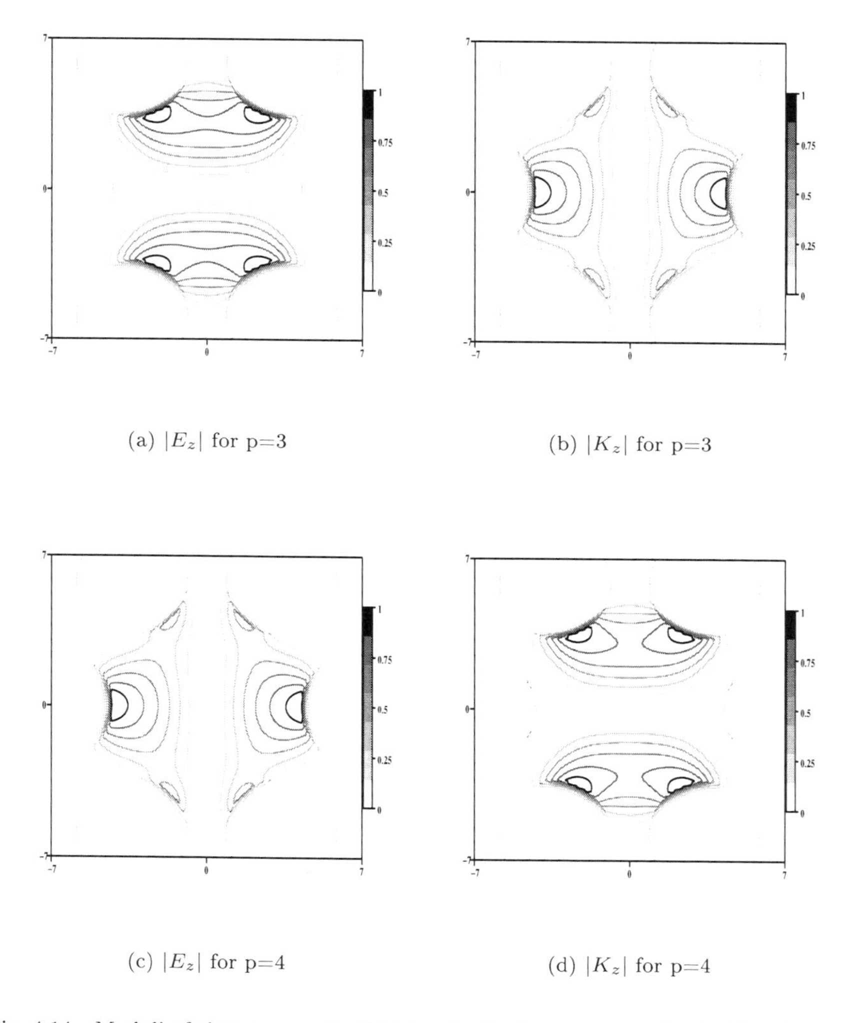

(a) $|E_z|$ for p=3 (b) $|K_z|$ for p=3

(c) $|E_z|$ for p=4 (d) $|K_z|$ for p=4

Fig. 4.14 Moduli of electromagnetic field longitudinal components for the C_{6v} six hole MOF example, in the core region, for the degenerate fundamental mode which belongs to the symmetry classes p=3 and p=4, $n_{\text{eff}} = 0.14387741\,10^1 + i\,0.43257457\,10^{-7}$. The field moduli are normalized to unity.

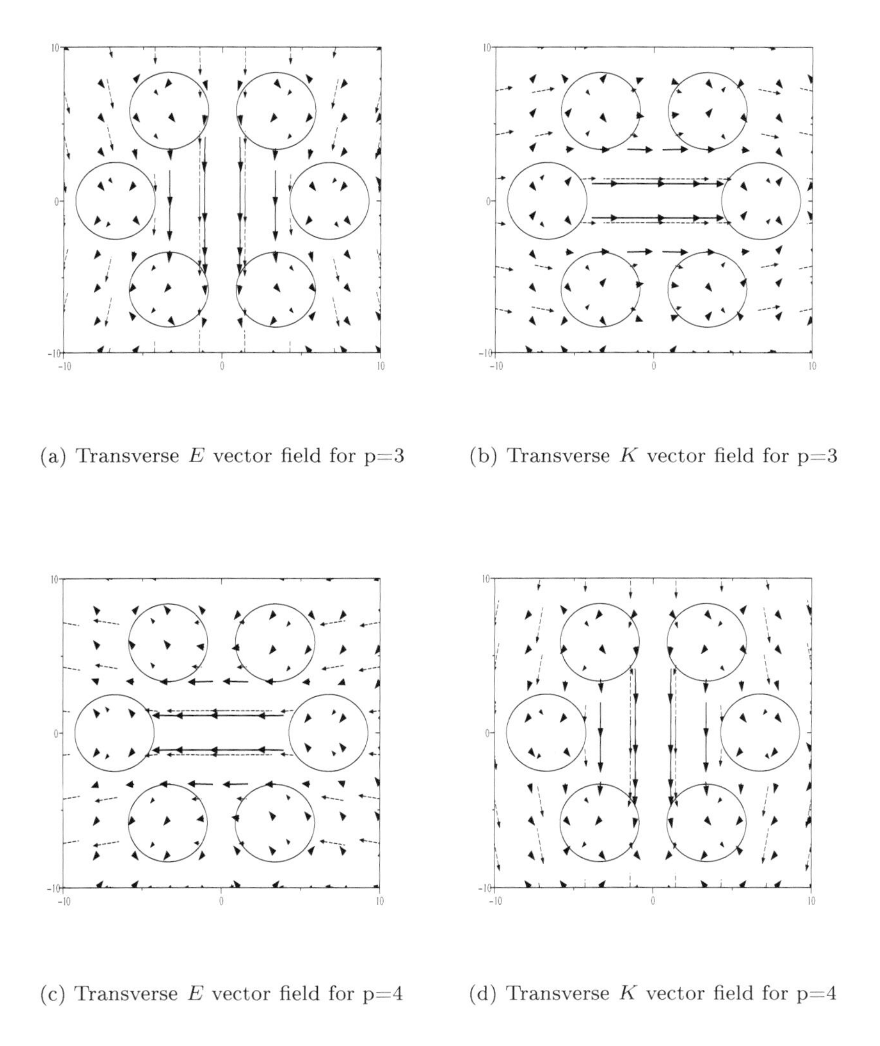

(a) Transverse E vector field for p=3

(b) Transverse K vector field for p=3

(c) Transverse E vector field for p=4

(d) Transverse K vector field for p=4

Fig. 4.15 Transverse electromagnetic vector fields for the degenerate fundamental mode of the C_{6v} six hole MOF example. The real part of the field is represented by plain thick vectors and the imaginary part is represented by dashed thin vectors.

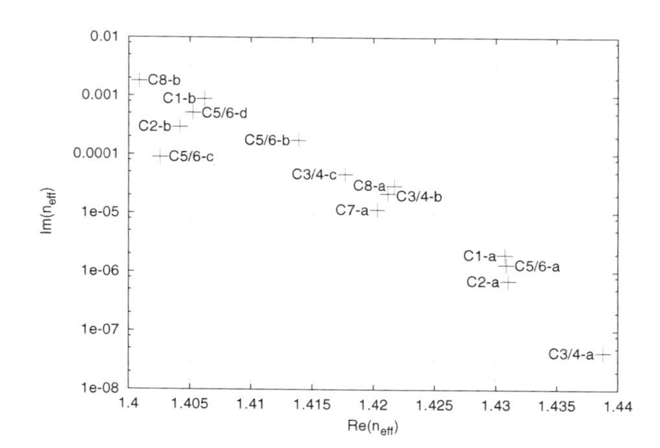

Fig. 4.16 Imaginary part of n_{eff} versus its real part for the main modes given in table 4.3 for the C_{6v} six hole MOF of section. 4.5.1. A log-scale is used for the imaginary part.

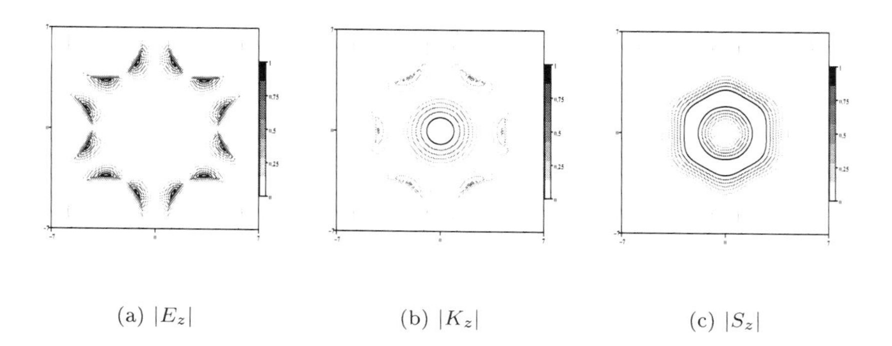

(a) $|E_z|$ (b) $|K_z|$ (c) $|S_z|$

Fig. 4.17 Moduli of electromagnetic fields and Poynting vector longitudinal components of the C_{6v} six hole MOF example, in the core region, for the higher order mode of symmetry class p=2 with the highest effective index real part, $n_{\text{eff}} = 0.14310182\,10^1 + i\,0.70048952\,10^{-6}$. The maxima of the field moduli are normalized to unity.

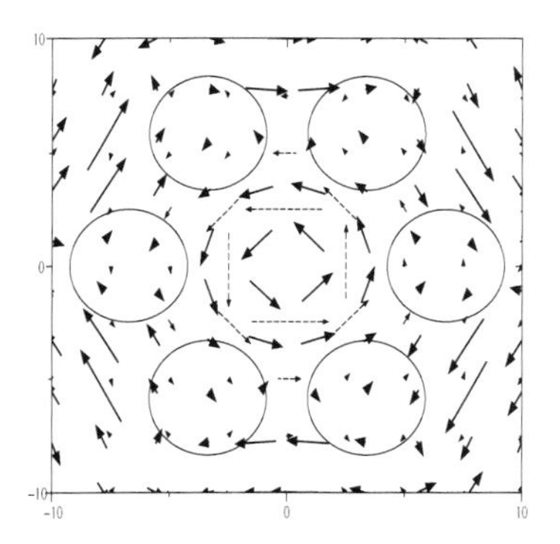

(a) Transverse E vector field

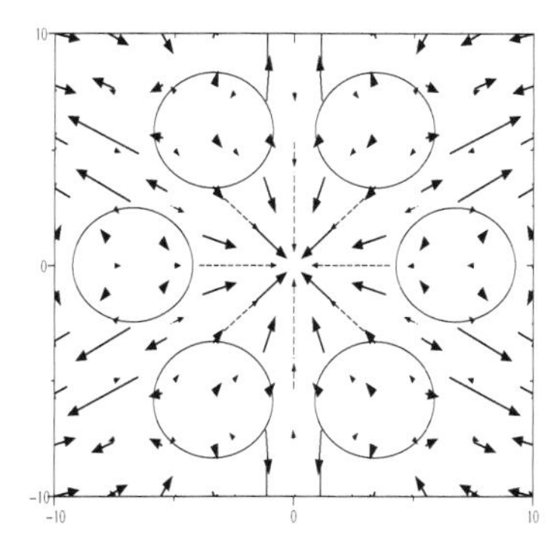

(b) Transverse K vector field

Fig. 4.18 Transverse electromagnetic vector fields of the C_{6v} six hole MOF example, for the higher order mode of symmetry class p=2 with the highest effective index real part. The real part of the field is represented by plain thick vectors and the imaginary part is represented by dashed thin vectors.

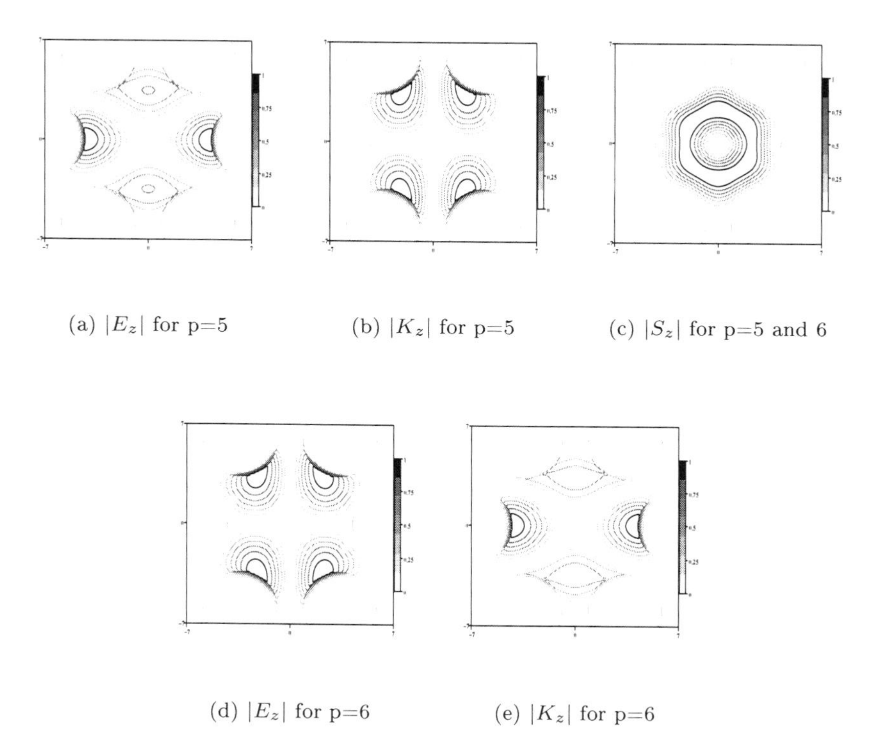

(a) $|E_z|$ for p=5 (b) $|K_z|$ for p=5 (c) $|S_z|$ for p=5 and 6

(d) $|E_z|$ for p=6 (e) $|K_z|$ for p=6

Fig. 4.19 Moduli of electromagnetic fields and Poynting vector longitudinal components of the C_{6v} six hole MOF example, in the core region, for the higher order modes of symmetry classes p=5 and p=6 with the highest effective index real part, $n_{\text{eff}} = 0.14308483\,10^{1} + i\,0.13214492\,10^{-5}$. The maxima of the field moduli are normalized to unity.

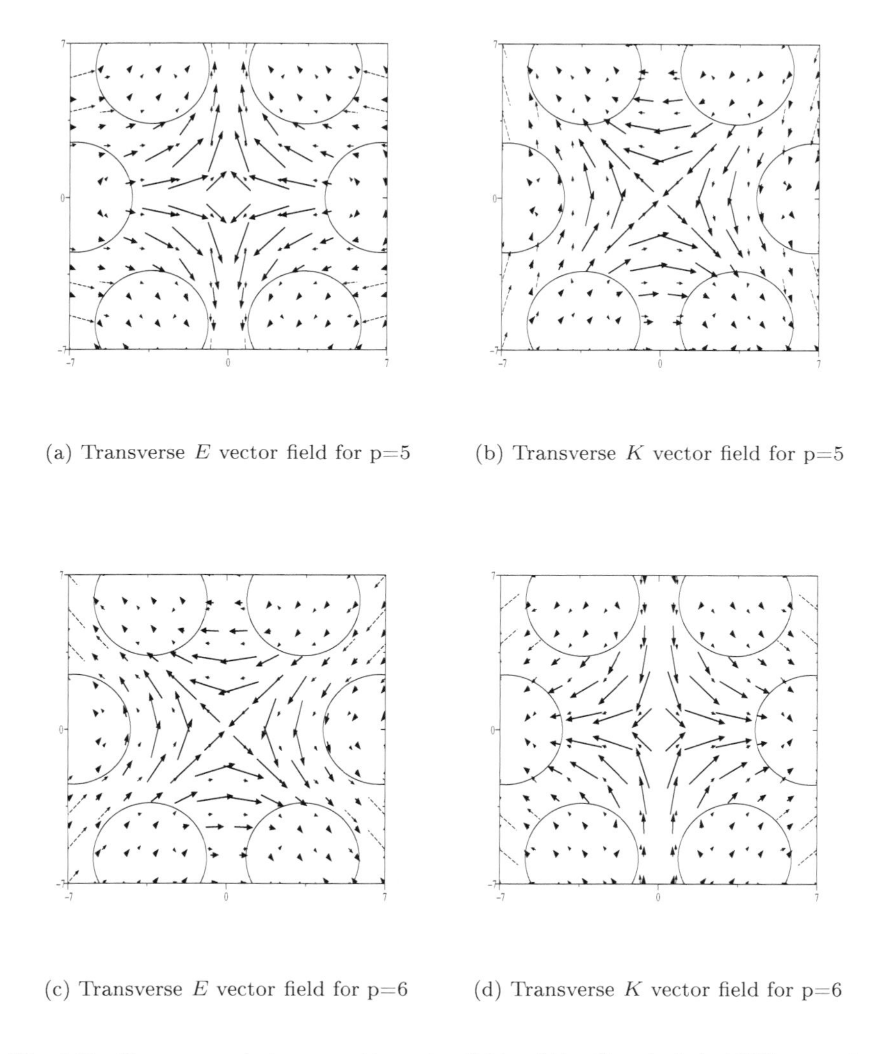

(a) Transverse E vector field for p=5

(b) Transverse K vector field for p=5

(c) Transverse E vector field for p=6

(d) Transverse K vector field for p=6

Fig. 4.20 Transverse electromagnetic vector fields of the C_{6v} six hole MOF example, for the degenerate mode of symmetry classes p=5 and p=6. The real part of the field is represented by plain thick vectors and the imaginary part is represented by dashed thin vectors.

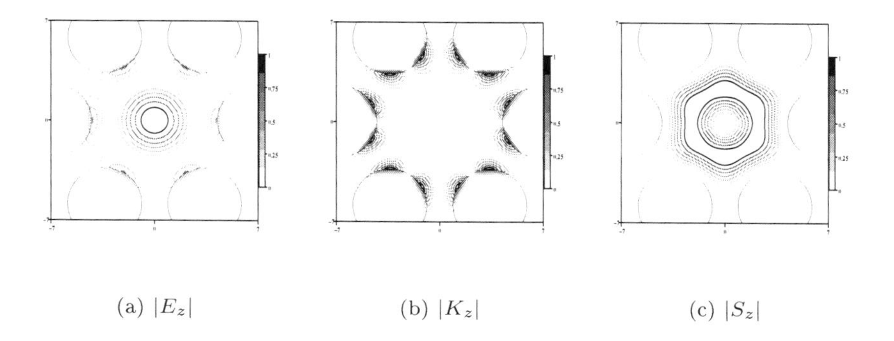

(a) $|E_z|$ (b) $|K_z|$ (c) $|S_z|$

Fig. 4.21 Moduli of electromagnetic fields and Poynting vector longitudinal components, in the core region, for the higher order mode of symmetry class p=1 with the highest effective index real part, $n_{\text{eff}} = 0.14307554\,10^{1} + i\,0.19457921\,10^{-5}$. The maxima of the field moduli are normalized to unity.

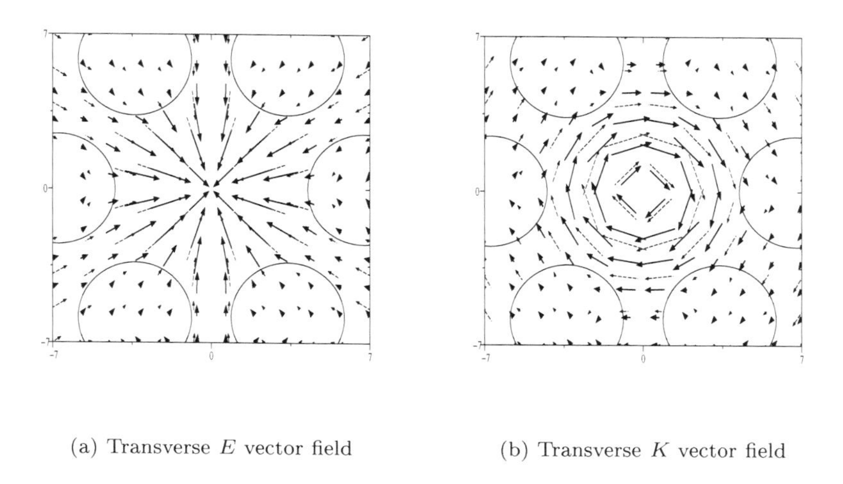

(a) Transverse E vector field (b) Transverse K vector field

Fig. 4.22 Transverse electromagnetic vector fields for the higher order mode of symmetry class p=1 with the highest effective index real part. The real part of the field is represented by plain thick vectors and the imaginary part is represented by dashed thin vectors.

4.6.2 A C_{2v} example: a birefringent MOF

MOF have already been used to realize highly birefringent optical fibers [OBKW+00] by means of the high index contrast and the numerous possible designs for the holes and their positions. In what follows we give a simple example which exibits a high birefringence: a C_{2v} six hole MOF with two types of hole. The MOF is nearly the same as that used in section 4.5.1: $\Lambda = 6.75\ \mu m$, the diameters of the four small holes is fixed at $5.0\ \mu m$, but two symmetrically positioned small holes are now big holes of diameters equal to $7.0\ \mu m$, and we choose $\lambda = 1.55\ \mu m$ and $\varepsilon_r = 2.0852042$. M is equal to 8 in all the numerical simulations of this section. We already know from the results described in section 4.3.1 that none of the modes are degenerate. This is exactly what we found in table 4.5 where we give the effective indices n_{eff} of the main modes of this C_{2v} MOF.

Table 4.5 Mode class, and effective index for the main modes of a C_{2v} six hole MOF. The results are obtained at $\lambda = 1.55\ \mu m$. According to McIsaac's theory none of the possible modes of this C_{2v} MOF can be degenerate.

Mode class	$\Re e(n_{\text{eff}})$	$\Im m(n_{\text{eff}})$	Label in Fig. 4.23
1	0.14245641E+01	0.26661425E-08	C1-a
	0.14302611E+01	0.16786253E-05	C1-b
	0.14059800E+01	0.44469672E-07	C1-c
	0.14096020E+01	0.45426880E-03	C1-d
	0.13969515E+01	0.30692507E-03	C1-e
2	0.14251917E+01	0.12531562E-08	C2-a
	0.14302577E+01	0.60522440E-06	C2-b
	0.13992741E+01	0.71576046E-06	C2-c
	0.14078141E+01	0.15989247E-03	C2-d
	0.13970789E+01	0.68785361E-04	C2-e
3	0.14373840E+01	0.21277878E-07	C3-a
	0.14163151E+01	0.56123303E-06	C3-b
	0.14189536E+01	0.15076718E-04	C3-c
	0.14053802E+01	0.16153919E-04	C3-d
	0.13946933E+01	0.10998837E-02	C3-e
4	0.14375326E+01	0.45399096E-07	C4-a
	0.14165924E+01	0.93230300E-06	C4-b
	0.14194844E+01	0.49669065E-04	C4-c
	0.14063350E+01	0.23955853E-04	C4-d
	0.13982127E+01	0.24979563E-02	C4-e

There are at least two interesting points concerning the modes in this C_{2v} si holes MOF. First of all, the determination of the fundamental mode is not straightforward: is it the mode labelled C2-a which has the lowest losses (see table 4.5 and Fig. 4.24) or the C4-a mode which have the high-

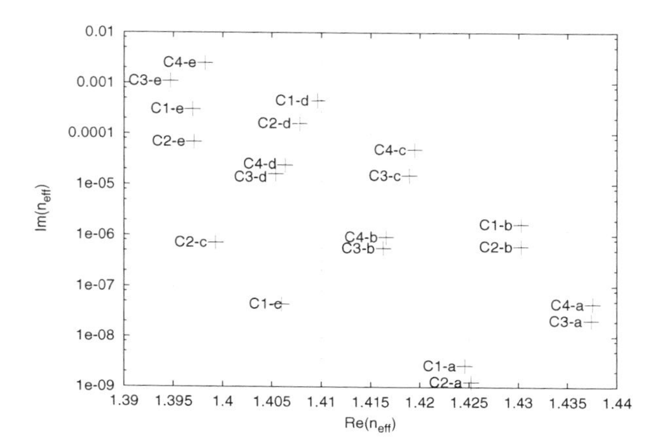

Fig. 4.23 Imaginary part of n_{eff} versus its real part for the modes given in table 4.5 for the C_{2v} six hole MOF. A log-scale is used for the imaginary part.

est real part of n_{eff}. We will describe a way to determine which one is the fundamental mode of the structure in the chapter 7, which is dedicated to MOF properties. The second point to consider is the lifting of the degeneracy of the C3/4-a mode described previously in the C_{6v} MOF into two distinct modes: C3-a and C4-b. The longitudinal components of these two modes are shown below in Fig. 4.25.

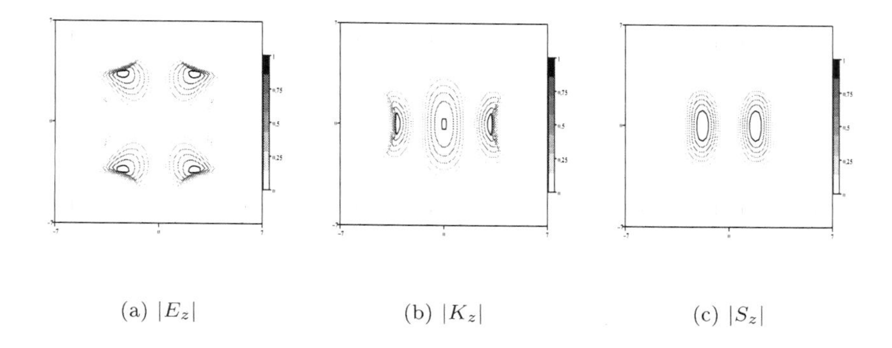

(a) $|E_z|$ (b) $|K_z|$ (c) $|S_z|$

Fig. 4.24 Moduli of electromagnetic fields and Poynting vector longitudinal components, in the core region of the C_{2v} six hole MOF exampke for the mode which belongs to the symmetry classes p=2, $n_{\text{eff}} = 0.14251917\,10^1 + i\,0.1253156210^{-8}$. The field moduli are normalized to unity.

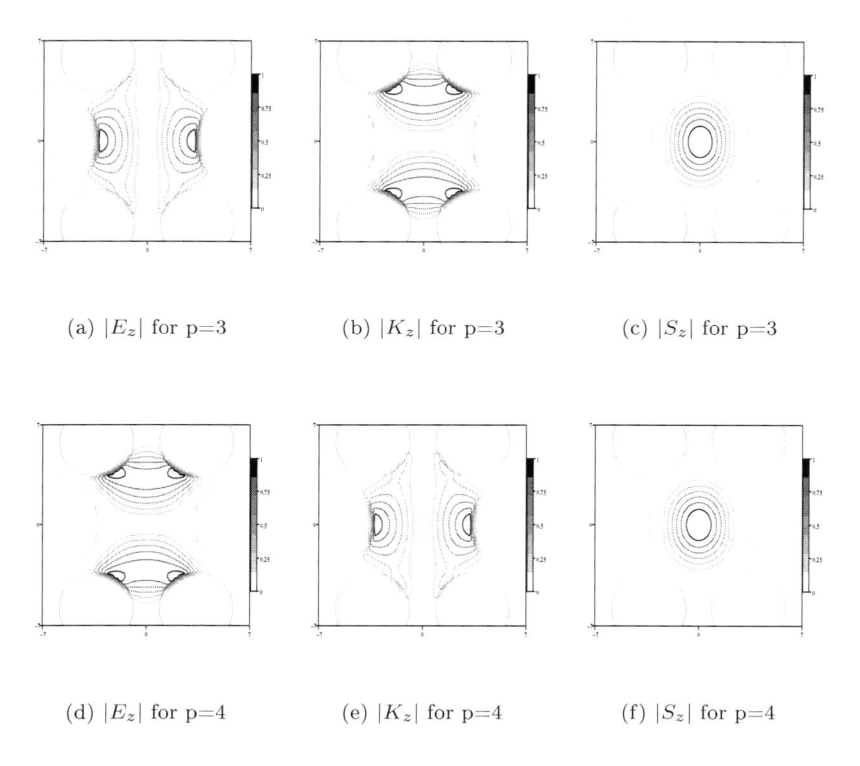

(a) $|E_z|$ for p=3 (b) $|K_z|$ for p=3 (c) $|S_z|$ for p=3

(d) $|E_z|$ for p=4 (e) $|K_z|$ for p=4 (f) $|S_z|$ for p=4

Fig. 4.25 Moduli of electromagnetic fields and Poynting vector longitudinal components, in the core region of the C_{2v} six hole MOF example for the modes of symmetry classes p=3 ($n_{\text{eff}} = 0.14373840\,10^1 + i\,0.21277878\,10^{-7}$) and p=4 ($n_{\text{eff}} = 0.14375326\,10^1 + i\,0.45399096\,10^{-7}$). The field moduli are normalized to unity.

4.6.3 *A C_{4v} example: a square MOF*

We conclude the illustration of the ordinary symmetric structures in MOF with a C_{4v} example (the fiber contains eight identical holes of diameter 5.0 μm, the pitch is 6.75 μm, and $\varepsilon_r = 2.0852042$). In this case, only the classes 3 and 4 are degenerate as expected from McIsaac's theory. The fundamental mode belongs to these classes, and it is the mode labelled C3/4-a: the correponding longitudinal components are plotted in Fig. 4.27. The first modes of this structure are given in Table 4.6 and in Fig. 4.26.

Table 4.6 Mode class, degeneracy, and effective index for the main modes of the square C_{4v} eight hole MOF. The results were computed at $\lambda = 1.55\,\mu m$.

Mode class	Degeneracy	$\Re e(n_{\text{eff}})$	$\Im m(n_{\text{eff}})$	Label in Fig. 4.26
1	1	0.1433960E+01	0.2984848E-06	C1-a
		0.1423173E+01	0.1387057E-04	C1-b
2	1	0.1433891E+01	0.2937135E-06	C2-a
		0.1423461E+01	0.1499999E-04	C2-b
		0.1410688E+01	0.6201385E-04	C2-c
3, 4	2	0.1439607E+01	0.8965050E-08	C3/4-a
		0.1430761E+01	0.1801941E-05	C3/4-b
		0.1427466E+01	0.5028845E-05	C3/4-c
		0.1419888E+01	0.2701725E-05	C3/4-d
		0.1415520E+01	0.8361918E-04	C3/4-e
		0.1413382E+01	0.8869551E-04	C3/4-f
5	1	0.1433642E+01	0.3221663E-06	C5-a
		0.1423361E+01	0.1371652E-04	C5-b
		0.1411219E+01	0.1878240E-03	C5-c
6	1	0.1434133E+01	0.2827584E-06	C6-a
		0.1423380E+01	0.1427653E-04	C6-b

4.7 Six Hole Plain Core MOF Example: Supercell Point of View

A newcomer in mathematical models for microstructured fibres may find that the literature on the supercell method is rather poor when it comes to explaining the links between PCF having finite and infinite periodic claddings. There is a feeling that people just skip over the difficulty by omitting issues as fundamental as the effect of leakage on Bloch wavevector diagrams. Such issues are of great practical importance and we thought that it would be wise to spend a few lines on this as yet virgin topic.

First, it is clear that one looses self-adjointness of the operator when dealing with lossy media, and hence the definition of Bloch waves has to be understood in a generalised sense (complex spectrum and evanescent states). From this, it is obvious that there is no one-to-one correspondence between propagating frequencies given by a traditional Bloch spectrum (infinite cladding) and those associated with the scattering spectrum (finite cladding and leakage). The work by Van der Lem and his colleagues [dLTM03] suggests that the perturbation induced by the introduction of a small amount of absorption in a periodic medium on Bloch frequencies is fairly small.

Indeed, the band structures of periodic arrays of cylinders consisting of

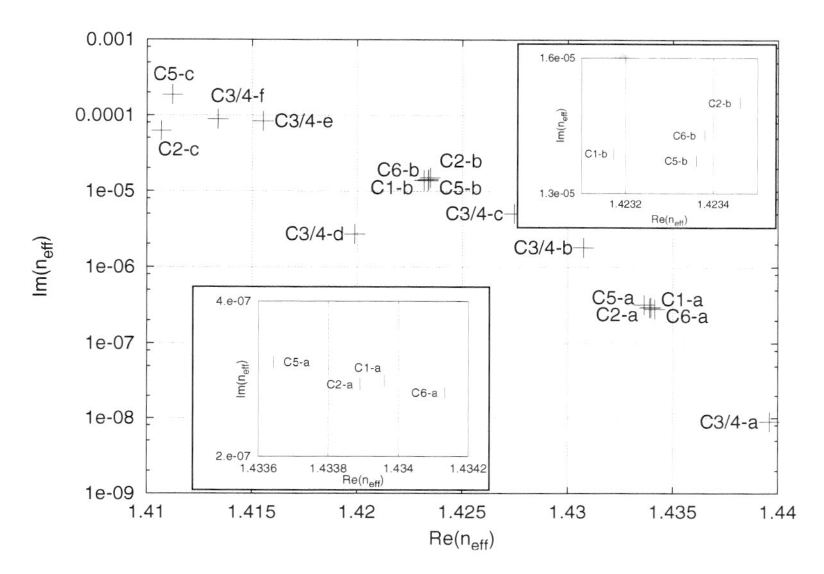

Fig. 4.26 Imaginary part of n_{eff} versus its real part for the modes given in table 4.6 for the C_{4v} eight hole MOF. A log-scale is used for the imaginary part.The upper right inset is a zoom around the positions of C1-a and C-6a, the lower left inset is a zoom around the positions of C1-b and C-6b.

Drude material *i.e.* with a relative permittivity

$$\varepsilon(\omega) = 1 - \frac{\omega_p^2}{\omega(\omega + i\gamma)} , \tag{4.53}$$

where ω_p stands for the plasmon frequency and γ stands for the loss, have been addressed in the transverse case in [dLTM03]. It was found that absorption causes the spectrum to become complex and to form islands in the negative complex half-plane. More importantly, the boundaries of these islands are not always formed by the eigenvalues calculated for Bloch vectors describing the edges of the first Brillouin zone and there are holes in those islands.[6] Nevertheless, for realistic physical parameters, the real part of the spectrum is hardly influenced by absorption, typically less than 0.25 per cent [dLTM03]. Therefore, a perturbation theory should prove very efficient for the analysis of lossy lattice of low index inclusions embedded in lossy matrix. This tends to explain why people observe good agreement between Bloch and scattering spectrums. A good way to check this could be

[6]This suggests that one should plot dispersion surfaces rather than dispersion curves (the Bloch vector should indeed describe the interior of the first Brillouin zone).

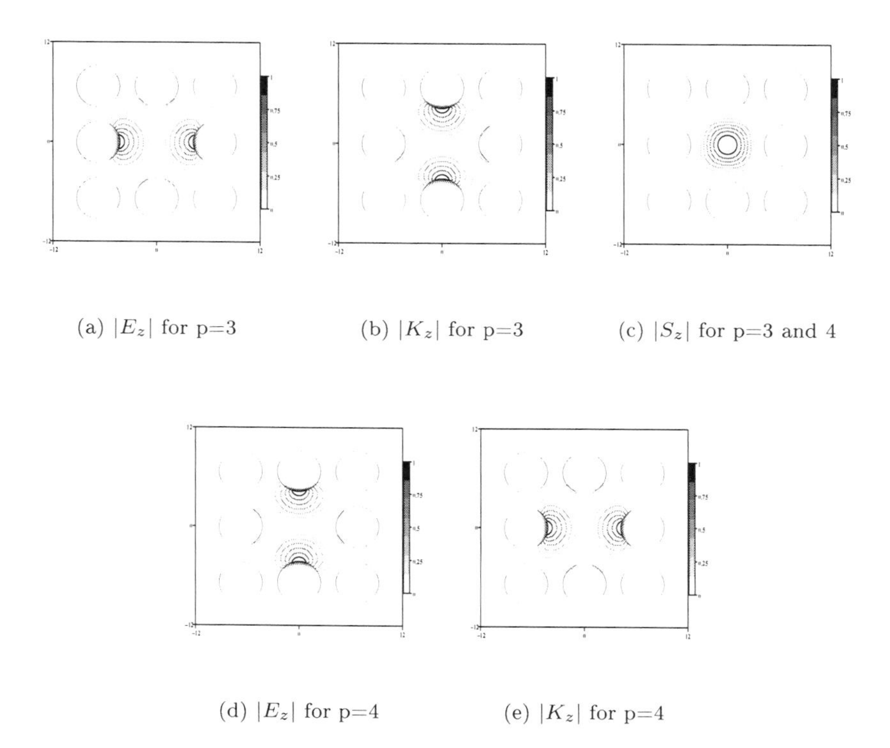

(a) $|E_z|$ for p=3 (b) $|K_z|$ for p=3 (c) $|S_z|$ for p=3 and 4

(d) $|E_z|$ for p=4 (e) $|K_z|$ for p=4

Fig. 4.27 Moduli of electromagnetic fields longitudinal components, in the core region for the degenerate fundamental mode which belongs to the symmetry classes p=3 and p=4, $n_{\text{eff}} = 0.14396074\,10^1 + i\,0.89650657\,10^{-8}$. In (c) modulus of the Poynting vector: it is the same for the two symmetry classes p=3 and p=4. The field moduli are normalized to unity.

to compare dispersion diagrams associated with an hexagonal cell containing six-air holes arranged in honeycomb structure against the six hole plain core MOF's case. When dealing with such structures, another point is to study the dispersion curves for a fixed β by varying the size of the supercell as shown in Fig. 4.28 for dispersion curves and in Figs. 4.29, 4.30 for associated eigenfields. In Fig. 4.28, we note that the inter-hole (center to center) spacing $\Lambda = 6.75\mu m$ provides us with the periodic counterpart of a six hole plain core MOF studied at the beginning of this chapter *i.e.* it consists of supercells containing six circular cylindrical air channels ($n = 1$) of normalised diameter $d/\Lambda = 0.74074$ drilled within a matrix of silica ($n = 1.45$).

The parameter L is introduced to specify the size of the hexagonal supercell: it is defined as the distance between the barycenter of the hexagon to any of its six vertices. On the vertical axis we represent the normalised frequency $\omega\Lambda/(2\pi c)$, where c is the speed of light in vacuum. On the horizontal axis we represent the normalised modulus $k\Lambda$ of the Bloch vector $\mathbf{k}$ when it describes the Brillouin zone ΓKM (its segments ΓK and KM are of respective normalised lengths $2\pi[\tan(\pi/6)+\tan^2(\pi/6)]/\Lambda \sim 5.71$ and $2\pi\sqrt{\tan(\pi/6)+\tan^2(\pi/6)}/\Lambda \sim 4.17$). We consider a normalised propagation constant $\beta\Lambda = 42.276816$ as in [NGZ+02]. We note that the frequencies of crossed lines are lower than those of dotted lines, in agreement with the fact that the fraction of silica becomes higher when the boundaries of the cell walk away. We find that the frequency of the fundamental mode equals $\omega_{1\Gamma}^{\bullet}\Lambda/(2\pi c) = 0.703$ for dotted lines and it equals $\omega_{1\Gamma}^{+}\Lambda/(2\pi c) = 0.699$ for crossed lines at point Γ. Its frequency equals $\omega_{1K}^{\bullet}\Lambda/(2\pi c) = 0.707$ for dotted lines and it equals $\omega_{1K}^{+}\Lambda/(2\pi c) = 0.702$ for crossed lines at point K (standing wave). If $\Lambda = 6.75\mu m$ (as in [NGZ+02]) this corresponds to respective (non normalised) wavelengths $\lambda_{1\Gamma}^{\bullet} = 2\pi c/\omega_{1\Gamma}^{\bullet} = 1.423\mu m$, $\lambda_{1\Gamma}^{+} = 2\pi c/\omega_{1\Gamma}^{+} = 1.431\mu m$, $\lambda_{1K}^{\bullet} = 2\pi c/\omega_{1K}^{\bullet} = 1.417\mu m$ and $\lambda_{1K}^{+} = 2\pi c/\omega_{1K}^{+} = 1.425\mu m$. If the wavelength $\lambda = 1.45\mu m$ is set up in the multipole method, we obtain an effective index (as defined previously in this chapter) $n_{\text{eff}} = 1.445395345 + i\,3.1510^{-8}$ [NGZ+02] which provides a normalised propagation constant $\beta\Lambda = 2\pi\Re e(n_{\text{eff}})\Lambda/\lambda = 42.27681784$ that closely matches the normalised propagation constant 42.276816 considered in the Finite Element Algorithm. The slight discrepancies between λ and $\lambda_{1\Gamma}^{\bullet}$, $\lambda_{1K}^{\bullet}$, $\lambda_{1\Gamma}^{+}$ and λ_{1K}^{+} can be attributed to the "philosophy" of the supercell's method: The structure is artificially periodised.

4.8 Conclusion

In this Chapter, the Multipole Method has been explain in some detail. It appears that the symmetry properties of MOF modes obtained from McIsaac's theoretical work can be take into account naturally with this method. Only circular inclusions have been considered but more general inclusion shapes can be treated provided that the inclusions do not overlap. Thanks to its speed and accuracy and to the way in which it can deal with material dispersion, the Multipole Method is well suited to study finite size MOF properties including their losses. Illustrations of its possibilities are to be found in Chapter 7. The Multipole Method can also be extended to periodic structures, as shown in the next Chapter.

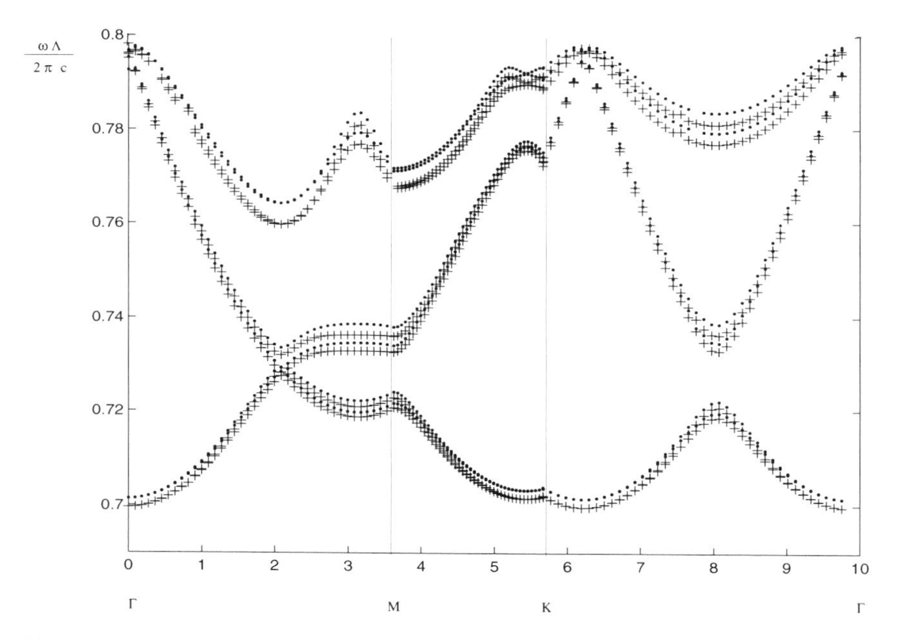

Fig. 4.28 Dispersion curves associated with a triangular array of cylinders of respective normalised pitches $\Lambda/L = 1$ (dotted lines) and $\Lambda/L = 2$ (crossed lines).

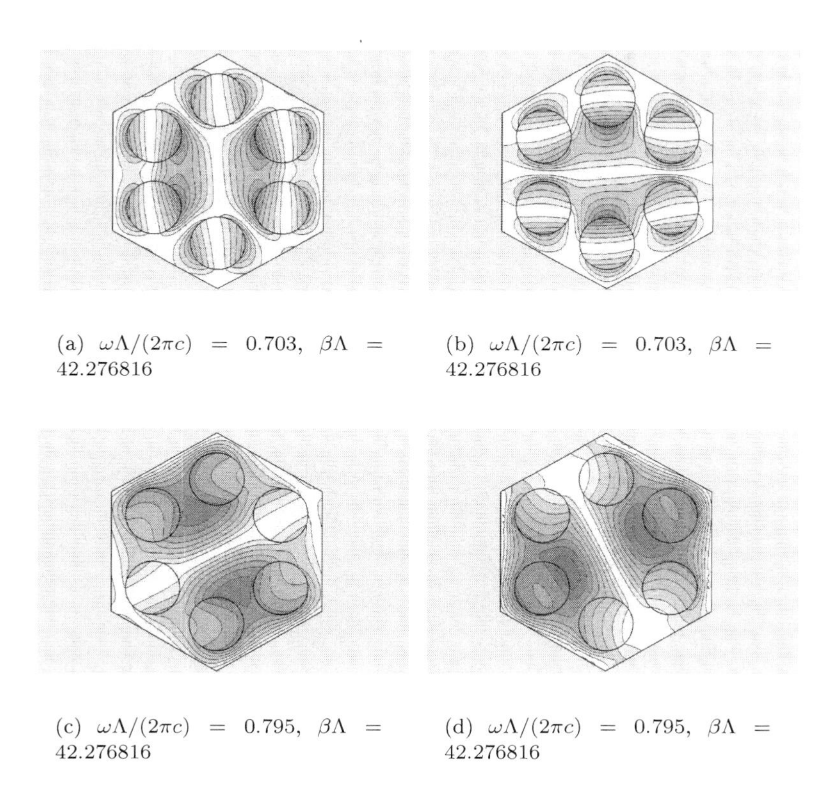

(a) $\omega\Lambda/(2\pi c) = 0.703$, $\beta\Lambda = 42.276816$

(b) $\omega\Lambda/(2\pi c) = 0.703$, $\beta\Lambda = 42.276816$

(c) $\omega\Lambda/(2\pi c) = 0.795$, $\beta\Lambda = 42.276816$

(d) $\omega\Lambda/(2\pi c) = 0.795$, $\beta\Lambda = 42.276816$

Fig. 4.29 On diagrams 4.29(a), 4.29(b), 4.29(c), 4.29(d), we depict the modulus of the longitudinal part of the magnetic field H_z within a supercell described itself by an hexagon containing the six circular cylindrical air channels of normalised diameter $d/L = 0.74074$ drilled within a matrix of silica. The Bloch vector is set to zero and the boundaries of the supercell correspond to the case $\Lambda/L = 1$ (see dotted lines on Fig. 4.28). The mesh used in the software GetDP was set to 7000 triangles. The weak asymmetry observed in above figures can be attributed to the mesh which does not preserve the C_{6v} symmetry. When the mesh is made finer, the correct symmetry is approached asymptotically.

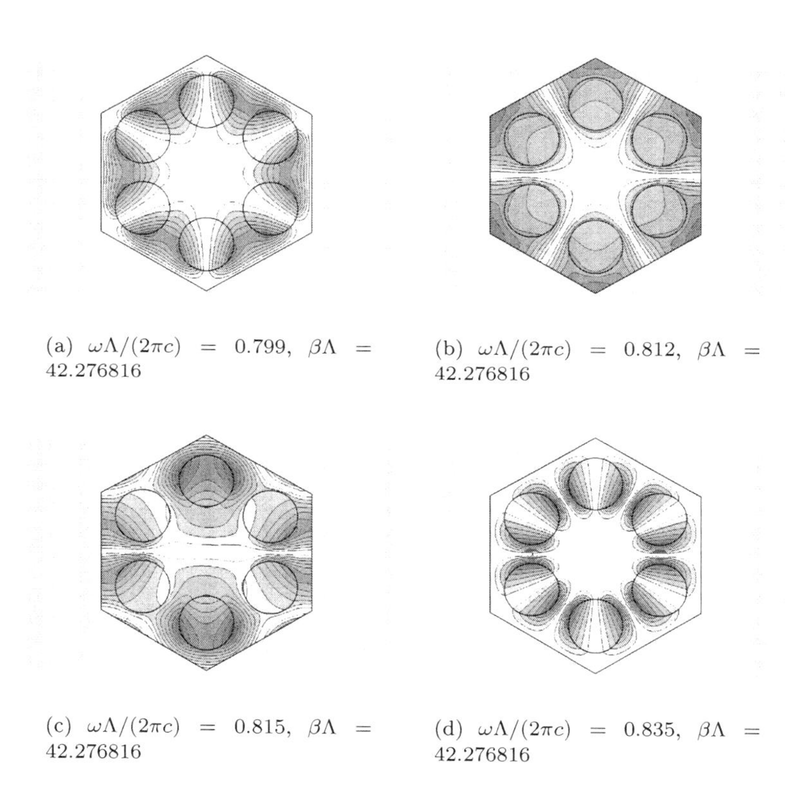

(a) $\omega\Lambda/(2\pi c) = 0.799$, $\beta\Lambda = 42.276816$

(b) $\omega\Lambda/(2\pi c) = 0.812$, $\beta\Lambda = 42.276816$

(c) $\omega\Lambda/(2\pi c) = 0.815$, $\beta\Lambda = 42.276816$

(d) $\omega\Lambda/(2\pi c) = 0.835$, $\beta\Lambda = 42.276816$

Fig. 4.30 On diagrams 4.30(a), 4.30(b), 4.30(c), 4.30(d), we depict the moduli of the longitudinal part of the magnetic field H_z within a supercell containing six circular cylindrical air channels of normalised diameter $d/\Lambda = 0.74074$ drilled within a matrix of silica. The Bloch vector is set to zero and the boundaries of the supercell correspond to the case $\Lambda/L = 1$ (see dotted lines on Fig. 4.28). The mesh used in the software GetDP was set to 7000 triangles. The lack of symmetry for the modes can be attributed to the presence of the supercell's boundaries. The remarks on lack of symmetries for the modes follow *mutatis mutandis* the ones of Figure caption 4.29.

Chapter 5

Rayleigh Method

5.1 Genesis of Baron Strutt's Algorithm

During the past decade, a great deal of attention has been payed to the analysis of filtering properties of Photonic Crystals [LBB$^+$03]. It is now well-known that in the propagation of electromagnetic waves through microstructured materials, many interesting and unusual phenomena may be observed. Such phenomena may include anomalous dispersion and so-called *photonic band gap effect*, where a range of frequencies is forbidden to propagate within the medium. The physics at work within these periodic structures has a lot to do with *electronic band gaps* whereby electrons in solids can only take quantified energy levels. Hence, the reader should be aware that most of the mathematical tools developed in this chapter take their roots in a classical book by J. B. Pendry published over 30 years ago [Pen74] (this tends to explain why condensed matter theorists – among whom J. Joannopoulos, S. John, J. B. Pendry and E. Yablonovitch – initiated and developed the field of Photonic Crystals [JMW95; Joh94; Pen94; Yab87]). But lots of features shared by Photonic Crystal Fibres (such as *effective index and photonic band gap guidance* [Gue01; Kuh03; GNZL03; BBB03]) are undoubtedly new and deserve a modern and specific approach.

A widely used method in wave propagation within periodic structures involves plane-wave expansions of fields and Fourier series representations of spatially varying material properties. Studies based on the plane-wave expansion method for the case of out-of-plane propagation were published by Maradudin and McGurn [MM94]. Although the plane-wave expansion is more flexible than the multipole-method (*e.g.* it can be applied to arrays of cylinders of arbitrary cross-section), it faces some problems of accuracy for high-contrast materials and is perhaps less well-suited for obtaining analytic results arising from questions of homogenisation, asymptotics of field structure and wave localisation [MMP02]. Until recently, the extension of the multipole method to a periodic array of cylinders – the so-called Rayleigh method– was carried out only for transversely propagating waves [NMB95]. Such important theoretical and practical issues as evolution of photonic band gaps with respect to the propagation constant β could not be addressed with enough accuracy in MOFs. For instance, the aforementioned photonic band gap guidance, whereby light propagates within a hollow-core MOF, requires first a thorough analysis of dispersion properties of Photonic Crystal Fibres (periodic counterparts of MOFs which possess, in actuality, finite cross-sections). As the proverb says, *Where there's a will, there's a way* (rooted from *A buona volontà, non manca facoltà*) and therefore close collaboration between Sydney and Liverpool Universities (the groups of R. C. McPhedran and A. B. Movchan) was maintained in order to broaden the Rayleigh method to the out-of-plane case [GPM03a].

5.2 Common Features Shared by Multipole and Rayleigh Methods

We now present the multipole method given in Chapter 4 p. 157 for the case of propagating waves

$$\boldsymbol{\mathcal{E}}(r,\theta,z,t) = \Re e\{\mathbf{E}(r,\theta)e^{i(\beta z-\omega t)}\} , \tag{5.1}$$

$$\boldsymbol{\mathcal{K}}(r,\theta,z,t) = \Re e\{\mathbf{K}(r,\theta)e^{i(\beta z-\omega t)}\} , \tag{5.2}$$

within a doubly periodic array of cylinders with circular cross-section of radius r_c. As in definition (4.2), ω denotes the angular frequency. As in Chapter 4, K_z denotes the rescaled magnetic field $K_z = \sqrt{\mu_0/\varepsilon_0}H_z = Z_0 H_z$. But the propagation constant β is now a strictly positive real number (no attenuation along the z-axis, unlike in Chapter 4).

It follows from Maxwell's equations (cf. Eq. 2.168) that either of the fields $u = E_z$ or $u = K_z$ satisfy the Helmholtz equation

$$(\Delta + (k_\perp^i)^2)u^{(i)} = 0 , \tag{5.3}$$

within the inclusions and

$$(\Delta + (k_\perp^M)^2)u^{(M)} = 0 , \tag{5.4}$$

within the bulk matrix. Here, the wave-numbers $k_\perp^i$ and $k_\perp^M$ are defined as $k_\perp^i = \sqrt{n_i^2 k_0^2 - \beta^2}$ and $k_\perp^M = \sqrt{n_M^2 k_0^2 - \beta^2}$ and the spectral parameter

$k_0 = \sqrt{\varepsilon_0 \mu_0}\omega = \omega/c$. We note that for refractive indices $n_M > n_i \geq 1$ (e.g. air-holes in a matrix of silica), there will be a range of frequencies ω for which $k_\perp^M$ is real whereas $k_\perp^i$ becomes purely imaginary. This corresponds to the physical situation of waves *propagating* within the matrix (by total internal reflection) and being *evanescent* within the cylinders.

Further, we have seen in Chapter 2 that the transverse components of **E** and **H** can be expressed in terms of their axial counterparts as

$$\mathbf{E}_t = \frac{i}{k^2 \varepsilon_r \mu_r - \beta^2} \left[\beta \mathbf{grad}_t E_z - Z_0 k_0 \mu_r R_{\pi/2} \mathbf{grad}_t H_z \right] , \quad (5.5)$$

$$\mathbf{H}_t = \frac{i}{k^2 \varepsilon_r \mu_r - \beta^2} \left[\beta \mathbf{grad}_t H_z + \frac{k_0 \varepsilon_r}{Z_0} R_{\pi/2} \mathbf{grad}_t E_z \right] , \quad (5.6)$$

where $R_{\pi/2}$ denotes the matrix of rotation by an angle of $\pi/2$. These two quantities should be conserved while crossing the interface between two media (as well as the axial fields E_z and H_z themselves). Hence, Eq. 5.3 and Eq. 5.4 couple through the boundary of each cylinder.

The physical principle behind the multipole method remains the same as in Chapter 4: It lies in an elegant identity between the non-singular parts of a *multipole expansion*

$$E_z^i = \sum_{m=-\infty}^{+\infty} c_m^E J_m(k_\perp^i r) e^{im\theta} , \quad (5.7)$$

$$K_z^i = \sum_{m=-\infty}^{+\infty} c_m^K J_m(k_\perp^i r) e^{im\theta} , \quad (5.8)$$

of the fields around a central cylinder *i.e.* for $r \leq r_c$ (keeping in mind that only non-singular terms can occur in the cylinders, for the field has to be finite) and

$$E_z^M = \sum_{m=-\infty}^{+\infty} \left[a_m^E J_m(k_\perp^M r) + b_m^E Y_m(k_\perp^M r) \right] e^{im\theta} , \quad (5.9)$$

$$K_z^M = \sum_{m=-\infty}^{+\infty} \left[a_m^K J_m(k_\perp^M r) + b_m^K Y_m(k_\perp^M r) \right] e^{im\theta} , \quad (5.10)$$

between the inclusions *i.e.* for $r \geq r_c$ (outside the cylinder in the unit cell Y to be repeated periodically in the transverse plane). Now the relations of continuity (See Eq. 2.170 p. 62 and Eq. 2.175 p. 63) can be written in a simple manner for cylinders with circular cross-section as per:

$$E_z^i(r_c, \theta) = E_z^M(r_c, \theta) \tag{5.11}$$

$$K_z^i(r_c, \theta) = K_z^M(r_c, \theta) \tag{5.12}$$

$$\begin{aligned} &\frac{\omega\varepsilon^i}{(k_\perp^i)^2}\frac{\partial E_z^i}{\partial r}(r_c, \theta) + \frac{\beta}{Z_0(k_\perp^i)^2}\frac{\partial K_z^i}{\partial \theta}(r_c, \theta) = \\ &\frac{\omega\varepsilon^M}{(k_\perp^M)^2}\frac{\partial E_z^M}{\partial r}(r_c, \theta) + \frac{\beta}{Z_0(k_\perp^M)^2}\frac{\partial K_z^M}{\partial \theta}(r_c, \theta) \end{aligned} \tag{5.13}$$

and

$$\begin{aligned} &-\frac{\omega\mu^i}{Z_0(k_\perp^i)^2}\frac{\partial K_z^i}{\partial r}(r_c, \theta) + \frac{\beta}{(k_\perp^i)^2}\frac{\partial E_z^i}{\partial \theta}(r_c, \theta) = \\ &-\frac{\omega\mu^M}{Z_0(k_\perp^M)^2}\frac{\partial K_z^M}{\partial r}(r_c, \theta) + \frac{\beta}{(k_\perp^M)^2}\frac{\partial E_z^M}{\partial \theta}(r_c, \theta) \end{aligned} \tag{5.14}$$

for any θ in $[0, 2\pi[$. This leads[1] to a matrix of boundary conditions as stated by Eq. 4.36. In brief, we have six sets of unknowns namely (a_n^E, b_n^E, c_n^E, a_n^K, b_n^K and c_n^K) for only four sets of equations! Now, the reader is entitled to ask himself the following question: *How can we find other equations for this problem?* To tell the truth, the problem of the propagation in a periodic medium has got an infinity of solutions. As already shown in sec. 2.6, the problem that is at hand amounts to looking for the so-called Bloch waves. For such particular solutions, a miracle happens: the balance between incoming and outgoing waves is perfectly enforced. This latter fact has mathematical consequences: there are two sets of equations that link a_n^E and b_n^E on the one hand, and a_n^K and b_n^K on the other hand.... The discovery of these equations is the topic of what follows.

[1] Eq. 5.9 and Eq. 5.10 are equivalent ways to write Eq. 4.19 since the Hankel functions $H_m^{(1)}$ can be expressed as a linear combination of Bessel functions of the first and second kinds. Indeed, one has $H_m^{(1)}(r) = J_m(r) + iY_m(r)$ for any integer m.

5.3 Specificity of Lord Rayleigh's Algorithm

We now assume some Floquet-Bloch boundary conditions as stated in Eq. 3.34 on the opposite sides of the basic cell Y, instead of matching the field decay outside the fiber cladding by a jacket condition. For oblique lattices, Eq. 3.34 can be equivalently written as

$$\mathbf{E}_{\mathbf{k}}(\mathbf{r}+\mathbf{R}_{\mathbf{p}}) = \mathbf{E}(\mathbf{r})e^{i\mathbf{k}\cdot\mathbf{R}_{\mathbf{p}}} \ , \tag{5.15}$$

$$\mathbf{K}_{\mathbf{k}}(\mathbf{r}+\mathbf{R}_{\mathbf{p}}) = \mathbf{K}(\mathbf{r})e^{i\mathbf{k}\cdot\mathbf{R}_{\mathbf{p}}} \ , \tag{5.16}$$

where $\mathbf{r}$ is the position vector, $\mathbf{k}$ is known as the Bloch wave-vector and $\mathbf{R}_{\mathbf{p}} = p^1\mathbf{a}_1 + p^2\mathbf{a}_2$ is the vector attached to the nodes $\mathbf{p} = (p^1, p^2) \in \mathbb{Z}^2$ of the lattice of translation vectors $\mathbf{a}_1$ and $\mathbf{a}_2$, which form the basis for the lattice as a whole (See Fig. 5.1). All of these aforementioned vectors lie in the transverse plane.

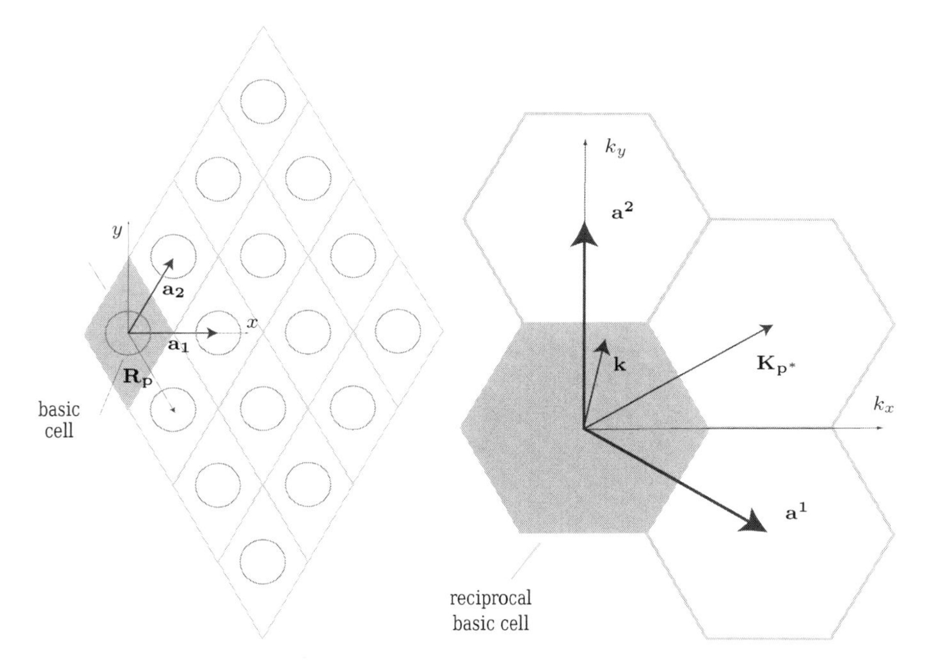

Fig. 5.1 An oblique array of cylinders in direct space and its associated reciprocal lattice.

It amounts to taking into account the superposed effect of singular sources arising from all the other cylinders within the array.[2]

5.4 Green's Function Associated with a Periodic Lattice

The contribution of the other cylinders within the array will depend on the geometry of the lattice. To consider the quasi-periodicity of the transverse field (electric or magnetic), we introduce a two-dimensional quasi-periodic *Green function* $G_{\mathbf{k}}$ which satisfies

$$\left(\Delta + (k_{\perp}^{M})^{2}\right) G_{\mathbf{k}}(\mathbf{r}, \mathbf{r}') = \sum_{\mathbf{p} \in \mathbb{Z}^2} \delta(\mathbf{r} - \mathbf{r}' - \mathbf{R_p}) e^{i\mathbf{k}\cdot(\mathbf{r}-\mathbf{r}')} , \tag{5.17}$$

where the sum stretches over the entire array of nodes $\mathbf{p}$ (locations of the centers of the cavities/inclusions). A solution for this differential equation is the sum of an infinite set of outgoing waves, spreading out from each point of the lattice

$$\begin{aligned} G_{\mathbf{k}}(\mathbf{r}, \mathbf{r}') &= -\frac{i}{4} \sum_{\mathbf{p} \in \mathbb{Z}^2} H_0^{(1)}(k_{\perp}^{M} \mid \mathbf{r} - \mathbf{r}' - \mathbf{R_p} \mid) e^{i\mathbf{k}\cdot\mathbf{R_p}} \\ &= -\frac{i}{4} \sum_{\mathbf{p} \in \mathbb{Z}^2} J_0(k_{\perp}^{M} \mid \mathbf{r} - \mathbf{r}' - \mathbf{R_p} \mid) e^{i\mathbf{k}\cdot\mathbf{R_p}} \\ &\quad + \frac{1}{4} \sum_{\mathbf{p} \in \mathbb{Z}^2} Y_0(k_{\perp}^{M} \mid \mathbf{r} - \mathbf{r}' - \mathbf{R_p} \mid) \Big) e^{i\mathbf{k}\cdot\mathbf{R_p}} . \end{aligned} \tag{5.18}$$

In fact, we need only include the singular Bessel functions Y_0 in Eq. 5.18, as shown for instance in [MMP02]. Defining $\boldsymbol{\xi} = \mathbf{r} - \mathbf{r}'$, we note that $\mid \mathbf{R_p} \mid > \mid \boldsymbol{\xi} \mid$, for $\mathbf{p} \in \mathbb{Z}^2 \setminus \{0, 0\}$. Hence, omitting the central point in the lattice,

[2] We note that this method is derived from the work of John William Strutt, the third Baron Rayleigh, back in 1892 [Ray92], and has been previously applied in scalar transport problems and in the transverse electromagnetic case [NMB95], which are both characterised by a single potential function (for TE or TM cases).

Graf's addition theorem for Bessel functions [Wat44] ensures that

$$\begin{aligned}
&\sum_{\mathbf{p}\in\mathbb{Z}^2\setminus\{0,0\}} Y_0(k_\perp^M \mid \boldsymbol{\xi} - \boldsymbol{R_p} \mid) e^{i\mathbf{k}\cdot\mathbf{R_p}} \\
&= \sum_{\mathbf{p}\in\mathbb{Z}^2\setminus\{0,0\}} \left[\sum_{l=-\infty}^{+\infty} Y_l(k_\perp^M \mid \mathbf{R_p} \mid) J_l(k_\perp^M \mid \boldsymbol{\xi} \mid) e^{il(\Phi_\mathbf{p}-\phi)} \right] e^{i\mathbf{k}\cdot\mathbf{R_p}} \\
&= \sum_{l=-\infty}^{+\infty} \left[\sum_{\mathbf{p}\in\mathbb{Z}^2\setminus\{0,0\}} Y_l(k_\perp^M \mid \mathbf{R_p} \mid) e^{il\Phi_\mathbf{p}} e^{i\mathbf{k}\cdot\mathbf{R_p}} \right] J_l(k_\perp^M \mid \boldsymbol{\xi} \mid) e^{-il\phi} \,, \quad (5.19)
\end{aligned}$$

where $\phi = arg(\boldsymbol{\xi})$ and $\Phi_p = arg(\mathbf{R_p})$.[3] We now introduce the so-called dynamic *lattice sums*

$$S_l^Y(k_\perp^M, \mathbf{k}) = \sum_{\mathbf{p}\in\mathbb{Z}^2\setminus\{0,0\}} Y_l(k_\perp^M \mid \mathbf{R_p} \mid) e^{il\Phi_\mathbf{p}} e^{i\mathbf{k}\cdot\mathbf{R_p}} \,, \quad (5.20)$$

so that

$$\begin{aligned}
&\sum_{\mathbf{p}\in\mathbb{Z}^2\setminus\{0,0\}} Y_0(k_\perp^M \mid \boldsymbol{\xi} - \mathbf{R_p} \mid) e^{i\mathbf{k}\cdot\mathbf{R_p}} \\
&= \sum_{l=-\infty}^{+\infty} S_l^Y(k_\perp^M, \mathbf{k}) J_l(k_\perp^M \mid \boldsymbol{\xi} \mid) e^{-il\phi} \,. \quad (5.21)
\end{aligned}$$

Green's function $G_\mathbf{k}$ can therefore be represented as a Neumann series within the central unit cell

$$\begin{aligned}
G_\mathbf{k}(\mathbf{r}, \mathbf{r}') &= \frac{1}{4} Y_0(k_\perp^M \mid \mathbf{r} - \mathbf{r}' \mid) \\
&+ \frac{1}{4} \sum_{l=-\infty}^{+\infty} S_l^Y(k_\perp^M, \mathbf{k}) J_l(k_\perp^M \mid \mathbf{r} - \mathbf{r}' \mid) e^{-il\phi}. \quad (5.22)
\end{aligned}$$

5.5 Some Absolutely Convergent Lattice Sums

The series in Eq. 5.20 is slowly convergent (the terms in the sum are of the order $O({R_p}^{-1/2})$ and therefore the sum is only semi-convergent). One way of accelerating this convergence is to employ the Poisson summation

[3] We note that shuffling the sum indices in the second equality of Eq. 5.19 requires the series to be positive or absolutely convergent for $\mid \mathbf{R_p} \mid > \mid \boldsymbol{\xi} \mid$. A justification for this is given *a posteriori* in section 5.5

formula. For this, we introduce the reciprocal lattice, depicted in Fig. 5.1, in the following way. We first define the area of the basic cell Y

$$A = |\mathbf{a}_1 \times \mathbf{a}_2| \,. \tag{5.23}$$

In a second step, we recall the vectors of the reciprocal basic cell Y^* as follows:

$$\begin{cases} \mathbf{a}^1 = \dfrac{2\pi}{A^2}\mathbf{a}_2 \times (\mathbf{a}_1 \times \mathbf{a}_2) & (5.24\text{a}) \\ \mathbf{a}^2 = \dfrac{2\pi}{A^2}\mathbf{a}_1 \times (\mathbf{a}_2 \times \mathbf{a}_1) \,. & (5.24\text{b}) \end{cases}$$

Analogously to $\mathbf{R}_\mathbf{p}$ we define the set of vectors $\mathbf{K}_{\mathbf{p}^*}$ in reciprocal space as $\mathbf{K}_{\mathbf{p}^*} = p_1\mathbf{a}^1 + p_2\mathbf{a}^2$, with $\mathbf{p}^* = (p_1, p_2) \in \mathbb{Z}^2$. We then define $\mathbf{Q}_{\mathbf{p}^*} = \mathbf{K}_{\mathbf{p}^*} + \mathbf{k}$ and $\theta_{\mathbf{p}^*} = arg(\mathbf{Q}_{\mathbf{p}^*})$. We are now in a position to rewrite Eq. 5.20 as an absolutely convergent expression for the lattice sums[4]

$$S_l^Y(k_\perp^M, \mathbf{k}) = \frac{1}{4}Y_0(k_\perp^M \mid \boldsymbol{\xi} \mid)\delta_{l,0} - \frac{1}{A}i^l \sum_{\mathbf{p}^* \in \mathbb{Z}^2} \frac{J_l(\mathbf{Q}_{\mathbf{p}^*}\boldsymbol{\xi})}{|\mathbf{Q}_{\mathbf{p}^*}|^2 - (k_\perp^M)^2} e^{il\theta_{\mathbf{p}^*}} \,. \tag{5.25}$$

Another remarkable fact is that the expression Eq. 5.25 for the lattice sums no longer involves the magnitude of the vector $\mathbf{R}_\mathbf{p}$ (*i.e.* the difference between the two vectors of Green's function in Eq. 5.18) contrarily to Eq. 5.20. This arbitrariness can be exploited by making use of the recurrence formulae

$$\int_0^\eta \mid \xi \mid^{l+1} J_l(k_\perp^M \mid \xi \mid) d\xi = \eta^{n+1} \frac{J_{l+1}(k_\perp^M \eta)}{k_\perp^M} \,, \tag{5.26}$$

$$\int_0^\eta \mid \xi \mid^{l+1} Y_l(k_\perp^M \mid \xi \mid) d\xi = \eta^{n+1} \frac{Y_{l+1}(k_\perp^M \eta)}{k_\perp^M} + \frac{2^{l+1} l!}{\pi (k_\perp^M)^{l+2}} \,. \tag{5.27}$$

After q successive integrations of Eq. 5.25 we obtain the following formula, originally derived in [MD92]:

$$S_l^Y(k_\perp^M, \mathbf{k}) J_{l+q}(k_\perp^M z) = -\delta_{l0}\left[Y_q(k_\perp^M z) + \frac{1}{\pi}\sum_{n=1}^{q} \frac{(q-n)!}{(n-1)!}\left(\frac{2}{k_\perp^M z}\right)^{q-2n+2} \right]$$

[4] We just recall the Poisson formula in the two-dimensional case. For any function $\psi \in \mathcal{S}$, if we denote by $\widehat{\psi}$ the Fourier transform of the function ψ, we have the folllowing identity: $\sum_{\mathbf{p} \in \mathbb{Z}^2} \psi(\mathbf{R}_\mathbf{p}) = \frac{1}{A} \sum_{\mathbf{p}^* \in \mathbb{Z}^2} \widehat{\psi}(\mathbf{K}_{\mathbf{p}^*})$.

$$-\frac{4i^l}{A}\sum_{\mathbf{p}^*\in\mathbb{Z}^2}\left(\frac{k_\perp^M}{\mid \mathbf{Q}_{\mathbf{p}^*}\mid}\right)^q\frac{J_{l+q}(\mid \mathbf{Q}_{\mathbf{p}^*}\mid)e^{il\theta_p}}{\mid \mathbf{Q}_{\mathbf{p}^*}\mid^2-(k_\perp^M)^2}\,. \tag{5.28}$$

From Eq. 5.20, we can also see that the lattice sums satisfy the identity

$$S_{-l}^Y(k_\perp^M,\mathbf{k})=\overline{S_l^Y}(k_\perp^M,\mathbf{k})\,, \tag{5.29}$$

and hence it is sufficient to calculate them only for non-negative values of l.

5.6 The Rayleigh Identities

Using the contour depicted in Figure 5.2, it follows from Green's theorem (for this we refer to our mathematical pot-pourri) that

$$\begin{aligned}&\int_{Y\setminus C}\Big(u(\mathbf{r}')\Delta' G_{\mathbf{k}}(\mathbf{r},\mathbf{r}')-G_{\mathbf{k}}(\mathbf{r},\mathbf{r}')\Delta' u(\mathbf{r}')\Big)\,d\mathbf{r}'\\&=\int_{\partial Y\cup\partial C}\left(u(\mathbf{r}')\frac{\partial G_{\mathbf{k}}(\mathbf{r},\mathbf{r}')}{\partial n'}-G_{\mathbf{k}}(\mathbf{r},\mathbf{r}')\frac{\partial u(\mathbf{r}')}{\partial n'}\right)dl'\,,\end{aligned} \tag{5.30}$$

where the vectors are located within the region $Y\setminus C$ as shown in Fig. 5.2 and the operator Δ' applies to primed variables. We can now take into

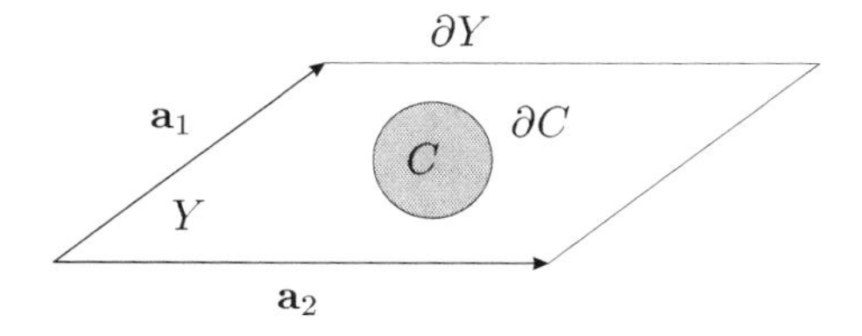

Fig. 5.2 The central unit cell, showing the contour within which Green's theorem is to be applied

account the fact that the function u and $G_{\mathbf{k}}$ satisfy a conjugate quasiperiodicity condition to Eq. 3.34, specifically:

$$u(\mathbf{r}+\mathbf{d})=u(\mathbf{r})e^{-i\mathbf{k}\cdot\mathbf{d}}. \tag{5.31}$$

This causes the integral around the boundary ∂Y to be zero in Eq. 5.30: The outward normal to ∂Y is anti-periodic. In addition, we know from the definition of the Green's function (See Eq. 5.17) that

$$\int_{Y\setminus C} \Big(u(\mathbf{r}')\Delta' G_{\mathbf{k}}(\mathbf{r},\mathbf{r}') - G_{\mathbf{k}}(\mathbf{r},\mathbf{r}')\Delta' u(\mathbf{r}') \Big)\, d\mathbf{r}' = u(\mathbf{r})\,, \tag{5.32}$$

for any $\mathbf{r}$ in $Y \setminus C$. Thus, applying Green's theorem provides us with the integral equation

$$u(\mathbf{r}) = \int_{\partial C} \left(u(\mathbf{r}')|_{\partial C} \frac{\partial G_{\mathbf{k}}(\mathbf{r},\mathbf{r}')}{\partial r'} - G_{\mathbf{k}}(\mathbf{r},\mathbf{r}')\, \frac{\partial u(\mathbf{r}')}{\partial r'}\bigg|_{\partial C} \right) dl'\,. \tag{5.33}$$

We now take into account the fact that ∂C is a circle of radius r_c. In such a situation $\frac{\partial u}{\partial n}$ is nothing but $\frac{\partial u}{\partial r}$. This expression can thus be evaluated by expanding both u and its radial derivative as Fourier series on the boundary ∂C:

$$u|_{\partial C} = \sum_{l=-\infty}^{\infty} \alpha_l e^{il\theta}\,,\ \frac{\partial u}{\partial r}\bigg|_{\partial C} = \sum_{l=-\infty}^{\infty} \beta_l e^{il\theta}\,, \tag{5.34}$$

where α_m and β_m are complex (as yet unknown) Fourier coefficients.

Using Graf's addition formula and some tedious computations,[5] we then expand Green's function in Eq. 5.22 in co-ordinates centered on the origin. Hence Eq. 5.33 leads to

$$\begin{aligned} u(\mathbf{r}) = &\sum_{n=-\infty}^{\infty} Y_n(k_\perp^M r)\Big(\frac{\pi r_c}{2}\Big)\Big[\beta_n J_n(k_\perp^M r_c) - \alpha_n k_\perp^M J_n'(k_\perp^M r_c)\Big] e^{in\theta} \\ &+ \sum_{n=-\infty}^{+\infty} \sum_{m=-\infty}^{+\infty} (-1)^{m+n} S_{m-n}^{Y}(k_\perp^M, \mathbf{k}) J_n(k_\perp^M r)\Big(\frac{\pi r_c}{2}\Big) \\ &\times \Big[\beta_m J_m(k_\perp^M r_c) - \alpha_m k_\perp^M J_m'(k_\perp^M r_c)\Big] e^{in\theta}\,. \end{aligned} \tag{5.35}$$

By identifying the multipole coefficients in the singular and non-singular parts of the expansions Eq. 5.9 and Eq. 5.10 together with Eq. 5.35, and

[5]The zealous reader can find a complete proof in [MMP02].

by making use of the uniqueness of Fourier decomposition, we obtain

$$a_n^{(E,K)} = \sum_{m=-\infty}^{+\infty} (-1)^{m+n} S^Y_{m-n}(k_\perp^M, \mathbf{k}) \left(\frac{\pi r_c}{2}\right) \times \left[\beta_m J_m(k_\perp^M r_c) - \alpha_m k J'_m(k_\perp^M r_c)\right] , \tag{5.36}$$

$$b_n^{(E,K)} = \left(\frac{\pi r_c}{2}\right) \left[\beta_n J_n(k_\perp^M r_c) - \alpha_n k J'_n(k_\perp^M r_c)\right] . \tag{5.37}$$

Hence, we are led to the following identities:

$$a_l^E = \sum_{m=-\infty}^{+\infty} (-1)^{l+m} S^Y_{m-l}(k_\perp^M, \mathbf{k}) b_m^E , \tag{5.38}$$

$$a_l^K = \sum_{m=-\infty}^{+\infty} (-1)^{l+m} S^Y_{m-l}(k_\perp^M, \mathbf{k}) b_m^K . \tag{5.39}$$

We note that these so-called *Rayleigh identities* have been derived in previous instances for the case of transverse incidence [MD92]. Here, the lattice sums may have a pure imaginary argument $k_\perp^M$ corresponding to non-propagating modes (and therefore to the existence of a cut-off frequency), which is a new feature due to conical propagation.

5.7 The Rayleigh System

These two sets of equations are linked via the boundary conditions (4.36) and lead to the so-called *Rayleigh system*:

$$M_l^{(EE)}(k_0) b_l^E + M_l^{(EK)}(k_0) b_l^K - \sum_{m=-\infty}^{+\infty} (-1)^{l+m} S^Y_{m-l}(k_\perp^M, \mathbf{k}) b_m^E = 0 , \tag{5.40}$$

$$M_l^{(KE)}(k_0) b_l^E + M_l^{(KK)}(k_0) b_l^K - \sum_{m=-\infty}^{+\infty} (-1)^{l+m} S^Y_{m-l}(k_\perp^M, \mathbf{k}) b_m^K = 0 . \tag{5.41}$$

In matrix form, this system reduces to $\mathbf{RB} = \mathbf{0}$, with obvious notation, namely: $\mathbf{R}(k_0, \mathbf{k}) = (\mathbf{M} + \mathbf{S})$, where $\mathbf{R}$ is the so-called Rayleigh matrix,

$\mathbf{M}$ is the matrix of boundary conditions and $\mathbf{S}$ the matrix of lattice sums.[6] This system possesses the aesthetically pleasing quality of separating the properties of each cylindrical surface (contained in the entries of $\mathbf{M}$) from the effect of the geometry of the lattice (contained in the entries of $\mathbf{S}$). It is noteworthy that the coefficients $M_m^{(EE)}$ and $M_m^{(KK)}$ are real, whereas $M_m^{(EK)} = -M_m^{(KE)}$ are pure imaginary, making the matrix $\mathbf{M}$ hermitian. Since $\overline{S_{m-l}^Y(k_\perp^M, \mathbf{k})} = S_{l-m}^Y(k_\perp^M, \mathbf{k})$, the Rayleigh matrix $\mathbf{R}$ is itself Hermitian (and hence has a real positive spectrum), as one would expect for a system associated with non-dissipative rods. This infinite system can be truncated. Some justification for this is given in the next section. Zeros of the determinant of the truncated system correspond to non-trivial solutions of the Maxwell system subject to Bloch boundary conditions. These roots are therefore associated with allowed propagating modes (Bloch waves).

5.8 Normalization of the Rayleigh System

Using large order approximations of Bessel functions [AS65] in the boundary terms $M_n^{(\alpha\gamma)}$, α and $\gamma \in \{E, K\}$, one can show that, as $n \longrightarrow +\infty$ [GPM03a]

$$M_n^{(\alpha\gamma)} = O\left((n+1)!^2 n \left(\frac{1}{2} k_\perp^M r_c \right)^{-2n} \right) . \tag{5.42}$$

Taking asymptotic expansions in Eq. 5.28, it can also be shown that

$$S_l^Y(k_\perp^M, \mathbf{k}) = O\left((l+1)! \left(\frac{1}{2} k_\perp^M d \right)^{-l} \right) , \text{ as } l \longrightarrow +\infty . \tag{5.43}$$

This causes numerical difficulties when $k_\perp^M d \ll 1$, since the off-diagonal terms increase extremely rapidly with index l. One solution is to rescale the terms of the previous system, so as to put it into a more amenable form. In a similar way to that adopted in [PMM+00] for the transverse elastodynamic case, we can introduce new unknown coefficients x_n, y_n such

---

[6] It is noticed in [MNBKD97] that $\mathbf{S}$ is a Toeplitz matrix *i.e.* a matrix whose entries are identical along each (top-left to lower-right) diagonal. Its bands contain *lattice sums*. The inverse of such symmetric, positive-definite $n \times n$ Toeplitz matrix can be typically found in $O(n^2)$ operations which should be compared to the $O(n^2 \log(n))$ operations hardly achieved by fastest algorithms for general symmetric positive-definite $n \times n$ matrices (see our mathematics pot-pourri for definition of Landau's notation).

that

$$x_n = \sqrt{\mid M_n^{(EE)} \mid}\left(b_n^E + \frac{M_n^{(EK)}}{M_n^{(EE)}} b_n^K\right), \tag{5.44}$$

$$y_n = \sqrt{\mid M_n^{(KK)} \mid}\left(\frac{M_n^{(KE)}}{M_n^{(KK)}} b_n^E + b_n^K\right). \tag{5.45}$$

Upon substituting this into the previous system (5.40)-(5.41), one can recast the Rayleigh system as[7]

$$x_l + \sum_{m=-\infty}^{+\infty}\left(D_{lm}^{(EE)} x_m + D_{lm}^{(EK)} y_m\right) = 0\,, \tag{5.46}$$

and

$$y_l + \sum_{m=-\infty}^{+\infty}\left(D_{lm}^{(KE)} x_m + D_{lm}^{(KK)} y_m\right) = 0\,, \tag{5.47}$$

where the new coefficients $D_{lm}^{(\alpha\gamma)}$, $(\alpha, \gamma) \in \{E, K\}$, can be found in [GPM03a]. Also, it is shown in [GPM03a] that if l is fixed as $m \to +\infty$,

$$D_{lm}^{(\alpha\gamma)} = O\left(\frac{m^{-l}}{\sqrt{m}}\left(\frac{r_c}{d}\right)^m\right), \tag{5.48}$$

and the other way around if m is fixed while $l \to +\infty$ (by interchanging the roles of l and m).

This normalised Rayleigh system (5.46)–(5.47) has matrix elements that decay exponentially away from the main diagonal, giving rise to higher-order multipole coefficients that decay similarly quickly.

5.9 Convergence of the Multipole Method

The expansions in Eq. 5.9 and Eq. 5.10 near r_m are dominated by the singular part of

$$\sum_{l=-\infty}^{+\infty} b_l^\alpha Y_l(k_\perp^M r) e^{il\theta}\;,\; \alpha \in \{E, K\}\,. \tag{5.49}$$

[7]Note that the transformation (5.44)-(5.45) preserves the hermitian structure of the infinite linear system (5.40)-(5.41).

Since Eq. 5.49 converges for at least one pair (r_m, θ), we are assured that

$$\lim_{l \to +\infty} \mid b_l^\alpha Y_l(k_\perp^M r_m) \mid = 0 \, . \tag{5.50}$$

Thus, for sufficiently large values of l, we deduce that

$$\mid b_l^\alpha \mid < \frac{1}{\mid Y_l(k_\perp^M r_m) \mid} \, . \tag{5.51}$$

Finally, for any $r > r_m$ and for a large enough value of l, by means of large-order expansions for Bessel functions [AS65] we have

$$\mid b_l^\alpha Y_l(k_\perp^M r) e^{il\theta} \mid < \frac{\mid Y_l(k_\perp^M r) \mid}{\mid Y_l(k_\perp^M r_m) \mid} = O\left(\left(\frac{r_m}{r} \right)^l \right) \, . \tag{5.52}$$

It follows that in the region near the boundary, the multipole expansions converge absolutely as

$$\sum_{l=-\infty}^{+\infty} \left(\frac{r_m}{r} \right)^l , \tag{5.53}$$

i.e. the convergence is exponential. A similar result holds for expansions near the outer boundary r_M, with the exception that in this region the J_l Bessel functions dominate in the expansions of Eq. 5.9 and Eq. 5.10.

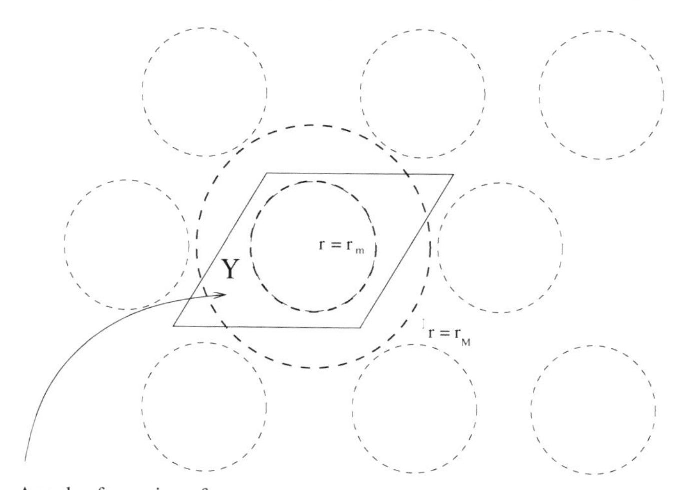

Fig. 5.3 The region of convergence for the multipole expansions

5.10 Higher-order Approximations, Photonic Band Gaps for Out-of-plane Propagation

In this section, we present some photonic band diagrams for hexagonal arrays of circular fibres. We give results for non dilute composites with circular inclusions of filling fraction $f = 0.45$ and $\beta\Lambda = 0.0$ (figure 5.4a), $\beta\Lambda = 0.7$ (figure 5.4b), $\beta\Lambda = 2.0$ (figure 5.4c) and $\beta\Lambda = 9.0$ (figure 5.4d). We only need solve the truncated Rayleigh system as far as the seventh multipole order, due to the normalisation. Computation of a complete dispersion diagram requires less than ten minutes on a PC with a 1Ghz processor and 256 Mb RAM. By making use of multipole expansions (5.8) and (5.10) we show on figure 5.5 the associated modulus of the eigenfield K_z for a normalised propagation constant $\beta\Lambda = 2.0$ and a null Bloch vector $\mathbf{k}$. We also give an illustration of the widths of the gaps (for the same filling fraction $f = 0.45$) versus the normalised propagation constant $\beta\Lambda \in [9.0; 11]$ (figure 5.6). We note that there is no anisotropy in the long-wave limit for the hexagonal lattice (the slopes or curvatures of the acoustic curves approaching the point Γ from the right and from the left are the same for each diagram). Also, figure 5.6 illustrates that new band gaps open up below the first band as the normalised propagation constant $\beta\Lambda$ increases, which is consistent with the work by Maradudin and McGurn [MM94]: On the diagram of Figure 5.6, the hatched regions correspond to frequencies for propagating modes, whereas the non hatched regions show the ranges of frequencies for which no wave is allowed to propagate within the photonic crystal fibre. The straight line passing through the point $(9.0, 9.0)$ corresponds to the "cut-off" frequency evaluated as a function of $\beta\Lambda$.

Finally, numerical comparisons between the Finite Element and Rayleigh Methods have been drawn in [NGZ+02; GNG+]. The fact that some very good matches were observed between those two radically different approaches provides a very sure footing for both methods. Indeed, the Finite Element approach is based on discrete variational formulation (a global analysis) leading to large sparse matrices, unlike the Rayleigh method which is rooted in local expansions (in the neighbourhood of each cylinder's boundary) leading to full small-size-systems. The price to pay with the Rayleigh method is the lack of flexibility in the geometry (*i.e.* a restriction to circular inclusions). But otherwise, it provides analytical results that can be used to derive various asymptotic formulae such as effective properties of microstructured optical fibres [GPM03b; GPM03a; PGM].

5.11 Conclusion and Perspectives

Throughout this chapter, we have concentrated on the filtering properties of periodic micro-structured optical fibres containing holes of circular cross-section. Perhaps the most obvious generalisation is to adapt the Rayleigh method to the case of elliptical cylinders [YMNB99]. This requires a new set of basis functions in order to allow the boundary conditions to be satisfied. Whereas the singularities associated with the circular cylinders all lie at the centres of the inclusions, the potentials for elliptical inclusions must exhibit a branch cut sitting between the two *foci* of each ellipse. Another approach is to perturb slightly the boundaries of the inclusions: a small parameter characterising the perturbation is introduced and this can be used to derive effective boundary conditions [PGMN03]. The asymptotic approach also allows the modelling of cylinders with a thin coating, leading to effective boundary conditions for perfectly bonded inclusions. Much of the foundation which underlies this chapter can thus be applied immediately without modification. Asymptotic analysis combined with the Rayleigh Method would enable one to analyse the effect of a weak twist on the photonic band gaps of PCF (cf. [NZG] for the setting of the spectral problem with Finite Elements). This study is of the foremost importance when one realises that photonic band gap guidance relies on gap's robustness [PBH$^+$03] with respect to the conical angle (need to preserve the existence of the gap when the conical parameter evolves). Shaping microstructured fibres onto hollow honeycomb PCF hence requires both knowledge of their band structure (*infinite periodic cladding*) and their *scattering spectrum* (hollow core surrounded by a *finite periodic cladding*). The Rayleigh Algorithm proves very handy in the first case, whereas Finite Elements are certainly best suited in the second case [GLNZ02].

In other work, not presented here, the Rayleigh algorithm has been generalised to the transverse elastodynamic waves in microstructured elastic composites, requiring a pair of potential functions for the coupled shear dilatation problem (solutions of two Helmholtz equations coupled by two boundary conditions) and a potential for the anti-plane shear scalar problem. We would like to draw an analogy between the results obtained for the TE and TM polarisations formulations for highly conducting inclusions and similar results for the boundary value problems that occur in continuum mechanics: the TE polarisation problem is equivalent to the problem of anti-plane shear for a medium containing very soft inclusions, whereas the TM polarisation problem is similar to that for the anti-plane shear of a region containing very heavy inclusions that each have the same shear modulus as the elastic matrix. Also, there is a formal equivalence between the conical mounting case in electromagnetism and the case of in-plane elastic waves: the latter can be readily tackled from our analysis [MMP02]. However, the elastic band properties of such structures for in-plane shear and dilatational waves are markedly different from their optical counterparts: it is known for example that a square array of cavities possesses an extremely low resistance to shear forces along the primary axes, in comparison to the same force applied to a hexagonal array. Given this, it was found that square arrays are more conducive than hexagonal ones to the formation of stop bands in elastodynamics (contrarily to electromagnetism).

In the case of conically propagating elastic waves, the study requires four potential functions, which are all solutions of Helmholtz equations and coupled by three boundary conditions together with a necessary gauge condition [GM04]. The problem of obliquely propagating elastic waves presents some new features such as different cut-off frequencies for pressure and shear waves (they propagate with different speeds) and requires some fundamental alterations to the algorithm presented in this chapter: the choice of an *adequate* gauge for the so-called *Lamé potentials* leads to a set of new basis functions [GM04]. The experimental and theoretical/numerical results discussed in [RMD+; GAMN] suggest that sonic band gaps can be used to manipulate sound with great precision and enhance its interaction with light. Such a photo-elastic coupling is slightly beyond the scope of the present book but we should undoubtedly leave a door open to such a generalisation of the material presented in the previous chapters.

Last, but not least, newly discovered *Left-Handed-Materials* within which light propagates in the wrong direction (due to a *Negative Refractive Index*) suggest brand new applications for Photonic Crystal Fibres [MG]. Generalisation of the Rayleigh Algorithm to analysis of evanescent states within periodic arrays of cylinders should prove very handy for such *meta-materials*. Design of these so-called *meta-waveguides* may also be addressed with Finite Elements fulfilling Bloch conditions [GGN+03; NGGZ]. The implementation of Finite Element Method for the analysis of electromagnetic waves propagating within microstructured fibres (whose periodic cladding may be finite or not) is the object of the preceding chapter.

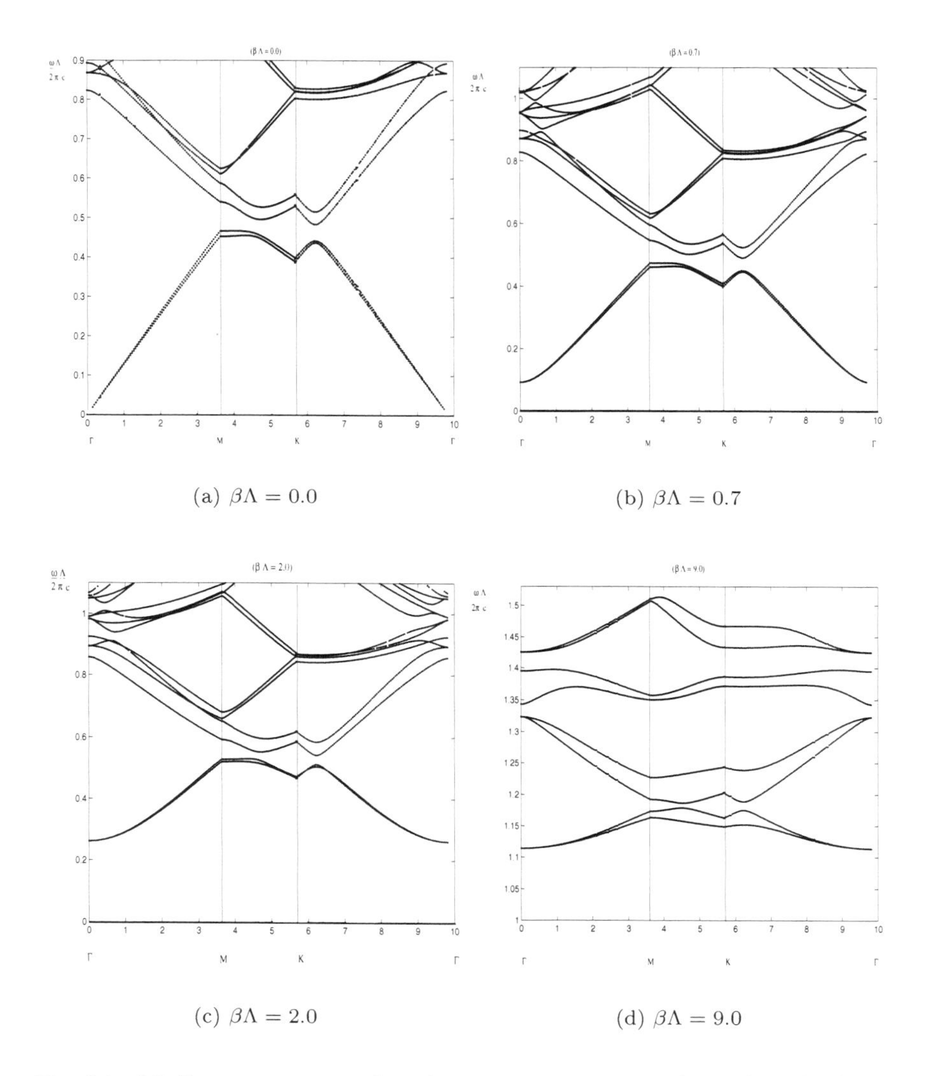

(a) $\beta\Lambda = 0.0$

(b) $\beta\Lambda = 0.7$

(c) $\beta\Lambda = 2.0$

(d) $\beta\Lambda = 9.0$

Fig. 5.4 All diagrams correspond to the same structure, namely a triangular lattice with a pitch Λ. This lattice is made of air channels of radius $r_c = 0.327\,\Lambda$, drilled within a matrix of silica (index $n_M = 1.45$). We show the evolution of the dispersion curves with respect to the normalised propagation constant $\beta\,\Lambda$.

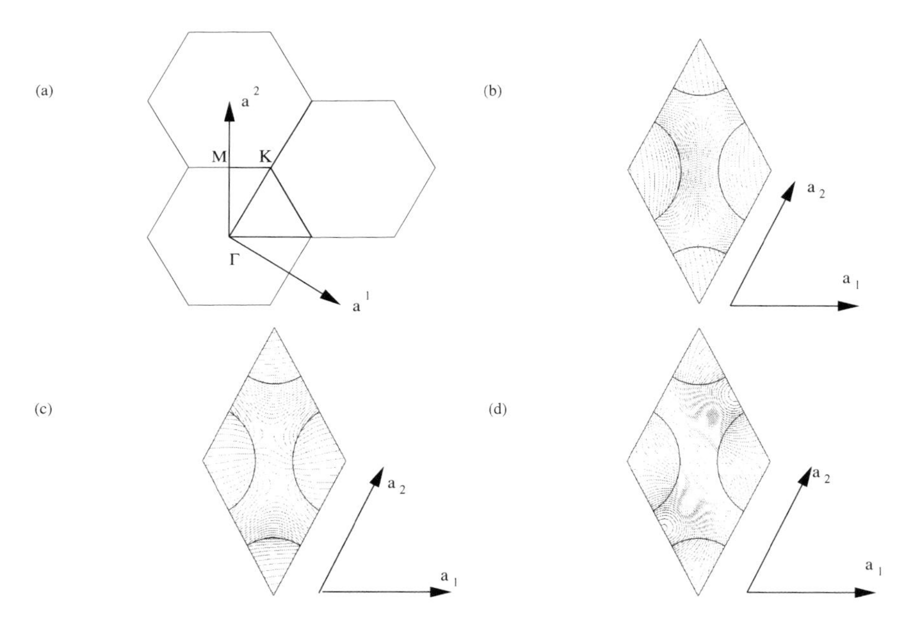

Fig. 5.5 Figure (a) shows the irreducible part of the first Brillouin zone represented by the triangle with vertices $\Gamma = (0, 0)$, $M = (0, \frac{2\pi}{\sqrt{3}\Lambda})$ and $K = (\frac{2\pi}{\sqrt{3}\Lambda}, \frac{2\pi}{\sqrt{3}\Lambda})$ which is attached to the lattice vectors $\mathbf{a}^1$ and $\mathbf{a}^2$ in reciprocal space. Figures (b), (c) and (d) each show the modulus of K_z within the basic cell generated by the vectors $\mathbf{a}^1$ and $\mathbf{a}^2$ (in physical space), when the Bloch vector is null. These correspond respectively to roots on Figure 5.4 associated with the first, third and fifth dispersion curves at point Γ. This is for a normalised conical parameter $\beta\Lambda = 2.0$.

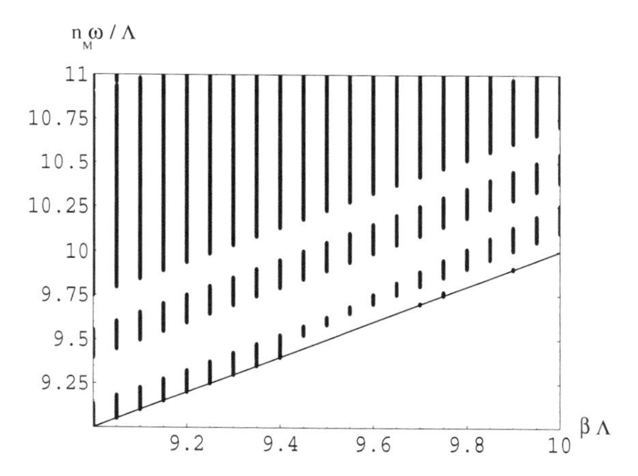

Fig. 5.6 Evolution of the widths and positions of the photonic band gaps (in normalised eigenfrequencies $n_M\omega/\Lambda$) with respect of the normalised propagation constant $\beta\Lambda$. We consider a hexagonal array of air cavities of normalized radius $r_c\,\Lambda = 0.327$ (air filling-fraction $f = 0.45$) within a matrix of silica (refractive index $n_M = 1.45$).

Chapter 6

À la Cauchy Path to Pole Finding

In the preceding chapters, it has been shown that the modes guided inside a photonic crystal fibre were associated with poles of the scattering matrix. The determination of the dispersion curves of these modes amounts to finding the poles of the scattering matrix. The numerical computation of poles of functions of a complex variable is by no means a simple problem. It cannot be considered as a simple extension of the same problem for functions of one real variable. Indeed, the description of a function of one real variable amounts, roughly speaking, to going along the real axis and to looking at the values of the function: it is just a one dimensional walk, along which the accidents of the function (its poles, asymptotes or zeros) are easily found. For complex variables, the walk should be two-dimensional: we can of course define a grid that serves to obtain a discrete version of the function, the poles of which are then searched for over this grid. Such an operation is an extremely time-consuming one. This is why one would like to have procedures to reduce as much as possible the number of numerical evaluations of the function at issue. There are two principal ways of finding the zeros or poles of a function of a complex variable. The first one is the iterative way: this has its roots in Newton methods for real-variable functions. Given an *a priori* estimate of the value of the zero, a sequence of successive approximations is generated that hopefully converges towards the correct value (such is for instance the Müller algorithm, cf. [PBTV86]). The main drawback of this method lies in its instability, and its incapacity to give the *a priori* estimate. The second way, which we shall take, is the use of Cauchy integrals for holomorphic functions. The starting point is precisely the Cauchy theorem, which we recall with no attempt at mathematical rigour (for the theory of functions of one complex variable, we refer the reader to [Rud86]). Let us consider a holomorphic function $z \rightarrow f(z)$ (that is, a function which is differentiable with respect to the complex variable z) and a path γ in the complex plane that goes round z_0, that is, an application from the interval $[0, 1]$ into $\mathbb{C}$ such that : $\gamma(0) = \gamma(1)$, and such that there are no other multiple points. We then have the following relations:

$$f(z_0) = \frac{1}{2i\pi} \int_\gamma \frac{f(z)}{z - z_0} dz, \tag{6.1}$$

$$\frac{d^n f}{dz^n}(z_0) = n! \frac{1}{2i\pi} \int_\gamma \frac{f(z)}{(z - z_0)^{n+1}} dz\,. \tag{6.2}$$

This is a major result: it says that in order to get the value of f at some point z_0, it suffices to make a walk around z_0. This property is a key for computing the value of the pole of a *meromorphic function* (a meromorphic function is a function that is holomorphic except for a discrete set of points, cf. [Rud86]). Indeed, let us assume that f has a pole of order 1 at z_0. We know then that f has a *Laurent expansion* of the form:

$$f(z) = \frac{a_{-1}}{z - z_0} + \sum_{n \geq 0} a_n (z - z_0)^n. \tag{6.3}$$

According to the Cauchy theorem, the following holds:

$$a_{-1} = \frac{1}{2i\pi} \int_\gamma f(z) dz, \tag{6.4}$$

and also that:

$$z_0 a_{-1} = \frac{1}{2i\pi} \int_\gamma z f(z) dz. \tag{6.5}$$

By comparing both results, we get the value of the pole z_0:

$$z_0 = \frac{\int_\gamma z f(z) dz}{\int_\gamma f(z) dz}. \tag{6.6}$$

It is to be stressed that the algorithm followed to obtain this value is completely different to that followed in an iterative approach. Indeed, we do not need to know an *a priori* value for z_0 here, all we have to know is an arbitrarily large region of the complex plane that contains the pole. This is by no means as severe a limitation as in the case of an iterative approach. From a computational point of view, the limitation lies in the evaluations of function f: these should be as few as possible. Hence the integration algorithm should be as efficient as possible (which of course is a truism).

In the evaluation of z_0 as described above, it should be noted that though there are two integrals to be computed, the values at which f is numerically evaluated can be used twice.

We have assumed above that the pole of f was of order 1; for the sake of completeness let us now consider the case of a function f that has a pole of order greater than 1. As a generic example, we choose $f(z) = \frac{a_{-2}}{(z-z_0)^2}$. As there is no term of order 1, the integral of f along γ is null: $\int_\gamma f(z)dz = 0$, but we have:

$$\int_\gamma zf(z)dz = \int_\gamma \frac{a_{-2}}{z-z_0}dz + \int_\gamma \frac{z_0 a_{-2}}{(z-z_0)^2}dz \tag{6.7}$$

$$= 2i\pi a_{-2}. \tag{6.8}$$

We need a second result in order to be able to get the value z_0: let us proceed to the computation of $\int_\gamma z^2\ f(z)dz$. Using the elementary transformation: $\frac{z^2}{(z-z_0)^2} = 1 - \frac{z_0^2}{(z-z_0)^2} + \frac{2zz_0}{(z-z_0)^2}$, we get the following equality:

$$\int_\gamma z^2 f(z)dz = a_{-2}\int_\gamma dz - z_0^2 \int_\gamma f(z)dz + 2z_0 \int_\gamma zf(z)dz \tag{6.9}$$

$$= 2z_0 \int_\gamma zf(z)dz, \tag{6.10}$$

from which we deduce immediately the value of z_0. In our example there is no term of order 1, although it could of course happen that there be such terms. Generally, for a pole of order n and hence a Laurent expansion of the type:

$$f(z) = \sum_{p=1}^{n} \frac{a_{-p}}{(z-z_0)^p} + \sum_{p\geq 0} a_p(z-z_0)^p\ , \tag{6.11}$$

one should evaluate the following $n+1$ integrals:

$$R_l = \frac{1}{2i\pi}\int_\gamma z^l\ f(z)dz,\ l = 0...n. \tag{6.12}$$

Indeed, let us define $r_{lp} = \frac{1}{2i\pi}\int_\gamma \frac{z^l}{(z-z_0)^p}dz, p = 1, \ldots, n$. Using relation (6.2), we get:

$$r_{lp} = \begin{cases} \binom{p-1}{l} z_0^{l-p+1} & \text{, if } l \geq p-1 \\ 0 & \text{, if } l < p-1 \end{cases} \tag{6.13}$$

where $\binom{p}{l} = \frac{l!}{p!(l-p)!}$. Finally, we obtain the following triangular linear system, with unknowns: $a_1, ..., a_{-n}$ and z_0:

$$R_l = \sum_{p=1}^{l+1} a_{-p} \binom{p-1}{l} z_0^{l-p+1}, l = 0...n, \tag{6.14}$$

or in matrix form:

$$\begin{pmatrix} 1 & 0 & \dots & \dots 0 \\ z_0 & 1 & \dots & \dots 0 \\ \vdots & \vdots & \ddots & \dots 0 \\ z_0^n & \dots & \binom{p}{n} z_0^{n-p} & \dots 1 \end{pmatrix} \begin{pmatrix} a_{-1} \\ \vdots \\ a_{-n} \\ 0 \end{pmatrix} = \begin{pmatrix} R_0 \\ \vdots \\ \vdots \\ R_n \end{pmatrix}. \tag{6.15}$$

There are $n+1$ equations and $n+1$ unknowns, so the system can be solved uniquely. However, in Electromagnetics such situations of higher order poles hardly ever happen: with a dielectric function that is constant outside a bounded domain, the poles are always of order one (cf. [Mel95]). Poles of order superior to one might appear in non-linear problems, though there are always poles of order one of some scattering matrix.

We have already said that problem at hand in the study of photonic crystal fibres was to determine the poles of an operator and not that of a meromorphic function. Our aim is now to show that the above procedure can be extended to deal with poles of operators instead of poles of functions.

6.1 A Simple Extension: *Poles of Matrices*

As a starting point we investigate the simplest extension, in terms of operators, of functions of a complex variable: we consider applications $A(z)$ of the complex variable z that take their values in the set of $n \times n$ matrices: $A : z \to A(z) \in M_n(\mathbb{C})$. The entries of $A(z)$ are meromorphic functions of z. We want to know if it is possible to compute the value of poles by means of Cauchy integrals. How can we compute *Cauchy integrals of matrix* $A(z)$ such as:

$$I_1 = \frac{1}{2i\pi} \int_\gamma A(z) dz \tag{6.16}$$

$$I_2 = \frac{1}{2i\pi} \int_\gamma z A(z) dz\,. \tag{6.17}$$

The obvious extension of the preceding section is to define the integral of $A(z)$ as being the integral of its entries, which are just functions. Let us first consider the even simpler case of 2×2 matrices. This shall help us in understanding the main features of pole finding for operators. We use the following matrix, which has a pole at $z = z_0$:

$$A_m(z) = \frac{1}{z - z_0} \begin{pmatrix} 1 & 0 \\ 0 & 1 \end{pmatrix}. \tag{6.18}$$

With this definition, it is easy to find that:

$$I_1 = \frac{1}{2i\pi} \int_\gamma A_m(z) dz = \begin{pmatrix} 1 & 0 \\ 0 & 1 \end{pmatrix} \tag{6.19}$$

and also that:

$$I_2 = \frac{1}{2i\pi} \int_\gamma z A_m(z) dz = \begin{pmatrix} z_0 & 0 \\ 0 & z_0 \end{pmatrix} = z_0 I_1, \tag{6.20}$$

where γ is a loop enclosing z_0 once in the direct sense. A direct comparaison shows that the eigenvalues of the second residue I_2 are obtained from those of the first one I_1 by multiplication by z_0.

All these operations are straightforwardly extendable to $n \times n$ matrices. The determination of the poles is reduced to the diagonalisation of the matrices I_1 and I_2. In our example, this is particularly easy because I_1 and I_2 are diagonal matrices already. In the general case, one should proceed to the determination of the eigenvalues of I_1, I_2. Indeed, if λ is an eigenvalue of I_1, then $z_0\lambda$ is an eigenvalue of I_2. The comparison between both sets of eigenvalues of course requires that they be ordered in the same way. This is done by considering the associated eigenvectors. Therefore, the complete determination of the poles requires the full diagonalisation, or at least the trigonalization, of only one of the matrices I_1 or I_2. Indeed, let us assume that P is a matrix that diagonalises I_1: there is a diagonal matrix D_1 such that: $I_1 = PD_1P^{-1}$. Then we obviously have: $I_2 = PD_2P^{-1}$, where $D_2 = z_0 D_1$. This gives immediately the value of z_0, the computation of D_1 and P giving $D_2 = P^{-1} I_2 P$.

This is representative of the purely algorithmic way of finding the poles of $A(z)$, but we also aim at understanding the underlying structure of the method. In particular, we may wonder about the meaning of the matrices I_1, I_2. To understand what we have just computed, let us go back to our

2×2 example matrix (6.18), and let us consider the inverse matrix of A_m:

$$\widetilde{A}_m(z) = (z - z_0) \begin{pmatrix} 1 & 0 \\ 0 & 1 \end{pmatrix} . \tag{6.21}$$

This is a regular matrix, and we note that $\widetilde{A}_m(z_0) = 0$, which means that the kernel of operator $\widetilde{A}_m(z_0)$ is the entire complex plane. The range of the residue is precisely this kernel. This is a general fact, which goes of course beyond our very particular example and the result applies to $n \times n$ matrices. However, $\widetilde{A}(z_0)$ need not be null in order for $A(z)$ to have a pole at $z = z_0$. It suffices that there exists some vector u_0 such that: $\widetilde{A}(z_0)u_0 = 0$, i.e. the null space of $\widetilde{A}(z_0)$ is not reduced to 0. Let us state this result with sufficient generality:

Theorem 6.1 *Let $\widetilde{A}(z)$ be a holomorphic $n \times n$ matrix function, and let $A(z)$ denote its inverse matrix whenever it exists. Let us further assume that z_0 is a pole of order 1 of $A(z)$. Then the range of the residue of $A(z)$ at z_0 is equal to the nullspace of $\widetilde{A}(z_0)$.*

Proof. We denote $u(z) = A(z)v$, where v is some constant vector. Expanding $\widetilde{A}(z)$ in a series around z_0, we get:

$$v = \left[\widetilde{A}(z_0) + (z - z_0)B(z)\right] u(z), \text{ with } B(z_0) \neq 0. \tag{6.22}$$

Then, if $u(z)$ has a pole of order 1, we get, by the Cauchy theorem (bearing in mind that v is constant):

$$0 = \frac{1}{2i\pi} \int_\gamma v dz = \widetilde{A}(z_0) Res(u, z_0) \tag{6.23}$$

where $Res(u, z_0)$ denotes the residue of $u(z)$ at $z = z_0$. Obviously, we have shown that the residue of $A(z)v$ at z_0 belongs to the kernel of $\widetilde{A}\,(z_0)$. Let us now prove the reciprocal inclusion. That is, the fact that for a given u_0 belonging to the kernel of $\widetilde{A}(z_0)$, there exists a vector w_0, such that, denoting by R_0 the residue of $A(z)$ at z_0, we have $R_0 w_0 = u_0$. Let us define: $w(z) = \frac{\widetilde{A}(z)u_0}{z - z_0}$. Using (6.22), we have immediately the relations:

$$A(z)w(z) = \frac{u_0}{z - z_0} \tag{6.24}$$

$$w(z) = B(z)u_0. \tag{6.25}$$

It remains for us to evaluate the residue of $A(z)w(z)$ at z_0. By using (6.24) and (6.25), we easily get:

$$\frac{1}{2i\pi}\int_{\gamma} A(z)w(z)dz = R_0 B(z_0)u_0 = u_0\,. \tag{6.26}$$

It suffices to set $w_0 = B(z_0)u_0$ to obtain the result. We conclude that the range of the residue operator equals exactly the kernel of $\widetilde{A}(z_0)$ provided that the pole is of order 1. □

A slightly subtler situation happens if the pole is of order greater than 1. In order to make this apparent, we consider the following simple example:

$$\widetilde{A}_m\,(z) = \begin{pmatrix} z - z_0 & 1 \\ 0 & z - z_0 \end{pmatrix} \tag{6.27}$$

the inverse of which, $A_m(z)$, can be written as:

$$A(z) = \frac{1}{z - z_0}\begin{pmatrix} 1\,0 \\ 0\,1 \end{pmatrix} + \frac{1}{(z - z_0)^2}\begin{pmatrix} 0\,1 \\ 0\,0 \end{pmatrix}. \tag{6.28}$$

The residue in that case is the identity matrix, whereas we have: $\widetilde{A}_m(z_0) = \begin{pmatrix} 0\,1 \\ 0\,0 \end{pmatrix}$ and therefore we did not get the kernel of $\widetilde{A}_m(z_0)$ by computing the residue of the inverse operator. What happened in this case? The point here is to realize that two different mathematical objects are to be considered: the first one is the eigenspace of the operator and the other one is the characteristic space. The link between both spaces is the multiplicity of the eigenvalue (we deal here with the null eigenvalue and hence the kernel). For the sake of generality, we turn to $n \times n$ matrices, and we consider $\widetilde{A}(z_0) \in M_n(\mathbb{C})$ and the successive powers of $\widetilde{A}(z_0)$: $\widetilde{A}(z_0), \widetilde{A}^2(z_0), \ldots$. For each power, we can denote the corresponding nullspace $E_n = \{u \in \mathbb{C}, \widetilde{A}^n(z_0)u = 0\}$. It is not difficult to see that each space is included in the next one: $E_n \subset E_{n+1}$. Indeed, if $u \in E_n$ then $\widetilde{A}^{n+1}(z_0)u = \widetilde{A} \times \widetilde{A}^n(z_0)u = 0$, and hence $u \in E_{n+1}$. The consequence is that the dimensions of theses spaces form an increasing sequence, but of course it should remain less than the dimension of the overall space. The only possibility is that the dimension be constant from a certain value n_0 of n. This means that for $n \geq n_0$, we have $E_n = E_{n_0}$. By definition, the space E_{n_0} is the characteristic space of $\widetilde{A}(z_0)$. The main point is that the residue ranges precisely over the characteristic space and not the null space. There is a difference between them however when the eigenvalue is degenerate in the mathematical sense.

In physics, it is common to say that an eigenvalue is degenerate whenever the associated eigenspace is not one dimensional. In mathematics, the term is used when the algebraic multiplicity of an eigenvalue is not equal to the dimension of the corresponding eigenspace. This is exactly what happens in the above example. For the matrix $\begin{pmatrix} 0 & 1 \\ 0 & 0 \end{pmatrix}$, the eigenvalue 0 has an algebraic multiplicity of 2, as a root of the characteristic polynomial. But the dimension of the kernel is just 1. On the contrary, a simple calculation shows that the square of this matrix is just the null matrix, whose kernel is the overall space. Hence the residue is the identity matrix, showing that any vector is in the characteristic space. Finally, let us remark that the algebraic multiplicity of an eigenvalue is equal to its order as a pole.

6.1.1 *Degenerate eigenvalues*

In spite of the preceding remark concerning the term "degenerate", we use it here in its usual meaning in physics: it qualifies an eigenvalue whose associated eigenspace has a dimension more than 1. We still deal with the kernel here, so that the situation to be addressed is the case of a kernel of dimension 2 or more. Let us give a simple example. We choose this time a 3×3 matrix (for 2×2 matrices, a kernel of dimension 2 implies the uninteresting case of the null operator):

$$A(z) = \begin{pmatrix} z - z_0 & 0 & 0 \\ 0 & z - z_0 & 0 \\ 0 & 0 & 1 \end{pmatrix} \tag{6.29}$$

For $z = z_0$, the kernel of this operator is generated by the two basis vectors :$(1, 0, 0)$ and $(0, 1, 0)$. For the inverse operator, we get:

$$\widetilde{A}(z) = \frac{1}{z - z_0} \begin{pmatrix} 1 & 0 & 0 \\ 0 & 1 & 0 \\ 0 & 0 & 0 \end{pmatrix} + \begin{pmatrix} 0 & 0 & 0 \\ 0 & 0 & 0 \\ 0 & 0 & 1 \end{pmatrix} \tag{6.30}$$

The residue at z_0 is $\begin{pmatrix} 1 & 0 & 0 \\ 0 & 1 & 0 \\ 0 & 0 & 0 \end{pmatrix}$, and therefore 1 is an eigenvalue of this matrix, which is twofold degenerate. This example can be easily extended to any $n \times n$ matrix. The conclusion is that when a pole is associated with a mode that is degenerate, in the physical sense, then so are the eigenvalues of the residue and with the same degree.

We now see that the method not only provides us with the value of the pole but also with its degree of degeneracy.

6.1.2 *Multiple poles inside the loop*

Let us now take a look at the situation in which more than one pole lies inside γ. In terms of the Laurent expansion, this entails the existence of two poles z_0, z_1 of order 1:

$$A(z) = \frac{R_0}{z - z_0} + \frac{R_1}{z - z_1} + A_0(z) \,. \tag{6.31}$$

The residues R_0 and R_1 are of course matrices and A_0 is regular near z_0 and z_1. Now, what happens when a Cauchy integral is computed with a loop γ that contains both z_0 and z_1? We have simply

$$I_1 = R_0 + R_1 = \frac{1}{2i\pi} \int_\gamma A(z) dz \tag{6.32}$$

and

$$I_2 = z_0 R_0 + z_1 R_1 = \frac{1}{2i\pi} \int_\gamma z A(z) dz \,. \tag{6.33}$$

The knowledge of both operators I_1 and I_2, obtained numerically, is sufficient to allow us to recover the values of the poles z_0 and z_1. It should also be noted that it is not at all necessary to compute the operators R_0 and R_1 in order to proceed to the determination of the poles. This point is a delicate one. Indeed, each pole is associated with a vector space that is the range of its associated residue. So, assuming that the vector spaces associated with the poles are one dimensional, we denote by u_0 and u_1 two vectors generating the ranges of R_0 and R_1 respectively. Each operator can then be written: $R_0 = \mu_0 P_{u_0}$ and $R_1 = \mu_1 P_{u_1}$, where P_u denotes the projector on the subspace generated by u: $P_u = u \otimes u$. Two different poles are associated with different eigenvectors and hence we have:

$$P_{u_1} P_{u_0} = P_{u_0} P_{u_1} = 0. \tag{6.34}$$

As a consequence, the non-null eigenvalues of $R_0 + R_1$ are (μ_0, μ_1) whereas those of $z_0 R_0 + z_1 R_1$ are $(z_0\mu_0, z_1\mu_1)$. Consequently, by diagonalizing both matrices (6.32,6.33), we recover both values of the poles z_0 and z_1. This situation is radically different to the case in which functions are considered: instead of being simple numbers, the residues are now operators. This is a major improvement: **we have, in principle, the possibility of**

computing the values of an arbitrary number of poles, and their associated modes, simply by evaluating an integral along a path in the complex plane.

Let us give an example, with our preferred operators: 2×2 matrices. We choose: $R_0 = (1,1) \otimes (1,1) = \begin{pmatrix} 1 & 1 \\ 1 & 1 \end{pmatrix}$ and $R_1 = (1,-1) \otimes (1,-1) = \begin{pmatrix} 1 & -1 \\ -1 & 1 \end{pmatrix}$. The non-null eigenvalue of R_0 is 2, that of R_1 is also 2, but of course the associated eigenvectors are different. We form the linear combination: $z_0 R_0 + z_1 R_1 = \begin{pmatrix} z_0 + z_1 & z_0 - z_1 \\ z_0 - z_1 & z_0 + z_1 \end{pmatrix}$, whose eigenvalues are easily seen to be $2z_0$ and $2z_1$.

6.1.3 *Miracles sometimes happen*

In this subsection, we investigate two surprising points, which are encountered in the numerical implementation of the method.

First, let us consider the matrix: $A(z) = \begin{pmatrix} 1 & 0 \\ 0 & 0 \end{pmatrix} \frac{1}{z - z_0} + \begin{pmatrix} 0 & 0 \\ 0 & 1 \end{pmatrix} \frac{1}{z - z_1}$, where we choose: $z_0 = 1 + i$, $z_1 = 2 + i$. Now, let us compute numerically the integral of $A(z)$ along the rectangular path γ_1 shown as a bold line in Figure 6.1. The integral is computed using the rectangle method.[1] For each side, we use $N = 15$ points of integration.

Keeping 6 figures, we obtain for the first integral (6.17):

$$I_1 = \begin{pmatrix} 0.90768 + i\,3.6517 \times 10^{-16} & 0 \\ 0 & 0.90768 + i\,3.475 \times 10^{-16} \end{pmatrix} \tag{6.35}$$

and for the second one (6.17):

$$I_2 = \begin{pmatrix} 0.90768 + i\,0.90768 & 0 \\ 0 & 1.8154 + i\,0.90768 \end{pmatrix}. \tag{6.36}$$

Of course the exact results are respectively:

$$\begin{pmatrix} 1 & 0 \\ 0 & 1 \end{pmatrix} \text{ and } \begin{pmatrix} 1+i & 0 \\ 0 & 2+i \end{pmatrix} \tag{6.37}$$

and therefore the numerical estimations are far from being good. However, let us take a look at the results for the numerical estimations of the poles.

[1] We acknowledge that we have not made the most clever choices, neither for the path, nor for the integration algorithm, but the discussion here stands at a pedagogical level, not on one of efficiency.

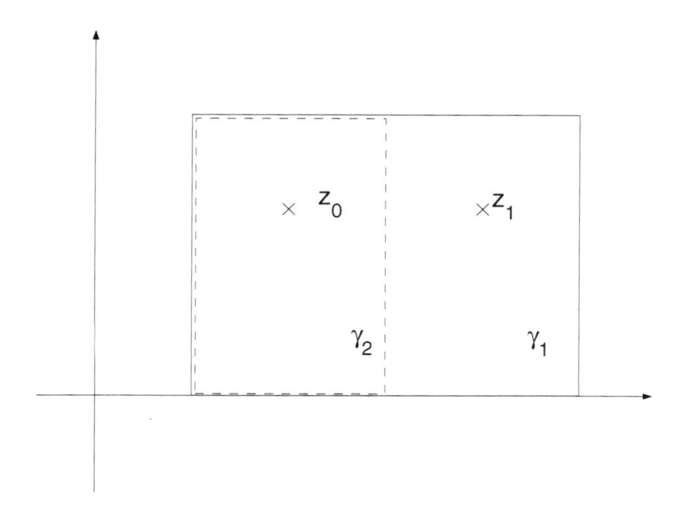

Fig. 6.1 Integration paths.

We get:

$$z_0 = \frac{0.90768 + i\,0.90768}{0.90768 + i\,3.6517 \times 10^{-16}} = 1 + i \tag{6.38}$$

$$z_1 = \frac{1.8154 + i\,0.90768}{0.90768 + i\,3.4751 \times 10^{-16}} = 2.0001 + i \tag{6.39}$$

and the numerical approximations of the poles are excellent. This impressive numerical phenomenon can be understood in the following way. For the sake of simplicity, we need only consider one pole here, so let us first rewrite I_2, cf. 6.17, in the form:

$$\int_{\gamma_1} \frac{zR}{z - z_0} dz = \int_{\gamma_1} R dz + z_0 \int_{\gamma_1} \frac{R}{z - z_0} dz\,. \tag{6.40}$$

The first integral on the right-hand side is null so, numerically, this is a very small number, which we denote by η. Besides, the difference between I_1, cf. 6.17, and its numerical value is also a (hopefully) small number δ. Finally, the numerical value of I_2 can be written as: $\eta + z_0 \times (I_1 + \delta)$. The error η can be expected to be small (this is just the integral of 1 over the path multiplied by a constant matrix), so that the numerical ratio between both integrals reads: $z_0 + \eta/(I_1 + \delta)$. We arrive at the conclusion that the precision in the evaluation of (I_1, I_2) does not influence much the precision on z_0. The preceding explanation should not be considered as a proof because we made some mathematical manipulations that have no numerical counterparts, but it helps in understanding that, roughly speaking, the numerical value of the ratio of the integrals does not depend upon the approximation of the integrals.

The second strange event occurs when one uses the second path γ_2 (indicated with a dashed line in figure 6.1). The pole z_1 is not inside the loop. The numerical results for the integrals are

$$I_1 = \begin{pmatrix} 0.96348 - i\,7.7747 \times 10^{-17} & 0 \\ 0 & 0.017445 - i\,3.7106 \times 10^{-17} \end{pmatrix}, \tag{6.41}$$

and

$$I_2 = \begin{pmatrix} 0.96348 + i\,0.96348 & 0 \\ 0 & 0.034889 + i\,0.017445 \end{pmatrix}. \tag{6.42}$$

while this time the exact results are:

$$\begin{pmatrix} 1 & 0 \\ 0 & 0 \end{pmatrix}, \begin{pmatrix} 1+i & 0 \\ 0 & 0 \end{pmatrix} \tag{6.43}$$

and the ratios between the eigenvalues are:

$$z_0 = 1 + i \tag{6.44}$$

$$z_1 = 2 + i\,. \tag{6.45}$$

Therefore, we see that we get both values of the poles, while the loop goes round one pole only! What happened? In order to understand this numerical trick, let us denote by: η and δ the respective numerical values of the integrals: $\int_{\gamma_2} dz$ and $\int_{\gamma_2} \frac{dz}{z-z_0}$. We then have the following expression for the numerical approximation:

$$\int_{\gamma_2} \frac{zdz}{z - z_0} = \eta + z_0\delta \tag{6.46}$$

and the ratio between both integrals gives, numerically: $z_0 + \frac{\eta}{\delta}$. The value of the constant function along γ_2, i.e. η, is much smaller than δ, and therefore we obtain the value of the pole, even though the pole is not inside the loop. This is the numerical miracle!

6.2 Cauchy integrals for operators

Our ultimate aim, of course, is to be able to compute the dispersion curves of a photonic crystal fibre. As has already been said, the main mathematical object to be considered is the scattering matrix, that is, the operator $S(k)$ such that $U^d = S(k)U^i$ where U^i is the incoming part of the field, U^d is the outgoing part of the field and k is the wavenumber. The scattering matrix can be written $S(k) = I_d + T(k)$ where $T(k)$ is the so-called scattering amplitude. For a given propagation constant, the eigenfrequencies are poles of the scattering amplitude. So, let us assume that there exists some pole k_p of T in some neighbourhood $\mathcal{V}$ of the complex plane. Then, for $k \in \mathcal{V}$, the scattering amplitude is $T(k) = \frac{R_p}{k-k_p} + T_0(k)$ where R_p is a residue operator and T_0 is holomorphic in $\mathcal{V}$. This is formally completely equivalent to the preceding section, where matrices were considered. Moreover, the following results can be established, though the proofs are rather involved (cf. [RS78]):

(1) operator R_p is a finite rank operator,
(2) its range, i.e. the vector space in which R_p takes its values, is precisely the nullspace of $T^{-1}(k_p)$.

These points extend to "infinite" matrices the results of the above section. Operator R_p can be represented as a Cauchy integral:

$$R_p = \frac{1}{2i\pi}\int_\gamma T(z)\,dz, \tag{6.47}$$

where integration takes place on a loop oriented in the direct sense, enclosing once the only pole k_p. The meaning of the integral is not at all obvious given that $T(z)$ is a holomorphic operator function. It is defined by considering finite Riemann sums: $\frac{1}{N}\sum_{k=0}^{N-1} T(\gamma(t_k))\gamma'(t_k)\Delta t_k$, where t_k are points in the interval $[0,1]$. The integral is then the limit of the sum when N tends to infinity. Mathematically speaking, it is not that simple to take such a limit.[2] However, for our purpose, it suffices to remember that all that is used numerically is finite matrices, so that we will never have problems in computing the finite sums. It is merely necessary for us to know that increasing the truncation order increases the precision of the result. We can also define the following operator Q_p

$$Q_p = \frac{1}{2i\pi}\int_\gamma zT(z)\,dz\,. \tag{6.48}$$

Numerically, it has been explained in the preceding chapters that the scattering amplitude admits a representation as an operator on $\ell^2(\mathbb{Z})$, that is

[2] In that for operators the very notion of limit has to be carefully elaborated.

as an infinite matrix, acting on double complex sequences. Once this representation is given, the residue operator can be computed provided that a region of the complex plane containing only one pole can be isolated. The method is then completely equivalent to the case of finite matrices dealt with in the preceding sections.

6.3 Numerical Applications

Let us now turn to some numerical applications. We deal with the structure depicted in figure 6.2.

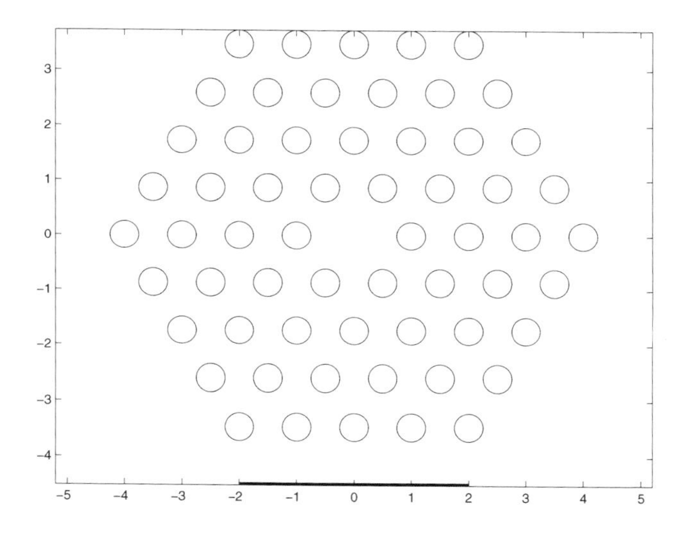

Fig. 6.2 The cross-section of the photonic crystal fibre.

This is a photonic crystal fibre with hexagonal symmetry and a hexagonal boundary. There are 60 air holes (relative permittivity $\varepsilon_r = 1$), embedded in an infinite medium of permittivity $\varepsilon_r = 9$. The ratio between the spacing and the radius of the rods is $R/\Lambda = 1/4$. We use the multiple scattering theory to compute the scattering matrix of this system (cf. chapter 4). All the numerical results have been obtained using a standard Personal Computer.

To keep things as simple as possible, we shall not look for the allowed propagation constants for one given frequency, but rather look for frequencies for which there are modes invariant along the axis of the fibre (and hence a null propagation constant). With such hypotheses, we may consider polarized waves only. Here, we shall only deal with p-polarized waves. As a consequence, the magnetic field component along z is the quantity of interest. The photonic fibre possesses a defect mode as can be seen in figure 6.3 where we have plotted the transmission spectrum:[3] this corresponds to the peak inside the low-transmission region.

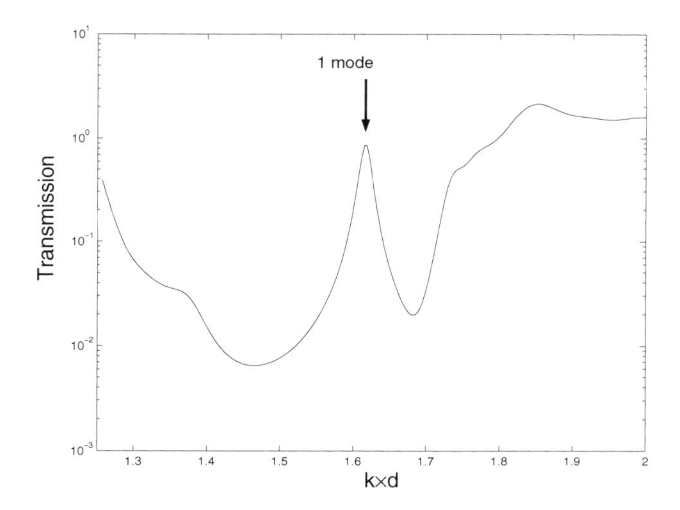

Fig. 6.3 Transmission spectrum for the fiber of figure 2. The bold line in fig. 6.1 indicates the segment through which the energy flux is computed.

This peak in the transmission spectrum is due to the existence of a pole k_p of the scattering matrix[4]. The reference value that we use for convergence comparisons is $k_p = 1.61620545141059 - i\,0.00883630644458$. This value has been obtained by using the Müller algorithm [PBTV86] and by minimizing the smallest eigenvalue of T^{-1}. For this given value of k_p the smallest eigenvalue has a modulus less than 10^{-14}. We then compute both Cauchy integrals (6.47,6.48). The integration path is a rectangle whose vertices have affixes: $(1.5, 2, 1.5 - i\,0.5, 2 - i\,0.5)$ and the integrals are computed with the integration algorithm described in [HMR98]. We denote by

[3]The transmission spectrum is obtained by illuminating the crystal fibre transversely, i.e. the wavevector of the incident plane wave is perpendicular to the axis of the fiber, and by collecting the transmitted light below the fiber, i.e. the transmission coefficient is defined as the energy flux through the segment depicted in figure 6.2.

[4]The reader is referred to [FZ03] and to chapter 4 of this book for a more complete exposition of the resonant properties of photonic crystals

$k_{p,N}$ the numerical value obtained by using N points of integration, which we compare with the above value of k_p which is the best numerical value that we can obtain. A good precision is rapidly obtained (see figure 6.4): for instance, using a discretisation of 20 points we obtain 4 exact figures though with such a rough discretization, we get operator R_p with a low precision.

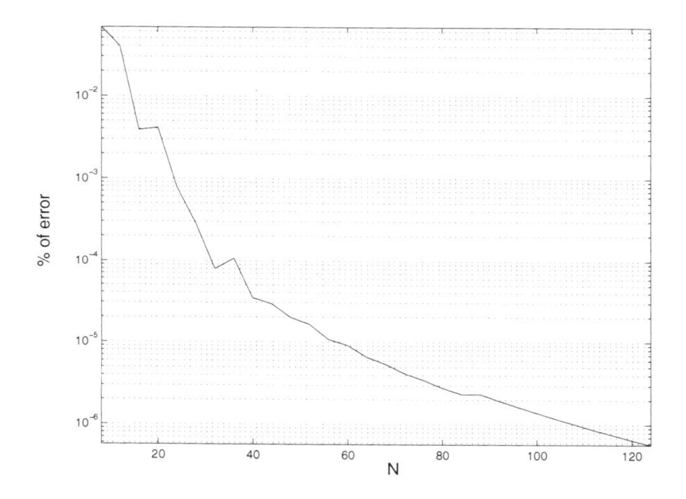

Fig. 6.4 Convergence of the numerical values of the pole.

The convergence towards the correct value of operator R_p can be checked by looking at the non-zero eigenvalue of R_p (figure 6.5). A much finer discretization than in the case of the pole is required to get a good representation of the defect mode. The ten first eigenvalues of R_p, sorted by modulus, are given in table (6.1), for various values of the number N of integration points.

These eigenvalues give two very specific pieces of information. First, we see that only two eigenvalues have a modulus that is not very small compared to 1: this indicates the precision reached in the computation of R_p. Indeed, operator R_p has a finite range, and hence an infinite null space: this means that, when computing the eigenvalues of $R_{p,N}$ we should find 0 infinitely many times. Numerically, we obtain "very small" eigenvalues. As a consequence, the precision of the computation of R_p is all the more delicate as the first of the "null" eigenvalues is small.

The second piece of information is the multiplicity of the non-zero eigen-

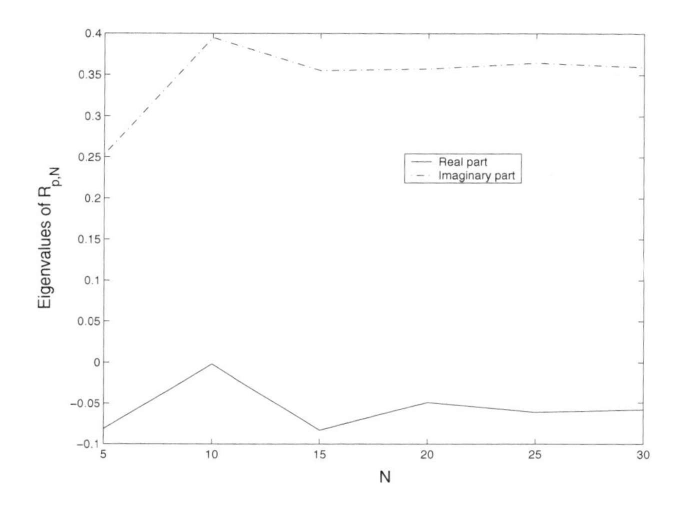

Fig. 6.5 Convergence of the non-zero eigenvalue of $R_{p,N}$.

Table 6.1 Eigenvalues of the residue $R_{p,N}$

N=20	N=40	N=60	N= 80
0.260229	0.419495	0.35039	0.367531
0.260229	0.419495	0.35039	0.367531
0.0177879	0.000468568	1.09213e-005	2.39925e-006
0.00407013	2.66898e-005	4.2348e-006	1.36528e-006
0.00472632	3.43616e-005	4.44443e-006	1.43643e-006
0.00472632	3.43616e-005	4.44443e-006	1.43643e-006
0.00241353	9.02527e-006	3.24191e-006	1.04219e-006
0.00241353	1.17358e-005	3.24191e-006	1.04219e-006
0.000709525	1.17358e-005	1.8887e-006	6.11417e-007
0.00023064	5.56785e-006	1.17332e-006	3.80356e-007

value: here its multiplicity is 2. As has been explained above, this shows that the underlying eigenspace is twofold degenerate. We have plotted in figures 6.6 and 6.7 the modulus of the electric fields of two linearly independent eigenmodes associated with k_p.

By looking at table (6.1), one can see that even at rather small N, the greatest eigenvalue is obtained twice, indicating the degree of degeneracy, although the numerical precision of the value of this eigenvalue is poor. Besides, it is clearly seen that the other eigenvalues quickly become very small as N increases. Given these results, it is now easy to

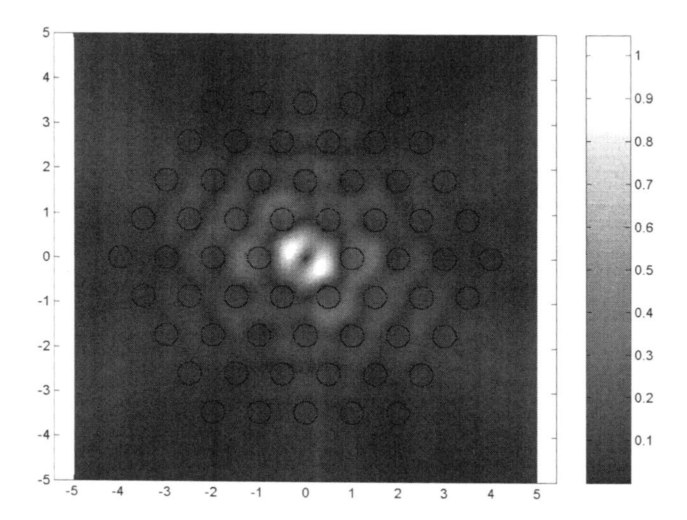

Fig. 6.6 Modulus of the electric field of an eigenmode.

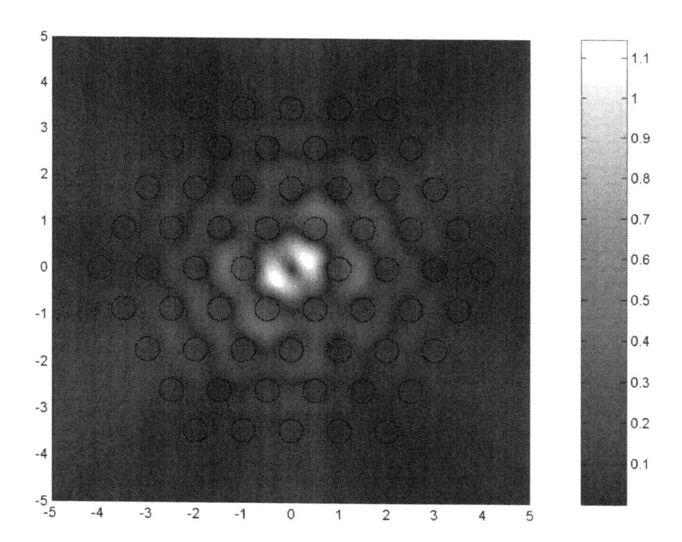

Fig. 6.7 Modulus of the electric field of an eigenmode.

get a very precise value of the pole by using a Müller-like algorithm. The results are shown in table (6.2). As a first guess, we use the value obtained from the residue method, by using $N = 8$ integration points:

$k_{p,8} = 1.6157112829714 - i\,0.00780320099663731$. For the last value in the

Table 6.2 Numerical evaluation of $k_{p,N}$ by a Müller algorithm, with the starting value given by the residue method

Number of numerical evaluations	k_p
0	1.6157112829714 - i 0.00780320099663731
3	1.61151683513391 - i 0.0118964383876807
6	1.6171584680062 - i 0.0135036466408934
9	1.6162056586867 - i 0.00883621128636636
12	1.61620545142902 - i 0.00883630648613322
15	1.61620545141059 - i 0.0088363064445728
18	1.61620545141059 - i 0.00883630644457503
21	1.61620545141059- i 0.00883630644457419
24	1.61620545141059 - i 0.00883630644457382
27	1.61620545141059 - i 0.00883630644457397
30	1.61620545141059 - i 0.00883630644457407
33	1.61620545141059 - i 0.00883630644457394
36	1.61620545141059 - i 0.00883630644457406

table, the smallest eigenvalues of $T^{-1}(k_p)$ each have a modulus less than 10^{-14}.

Let us now turn to a second set of numerical examples. We still consider the same basic structure as that in the preceding paragraph, but this time, the core is much larger (see figure 6.8).

The fibre is no longer monomode. This means that more than one pole lies in the band gap. However, if we plot the transmission spectrum (see figure 6.9), it cannot be guessed that there are in fact 3 poles: the curve does not exhibit three peaks, but rather it is smooth. This is due to the proximity of the poles to each other, which prevents the transmission curve from being correspondingly peaked.

Let us now apply the method of residues in the miracle regime. The path for integration is a rectangle with vertices $1, 2, 1 - .5i, 2 - .5i$. We give in table (6.3) the numerical values of the poles and also the moduli of the eigenvalues of R_p, for two values of N.

It can be immediately seen on this table that there are three poles inside the loop: two of them are twofold degenerate while the third one is not. It is noteworthy that these pieces of information are obtained with a small number of integration points.

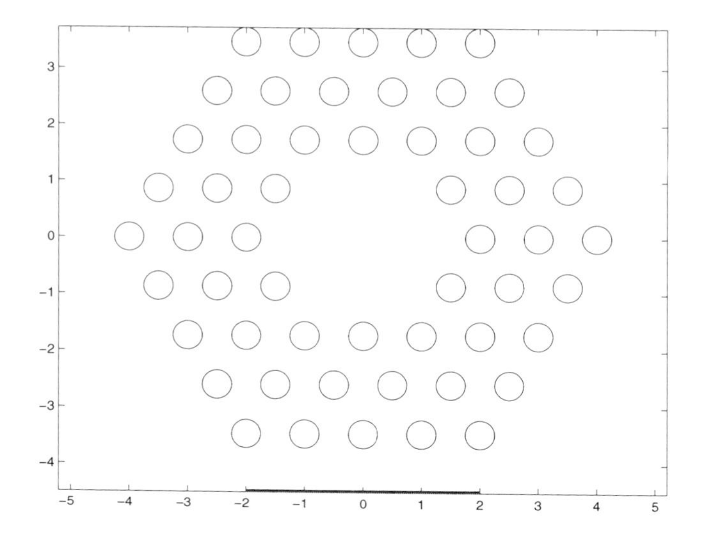

Fig. 6.8 A photonic crystal fibre with a larger core.

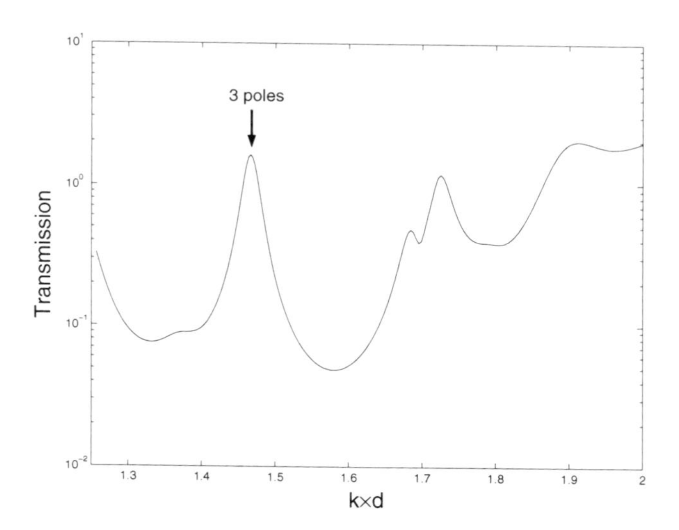

Fig. 6.9 Transmission spectrum for the larger core fibre of figure 6.8.

6.4 Conclusion

We have shown that it is possible to turn a rather abstract mathematical object into a useful numerical tool. This technique can be applied equally

Table 6.3 Eigenvalues and poles

N=40		N=60	
$\|\mu\|$	k_p	$\|\mu\|$	k_p
0.140584	1.46355 - i 0.0108734	0.158532	1.46355 - i 0.0108741
0.144985	1.46871 - i 0.0129151	0.151615	1.46871 - i 0.0129159
0.144985	1.46871 - i 0.0129151	0.151615	1.46871 - i 0.0129159
0.162388	1.53675 - i 0.0105272	0.152659	1.53676 - i 0.0105277
0.162388	1.53675 - i 0.0105272	0.152659	1.53676 - i 0.0105277
0.00011729	1.3677 -i 0.0334689	3.47592e-006	1.36909 -i 0.0230242
5.27993e-005	1.36228 -i 0.0349126	8.21936e-007	1.32383 -i 0.0125879
3.09168e-006	1.72023 - i 0.0139472	6.61699e-007	1.7202 - i 0.0139002
2.74738e-006	1.70914 - i 0.00972737	5.88712e-007	1.70921 - i 0.00980479
2.74738e-006	1.70914 - i 0.00972737	5.88712e-007	1.70921 - i 0.00980479

well to any situation in which a meromorphic operator with non essential poles is involved, which is the usual case. It has many special features that have been described in detail: it is possible to determine the value of several poles by computing only two integrals, and it gives immediately the degree of degeneracy a pole. However, one should not think that this is the only and ultimate method in pole searching. It is a powerful tool that gives mathematical information by numerical means: the number of poles in the loop and the dimensions of the eigenspaces. However, it is not a very efficient method to determine numerically a very precise value of the pole, if needed, simply because it then requires a precise computation of the integrals and hence a great number of evaluations of the scattering matrix. That is why it can be recommended for initial use, in order to determine a rough estimation of the location of poles, after which we can then use a Müller-like algorithm to get precise values. In this situation, like in many others, only a multiplicity of points of view allows a fine solution of the problem posed.

Chapter 7

Basic Properties of Microstructured Optical Fibres

In this chapter, we focus on the main properties of the fundamental mode of solid core MOF based on a C_{6v} triangular lattice of low index circular inclusions (the optical indices of these inclusions n_i will be fixed to 1.0 in all this chapter). The second mode of this structure will also be studied in order to clarify the notion of single-modeness in MOF. We define cutoff regions for these two modes, so that it will be possible to determine the different operation regimes of solid core C_{6v} MOF.

Finally, we study of the chromatic dispersion of the fundamental mode, and we show that chromatic dispersion in MOF can be fully managed in a useful way for practical applications using new MOF designs. We cannot end this chapter without an illustration of the most astonishing property of MOFs: the possibility of guidance in a hollow core.

7.1 Basic Properties of the Losses

All the modes of the MOF structure that we considered in chapter 4 are leaky modes. The *confinement losses* (See section 1.5.1 page 37) of these modes $\mathcal{L}$ (in decibel per meter) are related to the imaginary part of n_{eff} through the relation:

$$\mathcal{L} = \frac{40\pi}{\ln(10)\lambda} \Im m(n_{\text{eff}}) 10^6 \tag{7.1}$$

in which λ is given in micrometres. These losses are due to the finite number of air holes. It is worth mentioning that these losses occur even if the material absorption is fully neglected. The results given in this section concern the degenerate fundamental mode of solid core C_{6v} MOF such as that depicted in Fig.4.5(b). We limit the present study to the case of MOF cores composed of a single missing air hole in the finite size triangular lattice (See Fig. 7.1). The silica matrix is considered as infinite and its permittivity is taken from the Sellmeier expansion at the corresponding wavelength [Agr01] ($\varepsilon_r = 2.0852042$ for $\lambda = 1.55\,\mu m$).

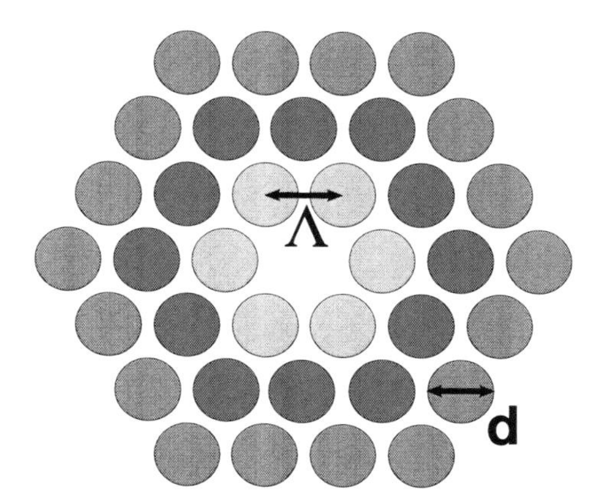

Fig. 7.1 Cross-section of the modelled MOF with 3 rings of holes (holes are shown coloured with different grey according to their ring number), $N_r = 3$. Λ is the pitch of the C_{6v} triangular lattice, and d is the hole diameter. The solid core is formed by one missing hole in the center of the structure.

In Fig. 7.2 we give the confinement losses versus the ratio[1] d/Λ (d is the hole diameter and Λ is the pitch of the triangular lattice of inclusions) and the number N_r of hole rings surrounding the MOF core. The losses decrease monotonically with both the number of low-index inclusion rings and the ratio d/Λ [WMdS+01]. We can notice that with the logarithmic scale for the losses the gap between successive values of N_r is nearly constant, and this means that for this kind of MOF the confinement losses decrease nearly exponentially with N_r[2]. In Fig. 7.3, we give these losses as a function of the pitch Λ for several values of the ratio d/Λ for a 3-ring MOF. For a fixed value of d/Λ, they decrease monotonically with the pitch Λ. It can also be seen in Fig. 7.2 and Fig. 7.3 that the MOF confinement losses can be made as low as 0.1 dB/km so the total losses in MOF can be limited by Rayleigh scattering and absorption rather that by the confinement losses [KRM03].

The real part of the effective index $n_{\text{eff}} = \beta/k$ as a function of N_r the number of hole rings for several values of d/Λ is shown in Fig. 7.4. We

[1]It is straightforward to see that this ratio d/Λ must be less than unity.

[2]The behaviour of the losses with N_r will be discussed in more detail in section 7.3.

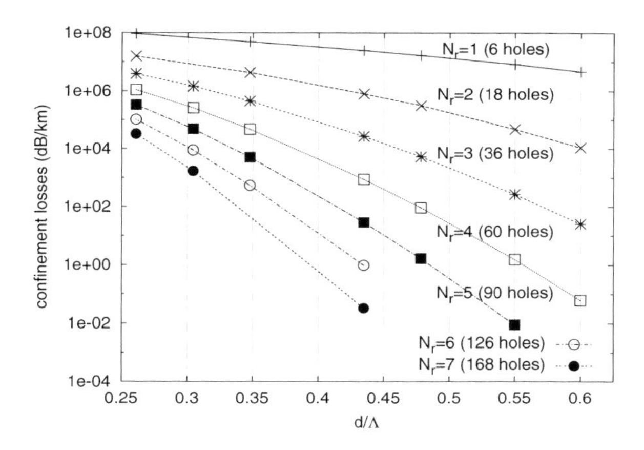

Fig. 7.2 Confinement losses (in dB/km with a log-scale) versus the ratio (d/Λ) and the number N_r of hole rings around the solid core. The MOF are based on a C_{6v} triangular lattice of air holes in silica (See the text). The pitch Λ is equal to 2.3 μm, and $\lambda = 1.55$ μm. N_r is the number of air hole rings surrounding the solid core.

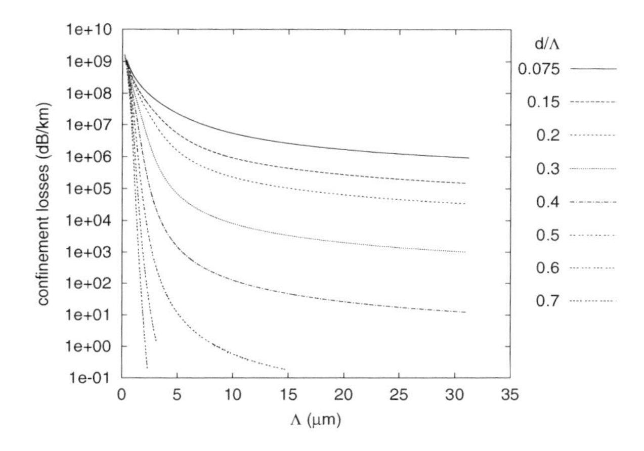

Fig. 7.3 Confinement losses (in dB/km with a log-scale) versus the pitch Λ for several values of the ratio d/Λ for a 3-ring MOF. The MOF are based on a C_{6v} triangular lattice of air holes in silica (See the text). The wavelength λ is equal to 1.55 μm.

see that $\Re e(n_{\text{eff}})$ increases with N_r and that the larger the ratio d/Λ the faster the limiting value of $\Re e(n_{\text{eff}})$, corresponding to an infinite number of hole rings, is reached. This behaviour in a solid core MOF can be easily understood as follows: the larger the ratio d/Λ, the better the confinement of the mode is improved by N_r, and consequently the faster the real part of the effective index reaches its limiting value.

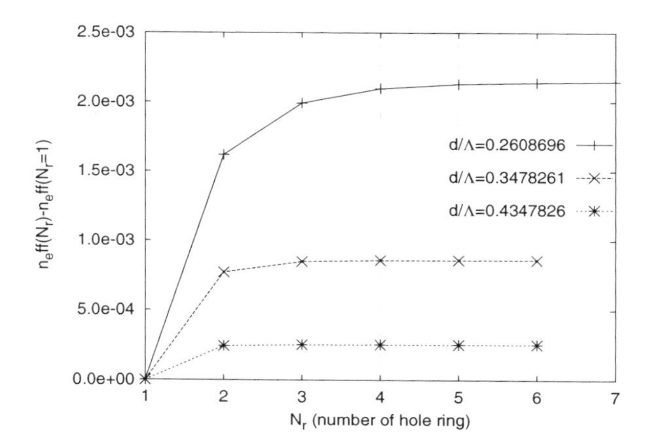

Fig. 7.4 Real part of the effective index $n_{\text{eff}} = \beta/k$ as a function of N_r the number of hole rings around the MOF core for several values of d/Λ. The wavelength λ is equal to 1.55 μm, and the pitch Λ is equal to 2.3 μm. The diameter of the inclusions d is successively equal to 0.6, 0.8 and 1.0 μm, whilst the corresponding real part of the effective index associated with $N_r = 1$ is succesively equal to 1.427698 ,1.424475, and 1.421159.

7.2 Single-Modedness of Solid Core C_{6v} MOF

One of the earliest known and most exciting properties of MOFs is that they can be *endlessly single-mode* [BKR97]. However, as mentioned earlier, a MOF in which a finite number of rings of holes is solely responsible for the confinement of light carries an infinite number of modes, all of which are leaky. By using Fig. 7.5, it is possible to consider the relative losses of the modes: If the losses of the different modes of a MOF are such that after a given length of propagation all modes except for one have faded away, the MOF can be considered to be single-mode for that length of propagation. Such a definition of single modedness is unsatisfactory in several ways: not only does it depend on the actual length of propagation, but it also depends on the pitch (see the losses of the three modes for small pitches in Fig. 7.5) and the number of rings of holes. Since the losses decrease with the number of rings of holes, for an infinite number of rings no MOF can be single-mode with such a definition.

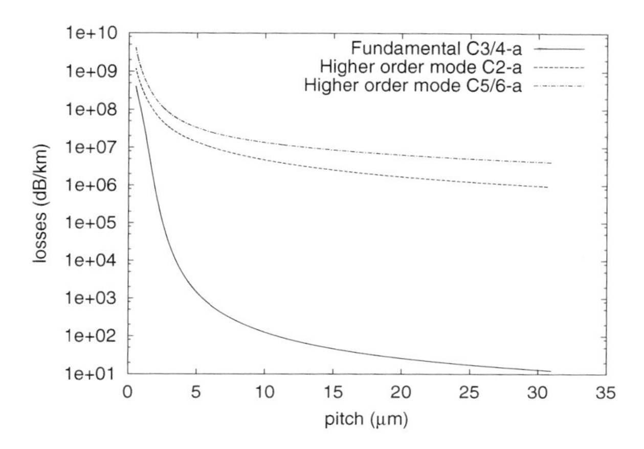

Fig. 7.5 Losses for the fundamental mode and for two higher order modes, for a 3 ring MOF, as a function of the pitch. The wavelength λ is equal to 1.55 μm, and $d/\Lambda = 0.4$ (see section 4.6.1 page 184 for the used mode labels).

7.2.1 *A cutoff for the second mode*

This definition through the relative losses of the main modes is not satisfactory. To obtain a more definitive characterization of single-modeness one approach is to study the second mode of a C_{6v} MOF[3] as a function of MOF parameters. This kind of study was realized for the leaky modes of W-fibres by Maeda and Yamada [MY77] in the late seventies. In the case of MOF, the chosen parameter is the ratio λ/Λ, *i.e.* the normalized wavelength. The second mode properties will be studied at fixed d/Λ [KMdS02], and the ratio λ/Λ will be varied by changing the pitch Λ and keeping constant the wavelength λ at 1.55 μm. In doing so the refractive index of the silica matrix keeps a constant value $n_M = \sqrt{\varepsilon_M}$. Since λ is constant the confinement losses are directly proportional to the imaginary part of n_{eff}.

In Fig. 7.6, we give the losses as a function of λ/Λ for a MOF such that $d/\Lambda = 0.55$ for several values of the number of rings N_r. A transition in the loss curves can be observed for all studied N_r values, and the higher the N_r value the steeper the transition.

In fact, several parameters[4] can be used to observe this transition:

- the normalized effective radius R_{eff}/Λ where R_{eff} is defined by the fol-

[3]We define the second mode as the mode with the nearest real part to that of the fundamental one.

[4]For the two first parameters, the integrals are taken over the cladding region of the MOF because, as already stated, the associated fields diverge at infinity.

lowing formula in which the function S_z is the real part of the longitudinal component of the Poynting vector:

$$R_{\text{eff}} = \frac{3}{2} \frac{\int S_z r^2 dr d\theta}{\int S_z r dr d\theta} \tag{7.2}$$

- the normalized effective area [Agr01] A_{eff} defined as:

$$A_{\text{eff}} = \frac{(\int |E_z|^2 dr d\theta)^2}{\int |E_z|^4 dr d\theta} \tag{7.3}$$

- the ratio of the magnetic field monopole coefficient (B_0^H) to the magnetic field dipole coefficient (B_1^H) for a cylinder in the first ring of the MOF.

$$\mathcal{M} = \frac{B_0^H}{B_1^H} \tag{7.4}$$

- the second derivative of the logarithm of the losses with respect to the logarithm of the pitch $\mathcal{Q}$ *i.e.* :

$$\mathcal{Q} = \frac{\partial^2 [\log \Im m(n_{\text{eff}})]}{\partial [\log \Lambda]^2} \tag{7.5}$$

The most sensitive quantity was found to be $\mathcal{Q}$ [KMdS02; KdSM$^+$02] (this quantity will be called loss transition parameter). This exhibits a sharp negative minimum giving an accurate value of the transition position (see Fig. 7.6). Using this quantity, it can be shown that the position of the transition converges with respect to N_r [Kuh03] and that the width of the transition defined as the width of the $\mathcal{Q}$ minimum at mid-height tends to zero as N_r tends to infinity as shown in Fig. 7.8. To illustrate the transition we give in Fig. 7.7 the fields for the mode above, during and after the transition. When the mode is well confined in the MOF core, the losses decrease exponentially with the number of hole rings like in Fig. 7.2. On the contrary above the transition (*i.e.* at long normalized wavelength) the mode spreads into all the cladding region defined by the hole lattice and the mode loss follows approximately a power law as a function of N_r. In this latter case the mode is well described as a space-filling cladding resonance [EWW$^+$00; MSS00; KBR98].

This second mode is in fact a defect mode, the observed transition defined by the observed minimum of $\mathcal{Q}$ constituting a change from a confined state for high λ/Λ values to an extended state for low λ/Λ values [Kuh03]. We define the locus of this transition as the cutoff of the second mode. This

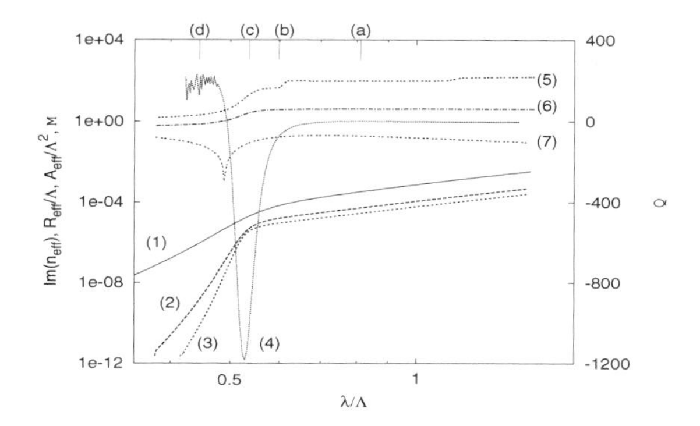

Fig. 7.6 Variation of different physical quantities during the transition, for a MOF with $d/\Lambda = 0.55$ used at $\lambda = 1.55$ μm. Curves (1) to (3) are $\Im m(n_{\text{eff}})$ for 4, 8 and 10 rings, curves (4) to (7) are $\mathcal{Q}$, A_{eff}/Λ^2, R_{eff}/Λ, and $\mathcal{M}$ as defined in the text, for $N_r = 8$. The points (a-d) indicate the positions of the field plots of Fig. 7.7.

cutoff is an intrinsic property of the mode and is not due to the finite size of the cladding region.

7.2.2 *A phase diagram for the second mode*

We have just defined the second mode cutoff in a MOF by varying λ/Λ at a fixed value of d/Λ. In Fig. 7.9 we give the loss curves as a function of λ/Λ for several values of the geometrical parameter d/Λ. As already mentioned, the position in λ/Λ of the cutoff converges with respect to N_r, and consequently we limit the study to an 8-ring MOF. The transition between a localized mode and an extended mode remains sharp for $d/\Lambda > 0.45$, whereas for $d/\Lambda < 0.45$ the transition becomes more and more gradual, disappearing entirely around $d/\Lambda \simeq 0.4$.

In the $(d/\Lambda, \lambda/\Lambda)$ plane, the loci of the cutoff define a curve which splits the MOF parameter space into two regions. Below this cutoff curve, the second mode is confined, and above it is extended or unconfined. The best fit of this limit curve obtained from finite size MOF studied with the Multipole Method is:

$$\frac{\lambda}{\Lambda} \simeq \alpha_{s.m.}\left(\frac{d}{\Lambda} - \left(\frac{d}{\Lambda}\right)_{s.m.}\right)^{\gamma_{s.m.}} \tag{7.6}$$

where $\alpha_{s.m.} = 2.8 \pm 0.12$, $\gamma_{s.m.} = 0.89 \pm 0.02$, and $(\frac{d}{\Lambda})_{s.m.} = 0.406 \pm 0.003$.

For $d/\Lambda < (d/\Lambda)_{s.m.}$, the second mode is always space filling, and consequently the fundamental mode is the only one to be potentially confined whatever the wavelength: the MOF is endlessly single-mode.

Using periodic boundary conditions in a plane wave basis and few points computed from the normalized effective area, Mortensen obtains for the parameter $(d/\Lambda)_{s.m.}$ an approximate value of 0.45 [Mor02]. From the experimental point of view, Folkenberg and his colleagues found that the above cutoff locus formula is in very good agreement with their experimental results obtained from high quality MOF [FMH+03]. On the theoretical side, Mortensen *et al.* exhibit an adapted V-parameter for microstructured optical fibres which follow nicely Eq. (7.6) [MFNH03].

7.3 Modal Cutoff of the Fundamental Mode

It is well known that the fundamental mode of conventional optical fibres does not undergo any cutoff, but for W-fibres this is no longer true [MY77]. Keeping in mind the useful analogies already drawn between these W-fibres and MOFs, it is a pertinent question to investigate the possible cutoff of the fundamental mode in MOF. In what follows we give some results concerning fundamental mode cutoff in MOF with a finite number N_r of hole inclusions.

7.3.1 *Existence of a new kind of cutoff*

The transition of the fundamental mode can be observed via a similar study as that conducted for the second mode. In Fig. 7.10, we give the variations of three physical quantities already described in the previous section, and once again these quantities are plotted as a function of λ/Λ. One can see that in a log-log plot the slope of the losses changes rapidly in a narrow region of the λ/Λ axis. As shown in Fig. 7.11, the change of $\Im m(n_{\text{eff}})$ is accompanied by a rapid variation in the field distributions. For large values of λ/Λ the mode has a large effective radius, *i.e.* it is cladding filling, and has high losses decreasing through a power law with respect to N_r. For small values of λ/Λ the mode is confined in the core and losses depend more strongly on the number of rings, they decrease exponentially with respect to N_r.

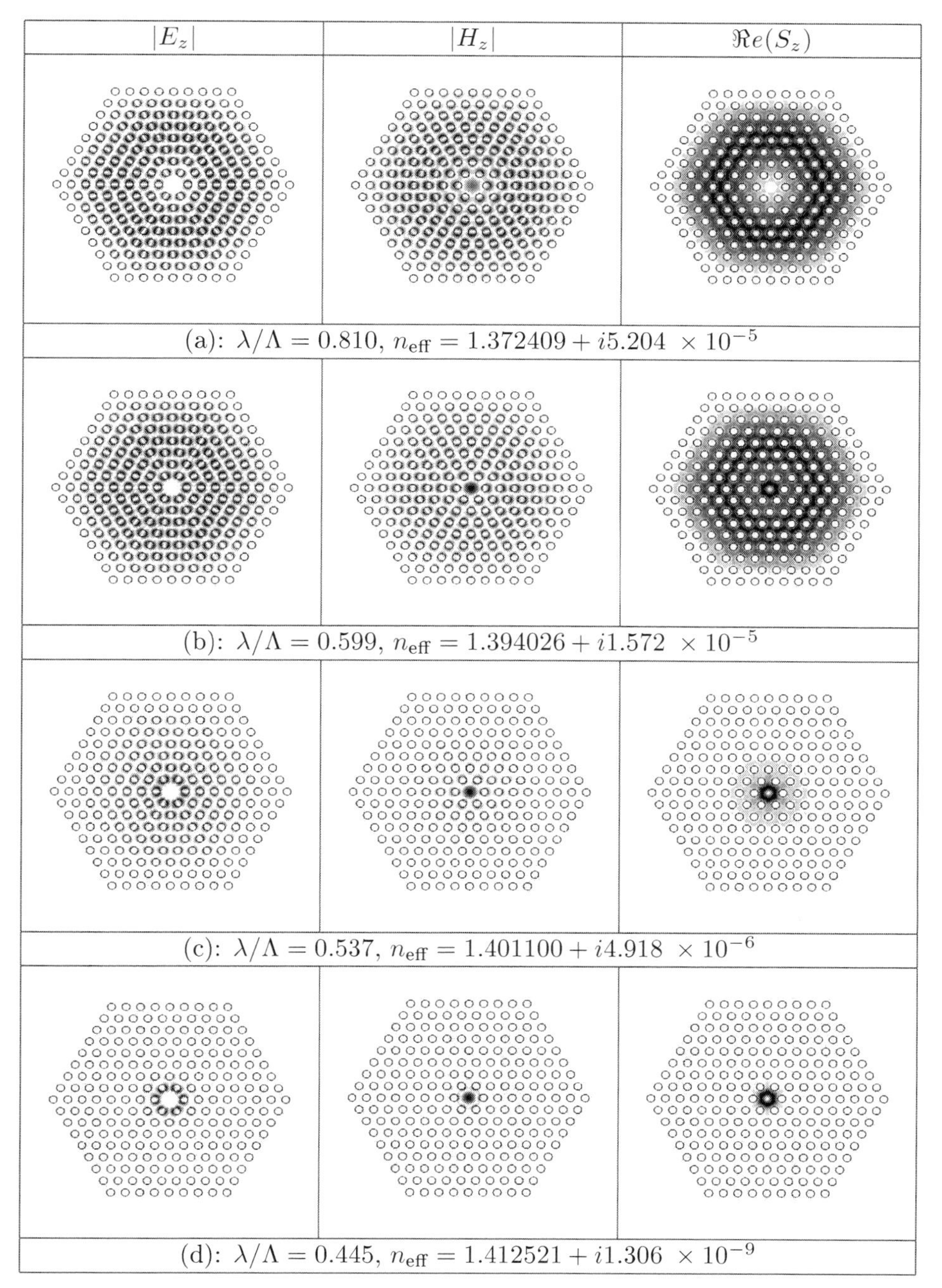

Fig. 7.7 Field distributions of the second mode across the transition. The letters in brackets refer to the points marked on Fig. 7.6. For all structures $d/\Lambda = 0.55$, $\lambda = 1.55\ \mu$m, and $N_r = 8$.

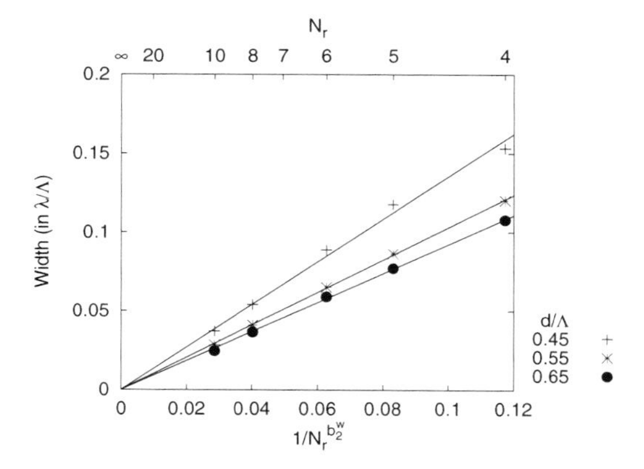

Fig. 7.8 Width of the cutoff of the second mode as a function of the number of rings for three different values of d/Λ. The depicted quantity is the half-width of the $\mathcal{Q}$ peak as a function of $N_r^{-b_2^w}$. Here $b_2^w \simeq 1.55$. The width of the cutoff transition clearly goes to zero as the number of rings increases.

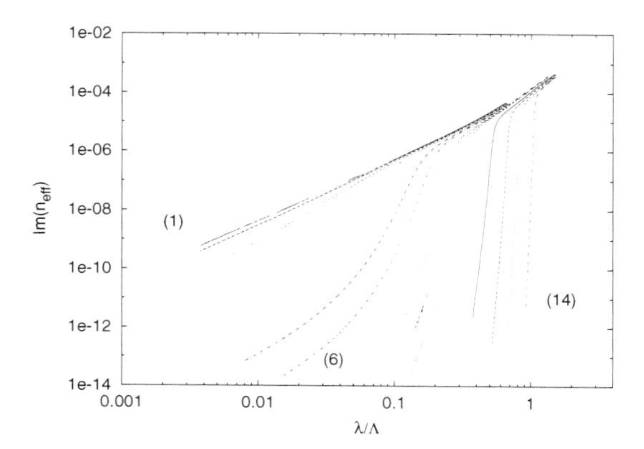

Fig. 7.9 $\Im m(n_{\text{eff}})$ as a function of wavelength/pitch, for an 8 ring MOF at a wavelength of $\lambda = 1.55$ μm for several diameter-to-pitch ratios. $\Im m(n_{\text{eff}})$ decreases monotonically with increasing d/Λ, as this parameter takes the values 0.40 (1), 0.41, 0.42, 0.43, 0.45, 0.46 (6), 0.48, 0.49, 0.50, 0.55, 0.60, 0.65, 0.70, 0.75 (14).

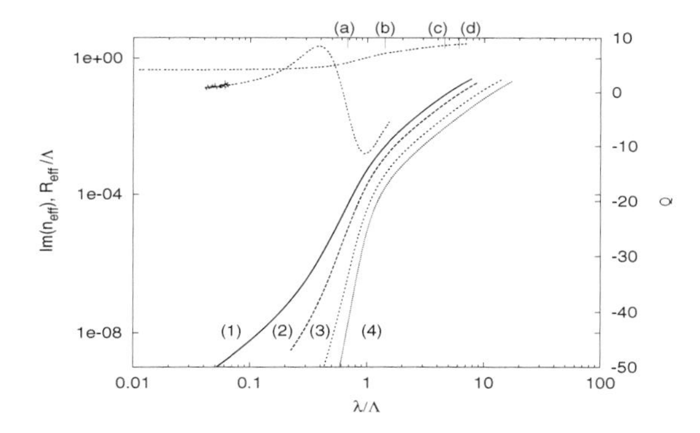

Fig. 7.10 Variation of different physical quantities during the transition of the fundamental mode, for a MOF with $d/\Lambda = 0.3$ used at $\lambda = 1.55\ \mu$m. Curves (1) to (4) are $\Im m(n_{\text{eff}})$ for 3, 4, 6 and 8 rings, curves (5) and (6) are R_{eff}/Λ and $\mathcal{Q}$ for $N_r = 3$ respectively. The points (a-d) indicate the positions of the field plots in Fig. 7.11.

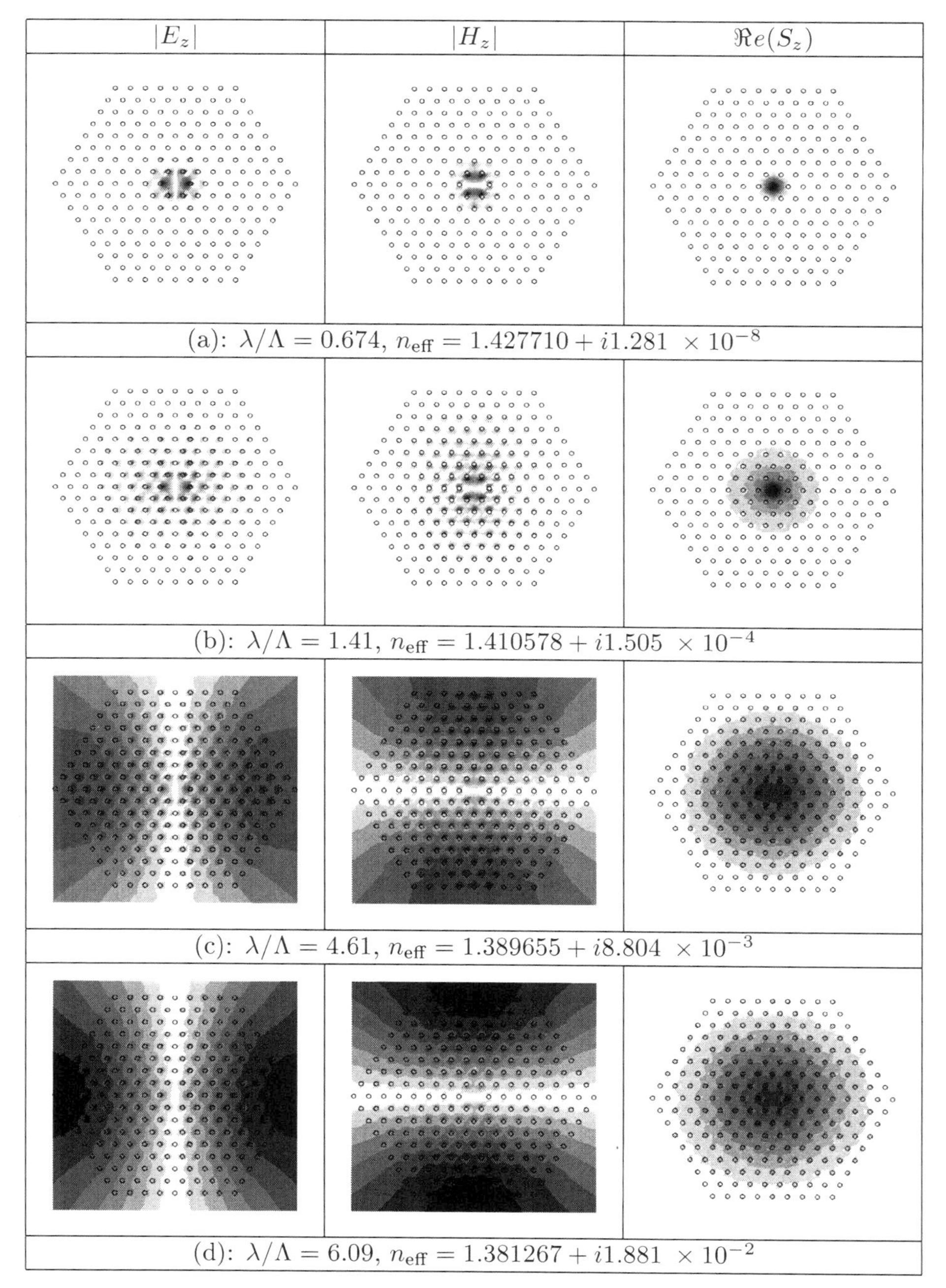

Fig. 7.11 Field distributions of the fundamental mode across the transition. The letters in brackets refer to the points marked on Fig. 7.10. For all structures $d/\Lambda = 0.55$, and $\lambda = 1.55$ μm, and N_r=8. In plots (c) and (d), the fields diverge due to high losses of the associated leaky modes.

Even though the fundamental mode transition seems at first sight very similar to that for the second mode, they are not equivalent. The dependence of the locus of the negative $\mathcal{Q}$ peak on the number of rings does not seem negligible, and its width does not seem to become infinitely sharp with increasing N_r [KdSM+02]. This second issue is adressed more quantitatively in Fig. 7.12: the half-width of the negative $\mathcal{Q}$ peak converges with respect to N_r to a finite value at least for low enough d/Λ ratio (See also Fig. 7.8 concerning the width of the second mode cutoff, in which the width decreases to zero with increasing N_r).

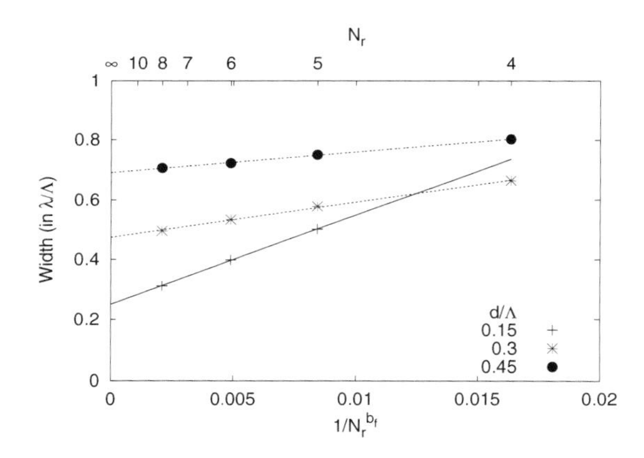

Fig. 7.12 Half-width of the negative $\mathcal{Q}$ peak of the fundamental mode as a function of $1/N_r^{b_f}$. The width of the peak converges to a non-vanishing value. Here $b_f \simeq 2.97$. This figure must be compared to Fig. 7.8 obtained for the second mode.

Besides this, the loss transition parameter $\mathcal{Q}$ has a positive peak before the negative one. Consequently, instead of a cutoff point we must define a cutoff region. We define this cutoff region as the interval between the positive and the negative peak observed in the $\mathcal{Q}$ curve as illustrated in Fig. 7.13.

7.3.2 *A phase diagram for the fundamental mode*

Using the same procedure as that described in section 7.2.2, one can establish the upper limit of the cutoff region (the $\mathcal{Q}$ minima in Fig 7.13). The associated curve (See the upper curve in Fig. 7.14) is obtained through the best fit of the computed $\mathcal{Q}$ minima values[5] for $N_r = 4$ using a similar

[5] N_r=4 is the largest number of rings for which we could directly extract with the Multipole Method the locus of the minimum of $\mathcal{Q}$ for values of d/Λ up to 0.75 , for higher

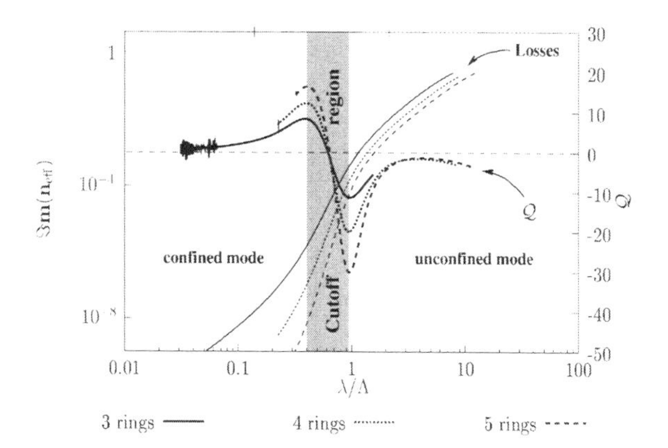

Fig. 7.13 The different operation regions of a solid core C_{6v} MOF. The curves show the fundamental mode losses and $\mathcal{Q}$ for MOF with $d/\Lambda = 0.3$ for $N_r = 3, 4$, and 5. The locus of $\mathcal{Q}$ extrema delimit the fundamental mode cutoff region (the grey region). The width of the cutoff region remains finite when $N_r \to \infty$.

function to that employed in Eq. 7.6.

$$\frac{\lambda}{\Lambda} \simeq \alpha_{f.u.}(\frac{d}{\Lambda} - (\frac{d}{\Lambda})_{f.u.})^{\gamma_{f.u.}} \tag{7.7}$$

where $\alpha_{f.u.} = 2.63 \pm 0.03$, $\gamma_{f.u.} = 0.83 \pm 0.02$, and $(\frac{d}{\Lambda})_{f.u.} \in]0; 0.06[$.

For λ/Λ above the limiting value defined by Eq. 7.7, the fundamental mode is always space filling. With increasing N_r, this upper limit of the cutoff region shifts slightly towards larger values of λ/Λ for $d/\Lambda \gtrsim 0.3$ and towards smaller values of λ/Λ for $d/\Lambda \lesssim 0.3$ [Kuh03].

The lower limit of the cutoff region represents the locus of the positive $\mathcal{Q}$ peaks (See the lower curve in Fig. 7.14). Due to numerical limitations relating to the smallest reachable $\Im m(n_{\rm eff})$ (limitations described in section 4.4 page 175), this curve can be obtained only for $N_r = 3$ in a wide range of d/Λ with the current numerical implementation of the Multipole Method. The upper and lower limits of the cutoff region split the MOF parameter space $(d/\Lambda, \lambda/\Lambda)$ in three regions: the unconfined fundamental mode region, the cutoff region, and the confined fundamental mode region.

N_r, $\Im m(n_{\rm eff})$ is smaller than the available numerical accuracy.

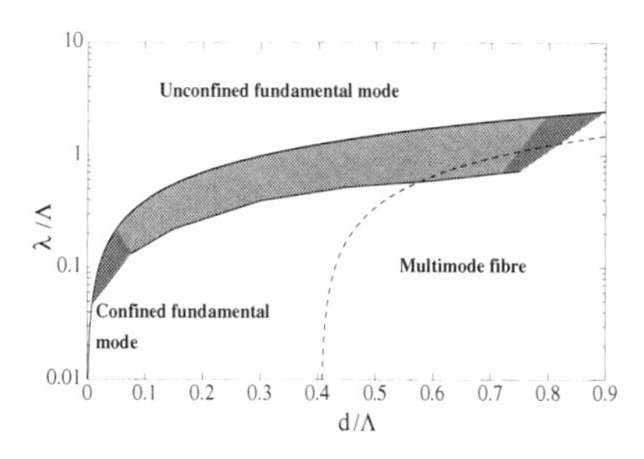

Fig. 7.14 Diagram of the operation regimes of solid core C_{6v} MOFs. The shaded grey region corresponds to the fundamental mode cutoff region. The dashed line corresponds to the second mode transition locus as computed via Eq. 7.6.

7.3.3 *Simple physical models below and above the transition region*

We must stress that the fundamental mode transition diagram presented here (Fig. 7.14) has been obtained with finite (and relatively small) values of N_r, but as can be seen in Fig. 7.13, the upper and lower limits of the cutoff region along the λ/Λ axis depend only slightly on N_r. Besides, it is possible that above a certain value of d/Λ the fundamental mode remains confined for $N_r \to \infty$ regardless of the wavelength. Nevertheless one must keep in mind that in all cases experimental MOF consist of a finite number of inclusion rings, so the results presented here retain their predictive value.

A detailed study of the fundamental mode cutoff and a deep asymptotic analysis of this phenomenon can be found in references [KdSM$^+$02; Kuh03]. We give below only the main qualitative results. MOFs have already been modeled for short wavelengths as step index fibres with varying cladding index [BKR97; KBR98; KAB$^+$00]. In the proposed approached models, both the short and long normalized wavelength limits are treated: a W-fibre model called CF2 is used for short wavelengths, and a step-index fibre model called CF1 is used for long wavelengths. These models are depicted in Fig. 7.15.

At long normalized wavelengths, we have observed that the fundamental mode fills the entire confining region, *i.e.* the core and the cladding so we can consider a homogenized fibre with a core radius ρ equal to $N_r\Lambda+d/2$ and an average effective index $[\bar{n}]$ embedded in the matrix. This effective index is computed by means of homogenization argument [Pou99] using the air filling fraction of the fibre [KdSM$^+$02]. The modal index is calculated using

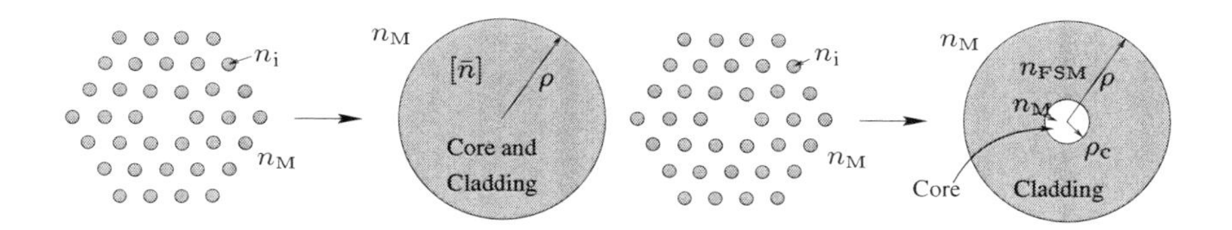

(a) CF1 model geometry: conventional step index fibre equivalent to a MOF at long wavelengths.

(b) CF2 model: Geometry of the conventional W-step index fibre equivalent to MOFs at short wavelengths.

Fig. 7.15 Models of conventional optical fibres used for the asymptotic analysis of MOF in the long (CF1) and short (CF2) normalized wavelength limits. n_M is the matrix index and n_i is the inclusion index (See the text for more details).

the analytical formula given in table 12-12 of reference [SL83] concerning a step-index uniaxial fibre with a core radius equal to $N_r\Lambda + d/2$ surrounded by a matrix of index n_M. The agreement between the results given by this step-index model fibre and those given by the Multipole Method for the MOF are excellent for $\lambda/\Lambda \geq 0.5$ [KdSM$^+$02]. Thus in this regime, the mode properties only depend on the total fibre radius and on d/Λ. MOF modes are fairly lossy in this regime, since the effective indices in the core region of size $\rho = N_r\Lambda + d/2$ are smaller than the matrix index. The loss decreases with increasing N_r following a power law. Consequently, N_r has to be impractically large to obtain MOFs with acceptable losses for applications. Besides, some important quantities such as the chromatic dispersion depend on N_r [KdSM$^+$02; KRM03]. At first sight, this regime of long normalized wavelength does not seem to be useful for practical applications.

In the short normalized wavelength region, we have observed that the fundamental mode is completely confined in the core region. Hence the fibre model is now a step-index W-fibre with a core made of silica of which the radius ρ_c is approximately[6] 0.64 Λ [BMPR00]. The intermediate region of the W-fibre is an annulus extending from the core radius to the end of the cladding region ($\rho = N_r\Lambda + d/2$). The index of this annulus is taken as n_{FSM}, the effective index of the fundamental space filling mode[7] [BKR97;

[6]This value of 0.64 Λ for the fibre core gives the best estimates for the actual value of the real part of the effective index of the fundamental mode given by more accurate methods (such as the finite element or multipole methods).

[7]n_{FSM} is the highest possible real effective index for a mode localized between the

MSS00]. Using this CF2 model, it can be shown that the losses vary as $(\lambda/\Lambda)^2$ for a fixed number of ring N_r, and decrease exponentially with increasing N_r at fixed λ/Λ. These behaviours are actually observed on the corresponding curves computed with the Multipole Method for the finite MOFs for $\lambda/\Lambda < 0.3$. In this regime, the real part of n_{eff} converges with increasing N_r and so does the chromatic dispersion.

Consequently, the small normalized wavelength region, or equivalently the regime in which the CF2 model is valid, is much more appropriate for practical MOF designs than the region associated with the CF1 model (long normalized wavelength). However, the further the CF2 region is penetrated toward short wavelength the stronger the analogy becomes between MOF and conventional optical fibres.

The region of the parameter space $(d/\Lambda, \lambda/\Lambda)$ between the regions in which the CF1 and CF2 models are valid is called the transition region. In this region, the properties of the fundamental mode change; they depend on the MOF parameters $(d/\Lambda, N_r)$ and they differ from those of conventional step-index or W-profile fibres. This transition region seems to be fruitful for applications.

7.4 Chromatic Dispersion

The chromatic dispersion, or equivalently the group velocity dispersion, plays a crucial role in conventional fibre optics [Oko82] both in linear and nonlinear [Agr01] phenomena. This key role is also present when microstructured optical fibres are considered [RWS00; DPG$^+$02; FZdC$^+$03]. Fairly early in MOF history, it became apparent that these new waveguides can exhibit peculiar and interesting dispersion properties and that they may be good candidates for dispersion management in optical systems [MBR98; FSMA00; KAB$^+$00].

The systematic study of chromatic dispersion in MOF [FSMA00] really began with the development of a vector simulation method with periodic boundary conditions [FSM$^+$00b]. Nevertheless we know from the previous section 7.3 that the MOF properties will converge with respect to N_r in the short normalized wavelength (λ/Λ) region of the operation regime diagram but that this is no longer true in the peculiar and promising transition regime. Therefore one must be able to take into account the finite cross-section of MOF to describe accurately the chromatic dispersion and to

low index inclusions constituting a lattice embedded in a high index infinite matrix.

compute their losses [KRM03].

In this section, mixing the contents of several articles [FSMA00; FSA+01; KRM03; RKM03] we describe how the chromatic dispersion can be managed in MOFs. We then show how it is possible using the great versatility of MOF design to get both ultraflattened chromatic dispersion and low losses.

7.4.1 *Material and waveguide chromatic dispersion*

The dispersion parameter D is computed through the usual formula from the real part of the effective index $\Re e(n_{\text{eff}})$ [Oko82]:

$$D = -\frac{\lambda}{c}\frac{\partial^2 \Re e(n_{\texttt{eff}})}{\partial \lambda^2} \tag{7.8}$$

D depends on λ, d, Λ, N_r, and also on n_M the matrix index.[8] Most of the time, for glasses, $n_M(\lambda)$ depends itself on λ (in dispersive media) [Fle78].

To explain the procedures used to study the chromatic dispersion we must first recall a few results obtained for step-index optical fibres [Oko82]. In a weakly guiding step index fibre, it can be shown that D is approximately composed of two components: one coming from the bulk material dispersion D_M and one coming from the waveguide itself D_W. So we have

$$D \simeq D_M + D_W \text{ with} \tag{7.9}$$

$$D_M = \frac{-\lambda}{c}\frac{d^2 n_M}{d\lambda^2} \tag{7.10}$$

$$D_W = -(\frac{\lambda}{c})\frac{\partial^2 [\Re e(n_{\texttt{eff}})|_{n_M(\lambda)=constant}]}{\partial \lambda^2} \tag{7.11}$$

D_M can be easily computed using a Sellmeier type expansion for n_M [Fle78; Agr01]. An explicit expression for D_W in terms of the usual step-index fibre parameters is available [Oko82] but this is useless in the present study which concerns MOF. We will assume that this splitting of the total chromatic dispersion D into two distinct terms is still approximately valid for MOF. This appromixation will be useful in understanding the global chromatic dispersion behaviour in MOF.[9] In the case in which no material dispersion is taken into account *i.e.* $n_M(\lambda) = constant$, the effective index n_{eff} of a

[8] In what follows we consider that the low index inclusions are simply holes of which the index is constant and equal to unity.

[9] We are not saying that this approximation will be used to compute the total chromatic dispersion in MOF (See section 4.4.2 page 177).

guided mode only explicitly depends on the pitch Λ of the inclusion lattice and the diameter d of these inclusions, and also on the wavelength λ. Hence as pointed out by Ferrando and his colleagues [FSMA00], that since n_{eff} is dimensionless, the dependence on the three above parameters can only occur through a dimensionless ratio. So, we can write the following when no material dispersion is involved:

$$n_{\text{eff}}(\lambda, d, \Lambda)|_{n_M(\lambda)=constant} = n_{\text{eff}}(d/\lambda, \Lambda/\lambda)\,. \tag{7.12}$$

We now have to introduce a geometric transformation of the triangular inclusion lattice region of the MOF. Let us consider a set of inclusions defined by Λ_{ref} and d. We can build from this a new cladding with the same filling ratio $f = d/\Lambda_{\text{ref}}$ simply by rescaling both these quantities: Λ and $d(\Lambda/\Lambda_{\text{ref}})$. If we now realize a scale transformation of the wavelength and look for the resulting scaling properties of D_W, it can be shown using Eq.7.12 that:

$$D_W(\lambda, \Lambda/\Lambda_{\text{ref}}, f) = \frac{\Lambda_{\text{ref}}}{\Lambda} D_W(\lambda\Lambda_{\text{ref}}/\Lambda, 1, f) \tag{7.13}$$

This scaling relation for waveguide dispersion allows us to compute straightforwardly the waveguide chromatic dispersion for all the rescaled MOF from the results obtained for the reference MOF: this is shown in Fig. 7.16. The scaling factor $\Lambda/\Lambda_{\text{ref}}$ can be used to modify the slopes of the chromatic dispersion curves: the higher the factor, the higher the absolute value of the slope. A $\Lambda/\Lambda_{\text{ref}}$ increase also induces an increase in the wavelength associated with zero D_W bounded by the first local maximum and the first local minimum of the waveguide dispersion (See Fig. 7.16).

On the other side, it is not possible to compute the D_W dependency on d/Λ using a similar argument: only numerical simulations will answer this question. Several curves are given for different filling ratios $f = d/\Lambda$ in Fig. 7.17. One can notice that the slope of the nearly linear part of these curves existing between the first local extrema is not really changed when the filling ratio is modified.

To achieve a specific total dispersion, one must compensate the material dispersion D_{mat} with D_W using the approximate relation Eq. 7.9. If the goal is to get flattened or ultraflattened chromatic dispersion in a target wavelength interval then one must control D_W to make it follow a trajectory parallel to that of $-D_{mat}$ in the target interval [FSMA00]. If this wavelength interval belongs to the interval in which the material dispersion behaves linearly (approximatively between 1.4 μm and 2.5 μm for silica)

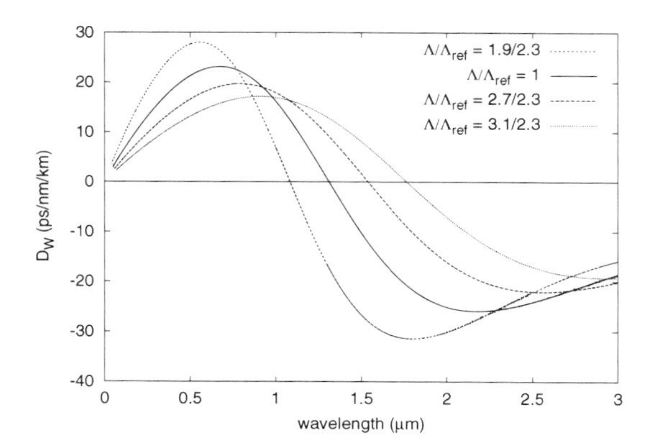

Fig. 7.16 Waveguide chromatic dispersion curves obtained for a non-dispersive matrix ($n_M = 1.45$) for several values of the pitch Λ for a fixed filling ratio $f = d/\Lambda = 0.6/2.3$, and $N_r = 3$. The reference curve is obtained for $\Lambda_{\rm ref} = 2.3\,\mu m$ with the Multipole Method and the other curves are obtained using the scaling law decribed by Eq. 7.13.

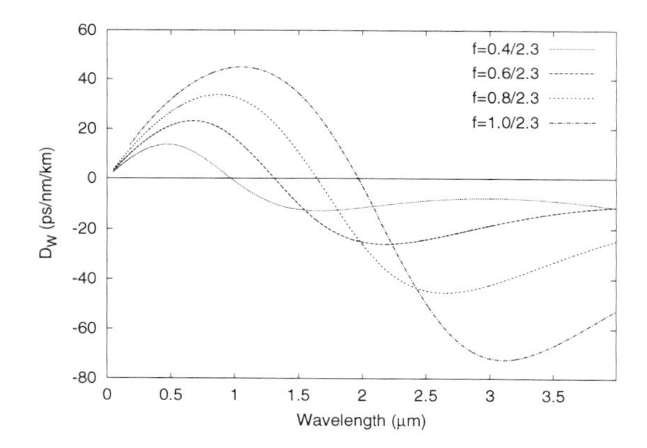

Fig. 7.17 Waveguide chromatic dispersion curves obtained for a non dispersive matrix ($n_M = 1.45$) for several values of the filling ratio $f = d/\Lambda$ for a fixed value of the pitch Λ, and $N_r = 3$. All the curves are obtained with the Multipole Method.

then a systematic approach can be used to obtain an initial MOF structure having an approximate ultraflattened dispersion curve.

We first have to adjust the slope of D_W using the scaling factor $\Lambda/\Lambda_{\rm ref}$ in order to make it parallel to $-D_{mat}$. Then we can play with d/Λ to modify the average value of the total chromatic dispersion, the co-linearity of D_W and $-D_{mat}$ being nearly conserved in the studied wavelength interval.

This scheme gives a first estimate MOF structure having the required total chromatic dispersion. Successive iterations of this process improve the quality of the results. To end the MOF design we must come back to the complete model: we have to take into account the material dispersion in our computation of chromatic dispersion curves. In some cases the gap between the complete model and the approximative one can be important compared to the required precision concerning the chromatic dispersion management.

This design scheme has also other important drawbacks. If the mean wavelength of the target interval is not included in the quasi-linear part of the waveguide dispersion then the scheme is no longer valid and only poor improvement can be obtained with sucessive iterations.

7.4.2 *The influence of the number of rings N_r on chromatic dispersion*

Another limitation of the design process described in the previous section is that the number of hole rings is not taken into account: we have worked with a fixed value for N_r, and as shown in section 7.3, depending on the position of the MOF in the phase diagram the MOF properties can depend strongly on N_r. This is exactly what we observe in the chromatic dispersion for several MOF configurations.

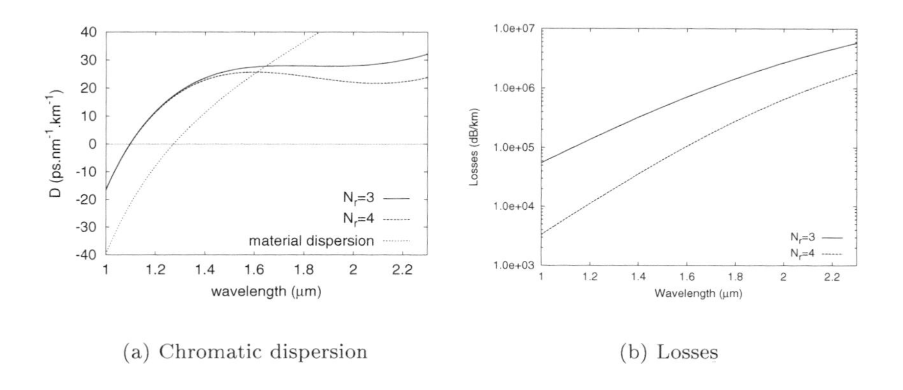

(a) Chromatic dispersion (b) Losses

Fig. 7.18 Example of the crucial influence of N_r on chromatic dispersion for some MOF configurations: the flat chromatic dispersion obtained for $N_r = 3$ is lost for higher N_r. Dispersion and losses as a function of wavelength for several values of N_r. The material dispersion is also shown. The hole diameter d is equal to $0.8\,\mu m$.

Directly linked to this issue are the losses. If one wants to practice chromatic dispersion management for applications using MOF the losses must be taken into account. A way to reduce the losses is to increase N_r as shown in the beginning of this chapter in section 7.1 and in Fig. 7.18(b), but we already know from section 7.3 that the behaviour of the losses with respect to N_r strongly depends on the values of λ/Λ and d/Λ.

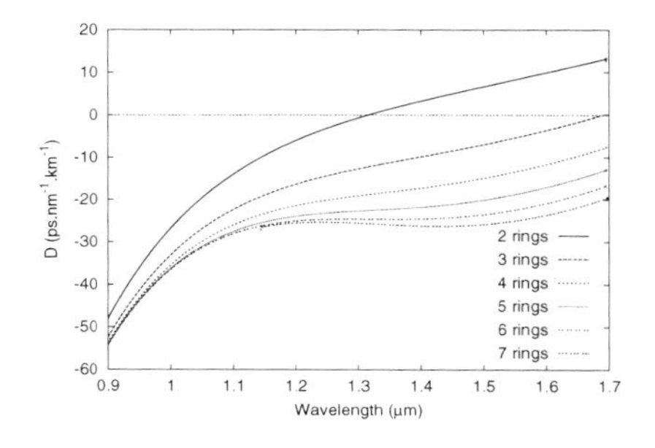

(a) Chromatic dispersion as a function of wavelength for several values of N_r

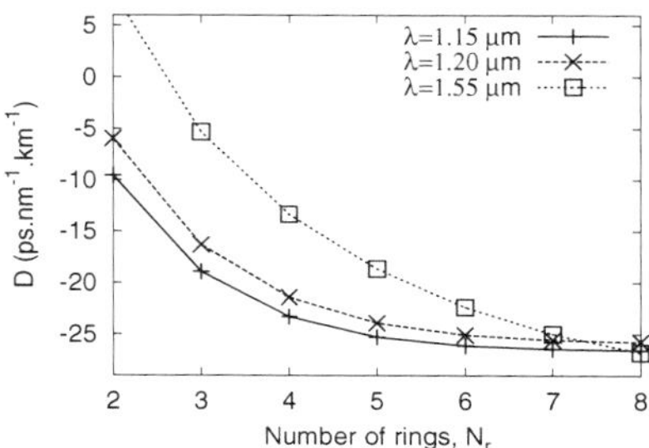

(b) Chromatic dispersion for three different wavelengths as a function of the number of rings N_r for the same MOF. For $\lambda = 1.55\,\mu m$ convergence is not obtained.

Fig. 7.19 Chromatic dispersion as a function of the number of rings N_r. Pitch $\Lambda = 2.0\,\mu m$; hole diameter $d = 0.5\,\mu m$.

7.4.3 *A more accurate MOF design procedure*

In Fig. 7.21, we show the variation of the total chromatic dispersion with respect to the numbers of rings for six different MOF geometries, all located in the parameter region in which the MOF properties converge with increasing N_r. All the curves show a simple variation with N_r, which can be modeled accurately by an exponential form $D_1 \exp(-\kappa N_r) + D_{\text{lim}}$. Such a fitting form has three parameters $(D_1, \kappa, D_{\text{lim}})$, which can be determined accurately from the results of N_r =3 to 6. We have established that the exponential fit thereafter accurately describes the dispersion of much larger structures and even the limiting parameter D_{lim}, the dispersion of a mode pinned by a single defect in an infinite lattice. In fact, using the limiting dispersion D_{lim} determined numerically for a set of wavelengths λ, we can also determine $S_{\text{lim}} = \partial D_{\text{lim}}/\partial\lambda$, the limiting chromatic dispersion slope.

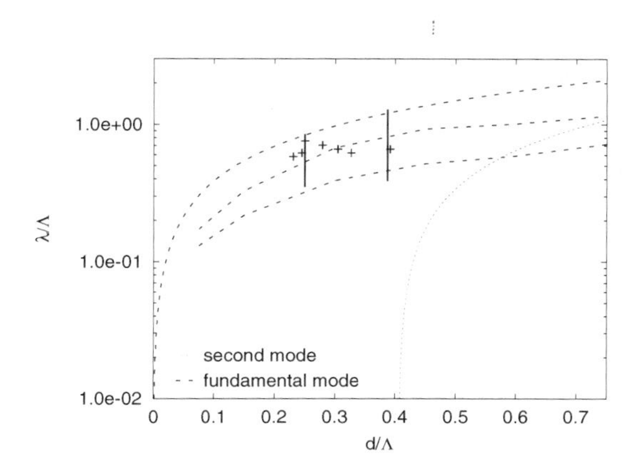

Fig. 7.20 Positions in the phase-diagram of the MOF parameters used in Fig. 7.19(a) (left black vertical segment). The dashed lines corresponds to the fundamental and second mode transitions. The crosses in the cutoff region correspond to the MOF configurations described in Fig. 7.21.

These two quantities describe the chromatic dispersion of the defect mode for the infinite lattice, and naturally tells us the chromatic variation of dispersion for large MOF structures.

This procedure has important advantages since MOFs with relatively small numbers of rings are relatively quickly modeled, and it directly takes into account material dispersion. Actually, this procedure can be used as the second step of MOF design for dispersion management based on the scheme previously described. Its main drawback is that it is much more time consuming compared to the first step.

In Fig. 7.22 we show the variations of these important parameters D_{lim} and S_{lim} as a function of the hole diameter d for different pitches Λ. This figure illustrates well how one can isolate a MOF exhibiting a target dispersion value for a sufficiently large number of rings N_r, which is flat over an interval containing the chosen wavelength value. Indeed such a MOF will have the desired value of D_{lim} and simultaneously a value of S_{lim} close to zero. Note that the pitches exhibited in Fig. 7.22 were chosen carefully to exemplify this desirable behaviour. We have also shown that, for the data of Fig. 7.22, the minima of S_{lim} as a function of d occur in the same diameter interval of $[0.65; 0.7]$ micrometres for all MOFs having $N_r \geq 6$. From Fig. 7.22, if one requires a positive nearly-zero flat chromatic dispersion then, using these curves, one should begin with its dispersion engineering with a MOF such that $\Lambda = 2.45\,\mu m$ and $d = 0.6\,\mu m$. Of course, Fig. 7.22

can be used to isolate MOF geometries having different characteristics, such as a prescribed slope with a fixed average value of dispersion over a given wavelength range. Note that, whereas the above designs have effectively a constant near zero chromatic dispersion for $N_r \geq 6$, their geometric losses impose much more stringent requirements on the number of rings, and the effective area of the fundamental mode $A_{\text{eff}} \simeq 36.5\,\mu m^2$ for $N_r = 6$: it requires $N_r \geq 18$ (1026 holes) to deliver losses below 1 dB/km at $\lambda = 1.55\mu$m. Even though some laboratories have already drawn 11-ring fibres [RKR02] (containing around 396 holes), there is clearly a technological issue. To overcome these limitations of conventional MOF, innovative MOF designs have been proposed [RKM03]. These consist of a solid core MOF in which the diameters of the inclusions increase with increasing distance from the fibre axis until the diameters reach a maximum value. With this new design and using three different hole diameters, it requires only seven rings (168 inclusions) to reach the $0.2\,dB/km$ level at $\lambda = 1.55\,\mu m$ with an amplitude of dispersion variation below $3.0\,10^{-2} ps.nm^{-1}.km^{-1}$ between $\lambda = 1.5\,\mu m$ and $\lambda = 1.6\,\mu m$.

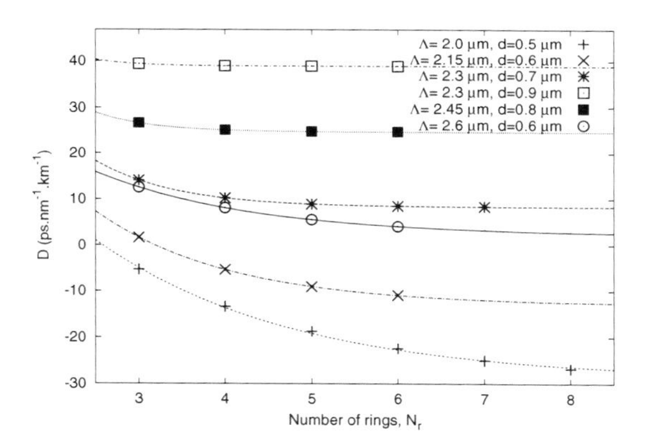

Fig. 7.21 Dispersion decay at $\lambda = 1.55\,\mu m$ as a function of the total number of hole rings N_r for different MOF structures. Λ is the pitch of the triangular lattice of holes, and d is the hole diameter. The points correspond to the computed numerical dispersion, and the lines to exponential based fits.

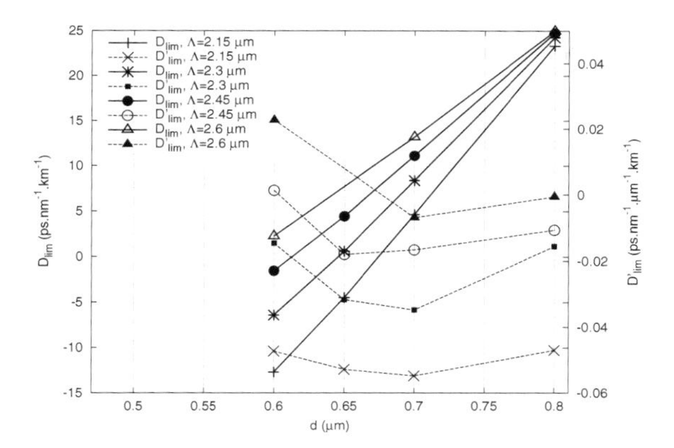

Fig. 7.22 Limit dispersion (plain thick lines, left y-scale) and limit dispersion slope (dashed lines, right y-scale) at $\lambda = 1.55\,\mu m$ as a function of the hole diameter d for several pitches. The chosen parameter values for Λ and d correspond to the small limit slope region.

7.5 A Hollow Core MOF with an Air-Guided Mode

We now illustrate one of the most striking possible properties of MOF: an hollow core MOF exhibiting an air-guided mode in the core [KBBR98; BBB99; BSB$^+$99; CMK$^+$99]. The MOF described in this section is the same as that used in reference [WMB$^+$01]. The pitch Λ is equal to 5.7816 μm, the diameter D_{co} of the center hole is equal to 13.1 μm, the diameter d of the circular inclusion in the cladding region is 4.026 μm, the matrix relative permittivity is equal to 1.9321 (*i.e.* $n_M = 1.39$) irrespective of the wavelength, and the inclusion relative permittivity is 1.0 (*i.e.* $n_i = 1$). We must mention that in our case we do not take into account material losses.[10]

7.5.1 *The photonic crystal cladding*

One way to obtain an air-guided mode in the hollow core MOF is to surround the core with a cladding region of inclusions having a complete photonic band gap in the transverse direction of the fibre around the desired wavelength (see section 1.3 page 12). In order to get this full band gap

[10]For realistic modeling, both matrix and inclusion material absorption must be taken into account. As will be shown later in this section, the geometrical losses due to the finite number of inclusion rings N_r around the core decrease nearly exponentially with respect to N_r so above a threshold value of N_r the main contribution to the losses comes from material absorption.

several rules can be followed [JMW95; LBB$^+$03; PBH$^+$03]. It is worth noting that the dispersion curves must computed in a conical (out-of-plane) propagation configuration in relation to the cylindrical inclusions[MM94]. These dispersion curves are associated with the Bloch waves (See sec. 2.6 page 72) of an infinite lattice. Figure 7.23 shows such curves for several values of the propagation constant β for the lattice associated with the cladding region of the hollow core MOF described above. Since for the study of these dispersion curves, we consider only lossless materials and an infinite lattice, β is real for the Bloch waves.

Computing these diagrams for several values of the normalized wave vector propagation constant $\beta\Lambda$, and determining their putative band gaps leads to Fig. 7.24 (such band gap diagrams are often called finger diagrams). As can be seen, several complete photonic band gaps exist for the studied lattice. Since we are looking for an air-guided mode we must have[11] $\Re e(\beta)/k_0 < 1$ (see section 1.5.5 page 25). This condition means that the dispersion curve of the air-guided mode of the finite MOF will be localized in the half-plane above the light line $\beta = k_0$ on the band gap diagram. This is indeed the case, as can be seen in Fig. 7.24.

One can see in Fig. 7.24 that the dispersion curve of the fundamental mode of the finite MOF crosses the fourth band gap instead of being inside it. This crossing can be understood from the fact that the modal dispersion curve has been computed for finite structures (N_r=3) while the band diagram comes from an infinite lattice.

7.5.2 *The finite structure*

A few results concerning finite size effects in hollow core MOF are now described. Due to the large diameter of the central hole we needed to use a fairly large value for the truncation order parameter M (see sections 4.4 and 4.5.1) to ensure a good convergence of the field expansions. Consequently we take M=19 for the numerical computations associated with the finite structures.

The finite MOFs studied in the following have the same parameters (Λ, d) as the infinite lattice used in the previous section. Nevertheless two other parameters are needed to described the finite structure: D_{co} the diameter of the hollow core which will be kept constant ($D_{co} = 13.1\,\mu m$), and the number of low index inclusions N_r which will be varied. The losses $\Im m(\beta)$

[11] We take the real part of β due to the expected leakiness of the mode for the finite MOF.

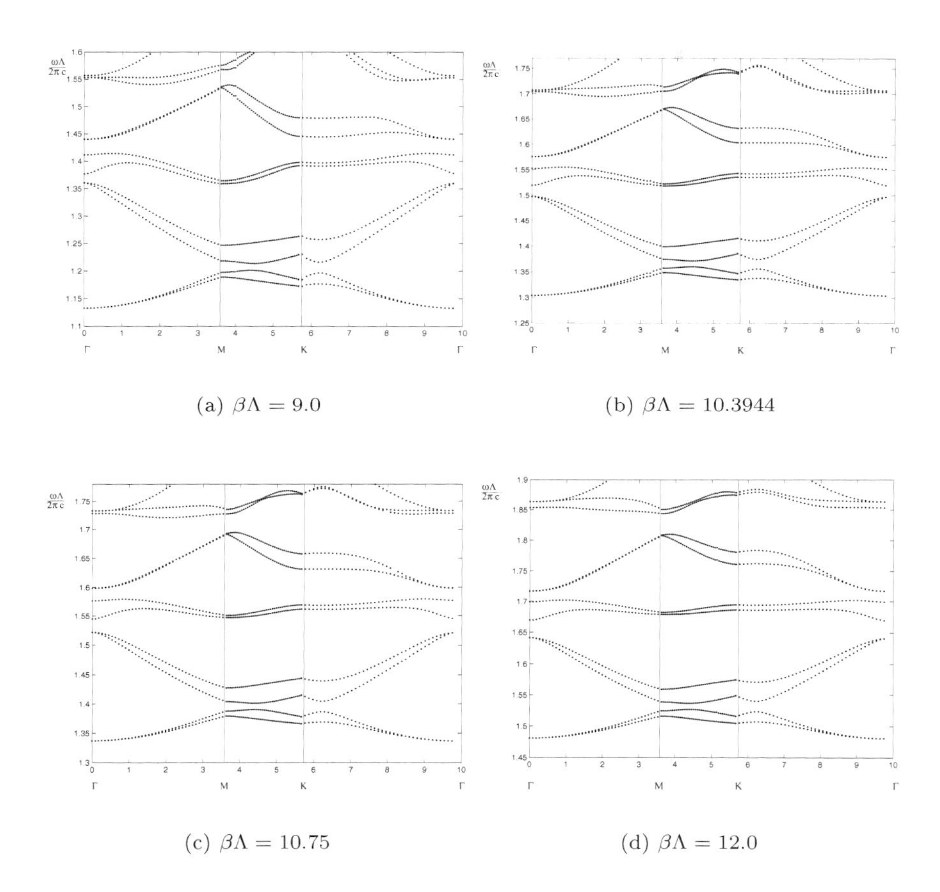

Fig. 7.23 Dispersion curves of the Bloch waves associated with a triangular lattice of circular cylindrical inclusions ($n_i = 1.0$) in an infinite silica matrix ($n_M = 1.39$). The geometrical parameters are: $\Lambda = 5.7816\,\mu m$, inclusion diameter $d = 4.026\,\mu m$. These curves have been obtained for an out-of-plane propagation, *i.e.* $\beta \neq 0$. The x-axis is associated with the route of the transverse wave vector along the edge of the irreducible Brillouin zone (see section 3.5.3 page 150). The y-axis represents the normalized frequency $\omega\Lambda/(2\pi c)$ in which ω is the frequency (angular) and c is the speed of light in vacuum.

of the fundamental mode of finite size MOFs versus the wavelength are shown in Fig. 7.25. These losses are computed for several values of N_r. As can be seen and as expected, the losses decrease with increasing N_r. More interesting is their wavelength dependence: there is an optimal wavelength value λ_{opt} for which the losses are minimal ($\lambda_{opt} = 3.416\,\mu m$ for $N_r = 2$,

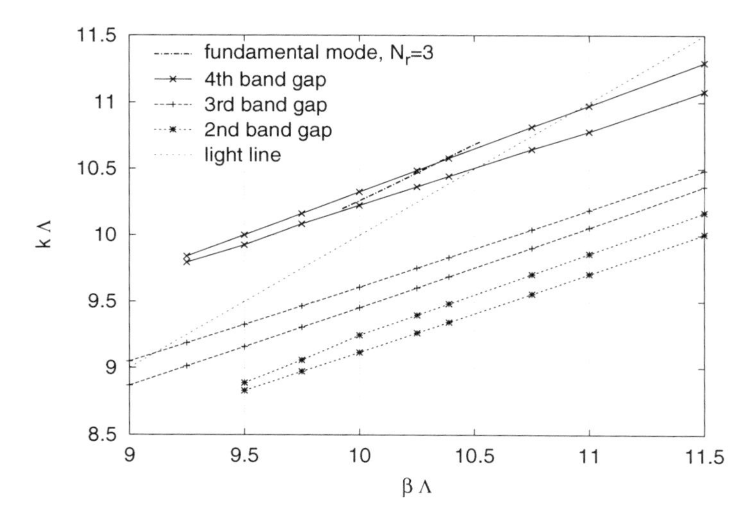

Fig. 7.24 Band gap diagrams associated with the dispersion curves of Fig. 7.23 (infinite lattice) and part of the dispersion curve of the fundamental mode for a three-ring finite size MOF. Only the 2nd, 3rd, and 4th band gaps are shown. Each band gap is localized between the two curves corresponding to the lower and upper boundaries of the bands (same line style). The thin straight dashed line is the light line $\beta = k_0$.

$\lambda_{opt} = 3.428\,\mu m$ for $N_r = 3$, $\lambda_{opt} = 3.440\,\mu m$ for $N_r = 4$, $\lambda_{opt} = 3.4413\,\mu m$ for $N_r = 5$, and $\lambda_{opt} = 3.4494\,\mu m$ for $N_r = 6$). This behaviour is due to the guidance mechanism in the hollow core: the photonic band gap effect.

It can also be noticed that the loss minima become sharper as N_r increases. This property is also related to the photonic band gap: the larger the value of N_r, the sharper the resonance.

The losses of the fundamental mode as a function of N_r are shown in Fig. 7.26. For a fixed wavelength, the losses decrease nearly exponentially with N_r. Otherwise if we consider the lowest loss values found for each N_r, *i.e.* the losses computed at λ_{opt}, the loss decreases is even better fitted by a decaying exponential. The exponential decay of the imaginary part of n_{eff} has already been found for solid core MOF in the CF2 regime (the regime in which the fundamental mode is confined, see section 7.3.2).

We now come back to the band diagram and to the dispersion curves of finite size MOF fundamental modes. Figure 7.27 is a zoom of Fig. 7.24 in which we add the dispersion curves of the fundamental mode computed for $N_r = 2, 3, 4$, and 5. It can be easily seen that the dispersion curves are very similar and that they cross the band gap even for $N_r = 5$. If we pick

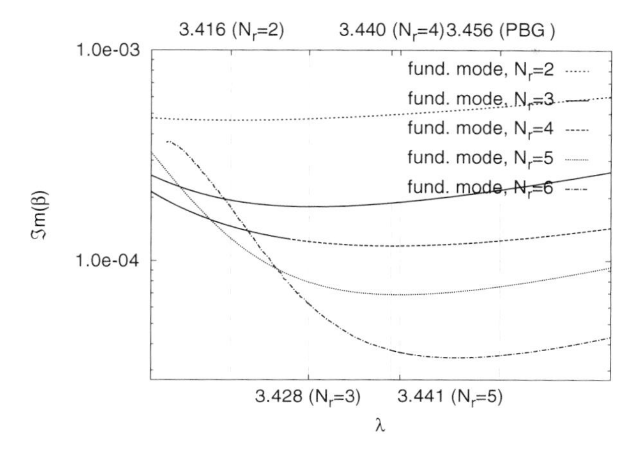

Fig. 7.25 Losses $\Im m(\beta)$ of the fundamental mode of several finite size MOFs ($N_r = 2, 3, 4,$ and 5) versus the wavelength (note the y log-scale). The wavelengths of the different finite structures associated with the lowest losses λ_{opt} are materialized with vertical lines. The wavelength corresponding to the middle of the photonic band gap (PBG) of the infinite lattice $\bar{\lambda}_{gap}$ is also shown.

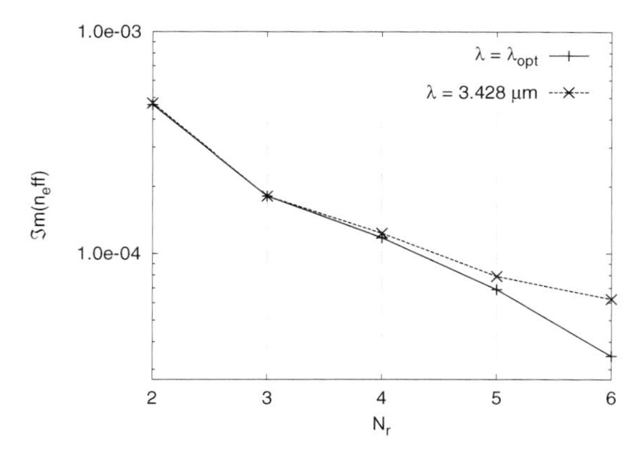

Fig. 7.26 Losses of the fundamental mode of finite MOFs versus the number of low index inclusion rings N_r for a fixed wavelength $\lambda = 3.428\,\mu m$ and for the optimal wavelength λ_{opt} in each case (see the text).

up the optimal value of β denoted by β_{opt} which is associated with λ_{opt} (using the results of the numerical simulations), it can be observed that even for $N_r = 5$, the value of β_{opt} is not yet in the band gap of the infinite structure. Nevertheless, it seems that as N_r increases, $\Re e(\beta_{opt})$ approaches the middle of the band gap. This comparison between the infinite lattice properties and those of finite MOFs clearly shows the crucial interest in being able to study finite structures when accurate results are needed as is the case for applications.

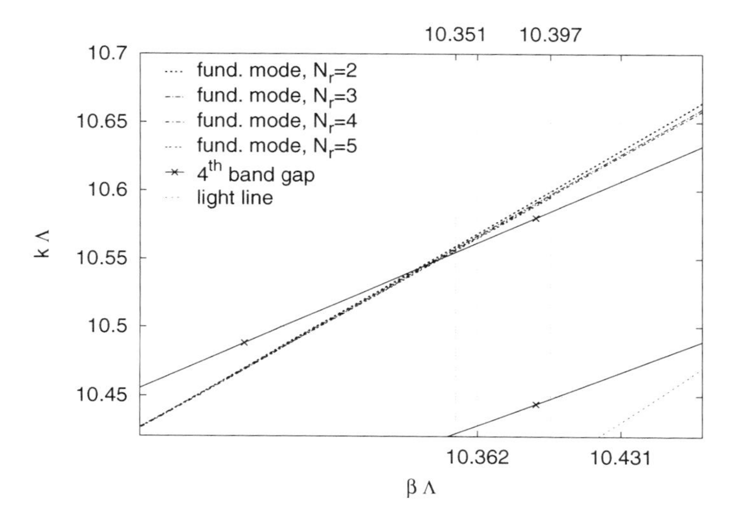

Fig. 7.27 Crossing of the dispersion curves of the fundamental mode of several finite size MOFs ($N_r = 2, 3, 4$,and 5) with the 4th band gap of the infinite lattice (See Fig. 7.24). The band gap is localized between the two plain curves corresponding to the lower and upper boundaries of the band (same line style). The $\Re e(\beta)$ values of the fundamental mode of the finite size MOFs, computed for the wavelength λ_{opt} associated with the lowest losses, are indicated by vertical lines ($N_r = 2$, $\Re e(\beta_{opt}) \simeq 10.431$; $N_r = 3$, $\Re e(\beta_{opt}) \simeq 10.397$; $N_r = 4$, $\Re e(\beta_{opt}) \simeq 10.362$; $N_r = 5$, $\Re e(\beta_{opt}) \simeq 10.3515$). The thin straight dashed line is the light line $\beta = k_0$.

We now describe in more detail the field properties obtained for the finite MOF fundamental mode at a fixed wavelength $\lambda = 3.428\,\mu m$. The computed effective index $n_{\text{eff}} = \beta/k_0$ is equal to $0.98111152 + i0.18086\,10^{-3}$ for $N_r = 3$ and $n_{\text{eff}} = 0.9808734 + i\,0.4724251\,10^{-3}$ for $N_r = 2$. This structure possesses the C_{6v} symmetry therefore we know that the fundamental mode is two-fold degenerate (see section 4.3 page 171). This property can be verified from Figs. 7.28 and 7.29. The vector fields of the fundamental mode are also given in Fig. 7.30: it can be seen that in the core the fields are nearly linearly polarized.

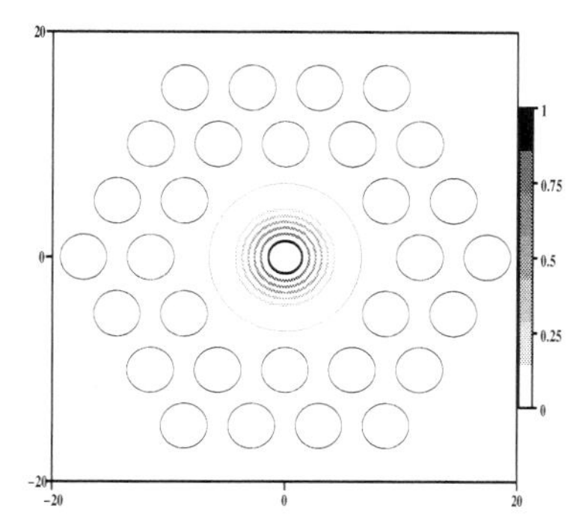

Fig. 7.28 Modulus of the Poynting vector for the fundamental mode of the hollow core fibre with $N_r = 2$: it is the same for the two symmetry classes p=3 and p=4. The modulus is normalized to unity.

7.6 Conclusion

In this Chapter, we have explained some of the main properties of solid and hollow core MOFs even though only linear phenomena were considered. The single-mode property of solid core MOF is now clearly defined, together with their operation regimes. We have focused on MOF based on subsets of triangular lattices but other kind of structures can be studied based on honeycomb or rectangular lattices. We have considered circular inclusions but elliptical inclusions or inclusions with even more complicated shapes could in principle be studied with the Multipole Method as long as these inclusions can be included in non overlapping circles. In order to extend the results shown in this chapter, other materials such as high index glasses or polymers should be studied and not only silica as we did.

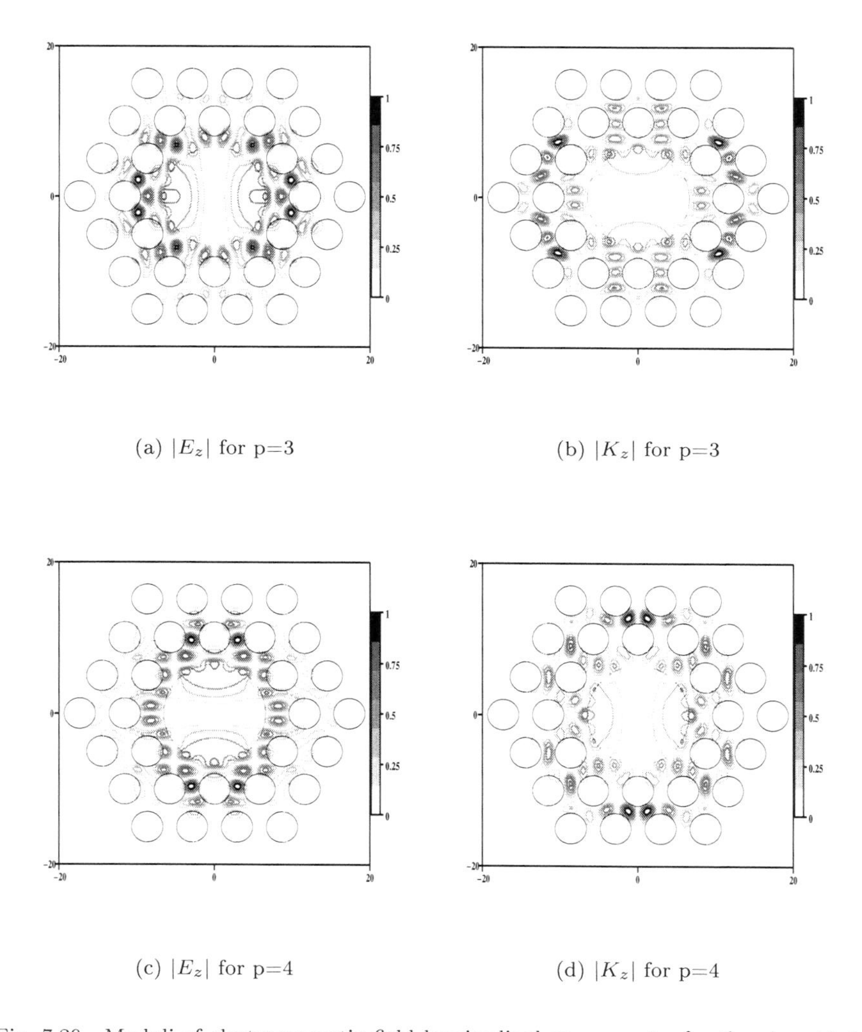

(a) $|E_z|$ for p=3

(b) $|K_z|$ for p=3

(c) $|E_z|$ for p=4

(d) $|K_z|$ for p=4

Fig. 7.29 Moduli of electromagnetic field longitudinal components, for the air-guided mode in the hollow core MOF ($N_r = 2$), this mode belongs to the symmetry classes p=3 and p=4. The field moduli are normalized to unity.

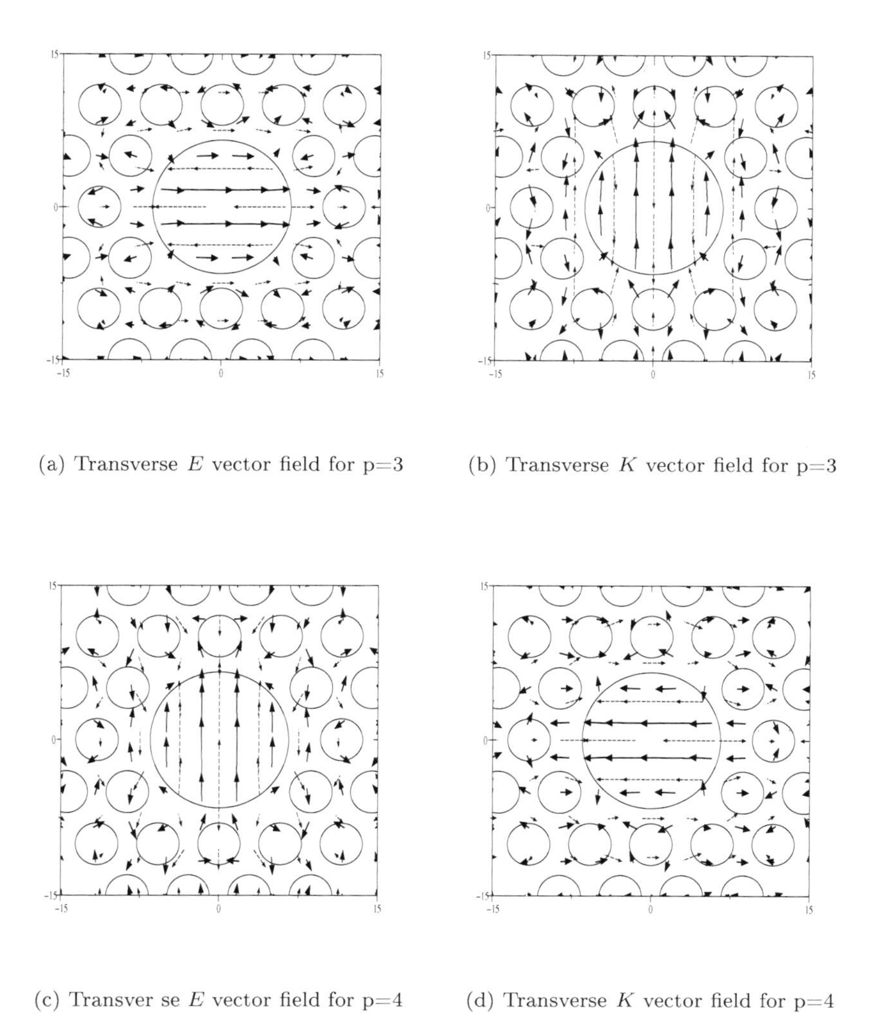

(a) Transverse E vector field for p=3

(b) Transverse K vector field for p=3

(c) Transver se E vector field for p=4

(d) Transverse K vector field for p=4

Fig. 7.30 Transverse electromagnetic vector field for the degenerate air-guided mode of the hollow core MOF ($N_r = 2$). The real part of the field is represented by plain thick vectors and the imaginary part is represented by dashed thin vectors.

Chapter 8

Conclusion

We believe that the terminology "new optical waveguides" is fully justified to qualify microstructured optical fibres (MOFs) even though leaky modes were already encountered in conventional W-fibres. MOF unique properties such as single-mode guidance (now clearly defined) and photonic band gap propagation (revisited for finite structures) make them much more than a simple generalization of conventional optical fibres.

In this book, special attention has been paid to accurate definitions of physical phenomena arising in MOFs. Contrarily to the case of propagating modes, the spectral theory of leaky modes is still inchoate. This theory is fundamental at least from a practical point of view *i.e.* in computing via a Finite Element Method, constants of propagation, losses and associated modes.

The keys of versatile and peculiar guiding properties of MOF are the high available index contrast between the matrix and the inclusions and the wide variety of available choices concerning their geometrical arrangement. The negative counterpart of these advantages is the impossibility of obtaining an accurate weak guiding theory similar to that which can be derived for conventional optical fibres, thus necessitating the development of powerful numerical tools.

Throughout this book, we have surveyed numerical tools which loosely range from transfer matrix methods, to recent finite element approach, and multipole methods. These two complementary numerical methods have been used to investigate MOF guiding properties with a reasonable success. Each method has its own advantages and weaknesses, but used together they allow a great variety of studies. As far as linear problems are concerned nearly all of the possible structures can be studied. Other numerical methods combining Finite Element Method versatility and Multipole

Method speed and accuracy may be implemented in the future allowing systematic investigation of complex MOF structures in a way similar to those presented in this book concerning MOFs which consist of circular inclusions. Actually, it seems that the Multipole Method looses its speed advantage whenever non circular inclusions are concerned. Nervertheless, new methods based on perturbative approaches may provide a useful tool to broaden the applicability of the Multipole Method. These would allow the study of MOFs in which the inclusions deviate slightly from circular cylinders using the results obtained for the circular reference inclusions.

In the topic of MOF, theoretical issues and practical designs are mixed: leaky modes and low-loss fibres, covariant formulation of the Maxwell operator and twist in fibres, transition regime in MOF and chromatic dispersion management....

Due to the huge potential of MOF for applications (as demonstrated by the increasing number of publications on practical aspects and the couple of several year-old start-up companies manufacturing and selling these fibres) we are fairly confident about the future of MOF even if this new kind of optical waveguides does not universally supplant single-mode fused silica conventional fibres in long-haul telecommunications. Hollow core MOFs are possibly the first optical components based on photonic band gaps available as a commercial product. It is worth mentioning that these theoretical and practical developments relating to MOFs have been realized within the last decade.

The issue of the optical and geometrical tolerances of MOF will have to be addressed in the next few years in the same way in which it had been treated in conventional optical fibres.

Most of the already published results concerning both solid or hollow core MOF were obtained for silica glass. We believe that only a short period will be required to get both numerical and outstanding experimental results with chalcogenide glasses or with other special glasses. The higher index contrast available in such structures between compared to that which can be obtained with silica glass should allow an even tighter control on waveguide properties.

Another promising area is the study of nonlinear properties in MOF. Some numerical modelings have already been done but a huge amount of work is still needed when we compare the available results for conventional optical fibres to those published regarding MOF. *But that's* ***not*** *all, folks!*

Appendix A

A Formal Framework for Mixed Finite Element Methods

The purpose of this appendix is to introduce the general framework for the mixed finite element methods [BF91; Bra97; BM94].

The following abstract problem is considered:

$$\begin{array}{lll}\text{Find } \sigma \in \Sigma \text{ and } u \in U \text{ such that :} & & \\ a(\sigma,\tau) \; + b(\tau,u) & = < g,\tau > & ,\forall \tau \in \Sigma \\ b(\sigma,v) & = < f,v > & ,\forall v \in U \end{array} \tag{A.1}$$

where Σ and U are two given Hilbert spaces, a and b are bilinear continuous forms on $\Sigma \times \Sigma$ and $\Sigma \times U$ respectively, and g and f provide linear continuous functionals on Σ and U respectively.

The *Brezzi theorem* asserts that necessary and sufficient conditions for the existence and uniqueness of a solution to problem A.1 are:

- The *ellipticity in the kernel* of a: $\exists\ \alpha > 0$ such that

$$a(\tau,\tau) \geq \alpha \|\tau\|_\Sigma^2 \;, \forall \tau \in K = \{\tau \in \Sigma : b(\tau,v) = 0 \;, \forall v \in U\}$$

- The *inf-sup condition*: $\exists\ \beta > 0$ such that

$$\inf_{v\in U} \sup_{\tau\in\Sigma} \frac{b(\tau,v)}{\|\tau\|_\Sigma \|v\|_U} \leq \beta$$

The inf-sup condition is also called the Babuška-Brezzi condition or the Ladyzhenskaja-Babuška-Brezzi (LBB for short) condition and can be written in various equivalent ways:

$$\begin{aligned}
&\inf_{v\in U}\sup_{\tau\in\Sigma}\frac{b(\tau,v)}{\|\tau\|_\Sigma\|v\|_U}\geq\beta && \Leftrightarrow\\
&\exists\tau\in\Sigma \text{ such that } \inf_{v\in U}\frac{b(\tau,v)}{\|v\|_U}\geq\beta\|\tau\|_\Sigma && \Leftrightarrow\\
&\sup_{\tau\in\Sigma}\frac{b(\tau,v)}{\|\tau\|_\Sigma}\geq\beta\|v\|_U \ ,\forall v\in U && \Leftrightarrow\\
&\exists\tau\in\Sigma \text{ such that } b(\tau,v)\geq\beta\|\tau\|_\Sigma\|v\|_U \ ,\forall v\in U\,.
\end{aligned} \tag{A.2}$$

The inf-sup condition looks strange and abstract but nevertheless let B be the linear operator from Σ to U associated with the bilinear form b, *i.e.* $b(\tau,v)=(B\tau,v)_U$,$\forall\tau\in\Sigma,\forall v\in U$. The inf-sup condition means that B is surjective *i.e.* that every $u\in U$ is the image of at least one $\tau\in\Sigma$. If this is not the case, U is the direct sum of the image $B\Sigma$ of Σ by B and of its orthogonal complement $B\Sigma^\perp$ (at least one dimensional): $U=B\Sigma\oplus B\Sigma^\perp$. In this case, there exists a nonzero $u\in B\Sigma^\perp$, which gives $(B\tau,u)_U=b(\tau,u)=0$,$\forall\tau\in\Sigma$ and therefore it is impossible to find a nonzero $\tau\in\Sigma$ such that $b(\tau,u)\geq\beta\|\tau\|_\Sigma\|u\|_U$ for this particular u and the inf-sup condition cannot be satisfied.

Now, by the Riesz theorem, to our u in $B\Sigma^\perp\subset U$ there corresponds an f in U' such that the second equation in A.1 cannot be satisfied hence the importance of the surjectivity of B.

The connection with problem 3.6 is straightforward if one sets $a(\sigma,\tau)=\int_\Omega\alpha^{-1}\sigma\tau d\mathbf{x}$, $b(\sigma,u)=\int_\Omega u\,\mathrm{div}(\sigma)d\mathbf{x}$, $g=0$, $\Sigma=H(\mathrm{div},\Omega)$, and $U=L^2(\Omega)$.

We go now to the discrete approximation problem in the finite dimensional spaces $U_h\subset U$ and $\Sigma_h\subset\Sigma$:

$$\begin{aligned}
&\text{Find } \sigma_h\in\Sigma_h \text{ and } u_h\in U_h \text{ such that :}\\
&a(\sigma_h,\tau)+b(\tau,u_h)=<g,\tau> \ ,\forall\tau\in\Sigma_h\\
&b(\sigma_h,v) \qquad\qquad\quad =<f,v> \ ,\forall v\in U_h\,.
\end{aligned} \tag{A.3}$$

Sufficient conditions for the existence and uniqueness of a solution to problem A.3 with an optimal error bound are:

- The *discrete ellipticity in the kernel* of a:
  $\exists\ \tilde\alpha>0$ **independent of** h such that

$$a(\tau,\tau)\geq\tilde\alpha\|\tau\|^2_\Sigma\ ,\forall\tau\in K_h=\{\tau\in\Sigma_h : b(\tau,v)=0\ ,\forall v\in U_h\}$$

- The *discrete inf-sup condition*:
  $\exists\, \tilde{\beta} > 0$ **independent of** h such that

$$\inf_{v \in U_h} \sup_{\tau \in \Sigma_h} \frac{b(\tau, v)}{\|\tau\|_\Sigma \|v\|_U} \leq \tilde{\beta}\,.$$

If these conditions are satisfied, a unique solution $(u_h, \sigma_h) \in U_h \times \Sigma_h$ exists with the following error bound:

$$\|\sigma - \sigma_h\|_\Sigma + \|u - u_h\|_U \leq \gamma(\inf_{\tau \in \Sigma_h} \|\sigma - \tau\|_\Sigma + \inf_{v \in U_h} \|u - v\|_U)$$

where (u, σ) is the solution of problem A.1 and γ is a constant **independent of** h.

The inf-sup condition is a fundamental tool in setting-up of mixed finite element formulations but unfortunately it is often difficult to satisfy or to impose. A practical tool which is useful when attempting to satisfy the condition is the notion of the *Fortin operator*, which is a projector with special properties.

The *projectors* are, in the present framework, operators that apply continuous spaces on discrete ones. In practice, they give the recipe for finding the element values of the discrete set of degrees of freedom of $u_h \in U_h \subset U$ from a given $u \in U$. By default, we will use the *orthogonal projectors* defined by

$$\Pi_h^U : U \to U_h \text{ such that } (u - \Pi_h^U u, v)_U = 0\ , \forall v \in U_h\,.$$

As an example, consider the L^2 orthogonal projector $\Pi_h^{L^2}$ from L^2 onto the discrete space $U_h \subset L^2$ of piecewise constant functions on a triangular mesh. A triangle T has a single degree of freedom *i.e.* the value $\overline{u}_\mathrm{T}$ of the function u_h of U_h inside T. The projection is defined by $\int_\mathrm{T} (u - \Pi_h^{L^2} u) v d\mathbf{x} = 0\ , \forall v \in U_h$. As the only $v \in U_h$ that are not null on T are constant on T, it is sufficient to satisfy the condition for $v = 1$, which gives: $\int_\mathrm{T} u d\mathbf{x} - \overline{u}_\mathrm{T} meas(\mathrm{T}) = 0$ and $\overline{u}_\mathrm{T}$ is simply the average value of u on T.

Now we state the following lemma:

Lemma A.1 *Fortin's criterion: Suppose that the bilinear form* b*:* $\Sigma \times U \to \mathbb{R}$ *satisfies the inf-sup condition. In addition, suppose that for the subspaces* Σ_h*,* U_h *there exists a bounded linear operator* $\Pi_h^\Sigma : \Sigma \to \Sigma_h$ *such that*

$$b(\tau - \Pi_h^\Sigma \tau, v_h) = 0\ , \forall v_h \in U_h$$

If $\|\Pi_h\| \leq c$ *for some constant* c *independent of* h*, then the finite element spaces* Σ_h*,* U_h *with the form* b *satisfy the inf-sup condition.* Π_h^Σ *is called a Fortin operator.*

A common way to satisfy the Fortin criterion is the following: if the projector Π_h^Σ is bounded independently of h, a sufficient condition is first that B, the linear operator associated with the bilinear form b as above, maps Σ_h into U_h: $B\Sigma_h \subset U_h$ and secondly that B commutes with the orthogonal projector Π_h^U from U onto U_h: $B\Pi_h^\Sigma\tau = \Pi_h^U B\tau \;, \forall\tau \in \Sigma$. In this case $b(\tau - \Pi_h^\Sigma\tau, v_h) = (B\tau - B\Pi_h^\Sigma\tau, v_h)_U = (B\tau - \Pi_h^U B\tau, v_h)_U = 0 \;, \forall v_h \in U_h$ using successively the very definition of B, its commutativity and the fact that Π_h^U is an orthogonal projector.

To sum up, a sufficient condition for satisfying the discrete inf-sup condition, together with the fact that the projector Π_h^Σ is bounded independently of h, is that the following diagram is commutative [BF91; Bra97]:

$$\begin{array}{ccc} \Sigma & \xrightarrow{B} & U \\ \Pi_h^\Sigma \big\downarrow & & \Pi_h^U \big\downarrow \\ \Sigma_h & \xrightarrow{B} & U_h \end{array}$$

The conclusion that can be drawn from the inf-sup criterion is slightly paradoxical at first sight: one could have believed that the larger the discrete spaces, the better since there is more flexibility to represent the approximation. This is partially true but the fact that B has to map Σ_h onto U_h implies that if you take a larger space $V_h \supset U_h$ that is not the image of Σ_h, it leads to an ill-posed discrete problem.

Appendix B

Some Details of the Multipole Method Derivation

B.1 Derivation of the Wijngaard Identity

To generalize the Wijngaard expansion of the field to MOF, we define a function $U(x, y)$ as

$$U(x,y) = \begin{cases} E_z, & r < R_0, \\ 0, & \text{elsewhere.} \end{cases} \tag{B.1}$$

Thus U is continuous inside the hole region, because of the continuity of the tangential field component, while its normal derivative is discontinuous at the boundaries of the inclusions. Both U and its normal derivative are discontinuous at the jacket boundary C ($r = R_0$). As a consequence, it can be deduced from Eqs. 4.3 and 4.4 that U satisfies, in the sense of distributions [Sch65],

$$\triangle U + k_\perp^2 U = s \tag{B.2}$$

where $k_\perp = k_\perp^i$ in inclusion i, and $k_\perp = k_\perp^e$ elsewhere. Source s is a singular distribution given by

$$s = \sum_{j=1}^{N_c} S_j \delta_{C_j} - T\delta_C - \mathrm{div}(\mathbf{n} Q \delta_C) \tag{B.3}$$

with S_j defined at the boundary C_j of the j-th hole as the jump of the normal derivative of U. Further, Q and T are, respectively, the limits of U and its normal derivative at $r = R_0$, where the normal $\mathbf{n}$ is outwardly oriented. The distribution $A\delta_C$ is defined by [Sch65]

$$\langle A\delta_C, \varphi \rangle = \int_C A(M)\varphi(M)dM \tag{B.4}$$

M being a point of C, dM the length of an elementary segment of C and φ an infinitely differentiable function with bounded support.

Equation B.2 can be rewritten as

$$\triangle U + (k_\perp^e)^2 U = \left[(k_\perp^e)^2 - (k_\perp)^2\right] U + s \tag{B.5}$$

and so in the hole region U follows from the convolution

$$U = G_e \star \left[s + \left((k_\perp^e)^2 - (k_\perp)^2\right) U\right], \tag{B.6}$$

where G_e is the Green function of the Helmholtz equation: $G_e = -iH_0^{(1)}(k_\perp^e r)/4$. From Eqs. B.3 and B.6 U can be re-expressed as

$$U = \sum_{j=1}^{N_c} G_e \star D_j + G_e \star D, \tag{B.7}$$

with

$$D_j = S_j \delta_{C_j} + \left[(k_\perp^e)^2 - (k_\perp^j)^2\right] U_j \tag{B.8}$$

$$D = -T\delta_C - \operatorname{div}(\mathbf{n} Q \delta_C), \tag{B.9}$$

$$U_j = \begin{cases} U \text{ in the } j\text{-th inclusion} \\ 0 \text{ elsewhere.} \end{cases} \tag{B.10}$$

Each term $V_j = G_e \star D_j$ of the sum on the right-hand side of Eq. B.7 is generated by sources placed inside or at the boundary of the j-th inclusion and satisfies a radiation condition outside this hole. It can be identified at any point outside this inclusion as the field scattered by it. Of course, since it satisfies the homogeneous Helmholtz equation outside the j-th inclusion, in the sense of distributions, it can be represented in the entire matrix region as a Fourier Bessel series

$$V_j = \sum_m B_m^{\mathrm{E}j} H_m^{(1)}(k_\perp^e r_j) e^{im\theta_j}. \tag{B.11}$$

The term $G_e \star D$ on the right-hand member of Eq. B.7 is generated by sources jacket boundary and it thus has no singularity inside this boundary. It can be identified as the incident field generated by the jacket and illuminating the matrix-inclusion region. It can also be represented in a Fourier Bessel expansion

$$V_{\text{inc}} = \sum_m A_m^{\mathrm{E}0} J_m(k_\perp^e r) e^{im\theta}. \tag{B.12}$$

From Eqs. B.7, B.11 and B.12, it can now be shown that in the entire matrix region, the field E_z can be represented by the Wijngaard expansion 4.20. The same argument can be used for the z component of the magnetic field H_z.

B.2 Change of Basis

Three changes of basis transformations are required: (i) conversion of outgoing fields sourced on one cylinder to regular fields in the basis of another cylinder; (ii) conversion of the regular field sourced on the jacket boundary to a regular field in the basis of each cylinder; and (iii) conversion of outgoing fields sourced at the cylinders to an outgoing field close to the jacket boundary. These are considered separately below.

B.2.1 *Cylinder to cylinder conversion*

Here we consider an outgoing cylindrical harmonic wave sourced from cylinder j and derive its regular representation in the coordinate system of cylinder l. From Graf's Theorem [AS65], we derive

$$H_m^{(1)}(k_\perp^e r_j)e^{im\arg(\mathbf{r}_j)} = \sum_{n=-\infty}^{\infty} J_n(k_\perp^e r_l)e^{in\arg(\mathbf{r}_l)} H_{n-m}^{(1)}(k_\perp^e c_{lj})e^{-i(n-m)\arg(\mathbf{c}_{lj})}, \tag{B.13}$$

so that the total field due to cylinder j is expressed as

$$\sum_{m=-\infty}^{\infty} B_m^j H_m^{(1)}(k_\perp^e r_j)e^{im\arg(\mathbf{r}_j)} = \sum_{n=-\infty}^{\infty} A_n^{lj} J_n(k_\perp^e r_l)e^{in\arg(\mathbf{r}_l)}, \tag{B.14}$$

where A_n^{lj}, defined in Eqs. 4.23 and 4.24, denotes the contribution to the nth multipole coefficient at cylinder l due to cylinder j.

B.2.2 *Jacket to cylinder conversion*

From Graf's Theorem [AS65] we now have

$$J_m(k_\perp^e r)e^{im\theta} = \sum_{n=-\infty}^{\infty} J_n(k_\perp^e r_l)e^{in\arg(\mathbf{r}_l)}(-1)^{n-m} J_{n-m}(k_\perp^e c_l)e^{-i(n-m)\arg(\mathbf{c}_l)}, \tag{B.15}$$

and from this, the change of basis transform is

$$\sum_{m=-\infty}^{\infty} A_m^0 J_m(k_\perp^e r) e^{im\theta} = \sum_{n=-\infty}^{\infty} A^{l0} J_n(k_\perp^e r_l) e^{in \arg(\mathbf{r}_l)}, \tag{B.16}$$

where A_n^{l0} denotes the multipole coefficient in the basis of cylinder l due to the regular field radiating from the jacket. Equation 4.27 is the matrix form B.16.

B.2.3 *Cylinder to jacket conversion*

The relevant transformation from Graf's theorem [AS65] is now

$$H_m^{(1)}(k_\perp^e r_l) e^{im \arg(\mathbf{r}_l)} = \sum_{n=-\infty}^{\infty} H_n^{(1)}(k_\perp^e r) e^{in\theta}$$
$$J_{n-m}(k_\perp^e c_l) e^{-i(n-m)\arg(\mathbf{c}_l)}. \tag{B.17}$$

The contribution from cylinder l to the outgoing field near the jacket boundary is

$$\sum_{m=-\infty}^{\infty} B_m^l H_m^{(1)}(k_\perp^e r_l)\, e^{im \arg(\mathbf{r}_l)} = \sum_{n=-\infty}^{\infty} B_n^{0l} H_n^{(1)}(k_\perp^e r)\, e^{in\theta}, \tag{B.18}$$

which can be written in the matrix notation 4.31.

B.3 Boundary Conditions: Reflection Matrices

We consider a cylinder centered at the origin of refractive index n_- and radius a embedded in a medium of refractive index n_+. To derive the reflection matrices of this cylinder we express the E_z and H_z fields in terms of Fourier-Bessel series in the local cylindrical coordinates (r, θ) inside and outside the cylinder,

$$E_z^\mp(r,\theta) = \sum_{m=-\infty}^{\infty} \left[A_m^{\mathrm{E}\mp} J_m(k_\perp^\mp r) + B_m^{\mathrm{E}\mp} H_m^{(1)}(k_\perp^\mp r) \right] e^{im\theta}, \tag{B.19}$$

for $r < a$ $(-)$ and $r > a$ $(+)$, with similar expressions for K_z. Here, $k_\perp^\pm = ({k^0}^2 n_\pm^2 - \beta^2)^{1/2}$ are the transverse wave numbers inside (outside) the cylinder. We introduce the vectors $\mathbf{A}^{\mathrm{E}\pm} = [A_m^{\mathrm{E}\pm}]$ and $\mathbf{B}^{\mathrm{E}\pm} = [B_m^{\mathrm{E}\pm}]$, as well as their K counterparts, and the condensed notation introduced in Eqs. 4.36 and 4.37 for $\tilde{\mathbf{A}}^\pm$ and $\tilde{\mathbf{B}}^\pm$. The interpretation of the J and H

terms was given in Section 4.2.3. At the cylinder boundary, reflection and transmission occurs and the waves mix with each other, which, because of the linearity of the Maxwell equations, can be expressed as a matrix relation between the various coefficients, as

$$\begin{cases} \tilde{\mathbf{A}}^- = \tilde{\mathbf{T}}^-\tilde{\mathbf{A}}^+ + \tilde{\mathbf{R}}^-\tilde{\mathbf{B}}^-, \\ \tilde{\mathbf{B}}^+ = \tilde{\mathbf{R}}^+\tilde{\mathbf{A}}^+ + \tilde{\mathbf{T}}^+\tilde{\mathbf{B}}^-. \end{cases} \tag{B.20}$$

Here, $\mathbf{R}^-$ and $\mathbf{R}^+$ are referred to as interior and exterior reflection matrices of the cylinder, over whereas $\mathbf{T}^+$ and $\mathbf{T}^-$ are transmission matrices, which do not matter in the analysis below. Note that the first equation of set (B.20) leads to Eq. 4.38, whereas the second leads to Eq. 4.37.

The $\mathbf{R}^\pm$ matrices can be derived from the continuity of the tangential field components at the cylinder boundary. To do this we need the expressions for the θ components of the fields, which can be expressed as a function of the z components as [SL83]

$$E_\theta(r,\theta) = \frac{i}{k_\perp^2}\left(\frac{\beta}{r}\frac{\partial E_z}{\partial\theta} - k\frac{\partial K_z}{\partial r}\right), \tag{B.21}$$

$$K_\theta(r,\theta) = \frac{i}{k_\perp^2}\left(\frac{\beta}{r}\frac{\partial K_z}{\partial\theta} + kn^2\frac{\partial E_z}{\partial r}\right), \tag{B.22}$$

where n is the refractive index. The partial derivatives that appear straightforwardly follow from Eqs. B.19

We can now write the continuity conditions for the z components by equating Eqs. B.19 on the boundary. Since the resulting equation is valid for all θ, terms with different m decouple and we find for each m

$$A_m^{\mathrm{E}-}J_m^- + B_m^{\mathrm{E}-}H_m^- = A_m^{\mathrm{E}+}J_m^+ + B_m^{\mathrm{E}+}H_m^+, \tag{B.23}$$

with the same result for K_z. Here we introduced the condensed notation $J_m^\pm = J_m(k_\perp^\pm a)$, *etc.*

In the same way we can equate the interior and exterior expressions for E_θ and K_θ. We then obtain two equalities of Fourier series in $e^{im\theta}$, in which, again, terms with different m decouple. These equations, which are not written out here, in combination with Eq. B.23 and its K_z counterpart are sufficient to obtain the $\mathbf{R}$ matrices.

We first concentrate on the interior reflection matrix $\tilde{\mathbf{R}}^-$; we obtain its coefficients by setting the exterior incoming field to zero: $\tilde{\mathbf{A}}^+ = \mathbf{0}$. It is now straightforward to solve, for a given m, the linear set of equation given to express $A_m^{\mathrm{E}-}$ and $A_m^{\mathrm{K}-}$ in terms of $B_m^{\mathrm{E}-}$ and $B_m^{\mathrm{K}-}$ by eliminating $B_m^{\mathrm{E}+}$

and $B_m^{\mathrm{K}+}$. We obtain

$$\begin{cases} A_m^{\mathrm{E}-} &= R_m^{\mathrm{EE}^-} B_m^{\mathrm{E}-} + R_m^{\mathrm{EK}^-} B_m^{\mathrm{K}-}, \\ A_m^{\mathrm{K}-} &= R_m^{\mathrm{KE}^-} B_m^{\mathrm{E}-} + R_m^{\mathrm{KK}^-} B_m^{\mathrm{K}-}, \end{cases} \tag{B.24}$$

with

$$\begin{aligned} R_m^{\mathrm{EE}-} &= \frac{1}{\delta_m}\Big\{ (\alpha^+_{J^- H^+} - \alpha^-_{H^+ J^-}) \times \\ &\quad (n_-^2 \alpha^+_{H^- H^+} - n_+^2 \alpha^-_{H^+ H^-}) - m^2 J_m^- H_m^- H_m^{+2} \tau^2 \Big\}, \\ R_m^{\mathrm{EK}-} &= \frac{1}{\delta_m}\left\{ \frac{2m\tau}{\pi k a} \frac{k_\perp^+}{k_\perp^-} H_m^{+2} \right\}, \\ R_m^{\mathrm{KE}-} &= -n_-^2 R_m^{\mathrm{EK}-}, \\ R_m^{\mathrm{KK}-} &= \frac{1}{\delta_m}\Big\{ (\alpha^+_{H^- H^+} - \alpha^-_{H^+ H^-}) \times \\ &\quad (n_-^2 \alpha^+_{J^- H^+} - n_+^2 \alpha^-_{H^+ J^-}) - m^2 J_m^- H_m^- H_m^{+2} \tau^2 \Big\}, \end{aligned} \tag{B.25}$$

where

$$\alpha^\pm_{J^\pm H^\pm} = \frac{k_\perp^\pm}{k} J_m'^{\pm} H_m^\pm \tag{B.26}$$

with other α coefficients defined analogously. Further

$$\begin{aligned} \delta_m = &(\alpha^-_{H^+ J^-} - \alpha^+_{J^- H^+})(n_-^2 \alpha^+_{J^- H^+} - n_+^2 \alpha^-_{H^+ J^-}) \\ &+ (m J_m^- H_m^+ \tau)^2 \end{aligned} \tag{B.27}$$

and

$$\tau = \frac{\beta}{a k_\perp^- k_\perp^+}(n_+^2 - n_-^2)\ . \tag{B.28}$$

To obtain the exterior reflection matrix $\tilde{\mathbf{R}}^+$ we set $\tilde{\mathbf{B}}^- = \mathbf{0}$, and eliminate the A_m^- coefficients. This yields

$$\begin{aligned}
R_m^{\mathrm{EE}+} &= \frac{1}{\delta_m}\Big\{ \left(\alpha^+_{J^- H^+} - \alpha^-_{H^+ J^-}\right) \times \\
&\quad \left(n_-^2 \alpha^+_{J^- J^+} - n_+^2 \alpha^-_{J^+ J^-}\right) - m^2 J_m^+ H_m^+ J_m^{-2} \tau^2 \Big\}, \\
R_m^{\mathrm{EK}+} &= \frac{1}{\delta_m}\left\{ \frac{2m\tau}{\pi k a} \frac{k_\perp^-}{k_\perp^+} J_m^{-2} \right\}, \\
R_m^{\mathrm{KE}+} &= -n_+^2 R_m^{\mathrm{EK}+}, \\
R_m^{\mathrm{KK}+} &= \frac{1}{\delta_m}\Big\{ \left(\alpha^+_{J^- J^+} - \alpha^-_{J^+ J^-}\right) \times \\
&\quad \left(n_-^2 \alpha^+_{J^- H^+} - n_+^2 \alpha^-_{H^+ J^-}\right) - m^2 J_m^+ H_m^+ J_m^{-2} \tau^2 \Big\}.
\end{aligned} \tag{B.29}$$

Appendix C

A Pot-Pourri of Mathematics

In this appendix, some mathematical vocabulary is introduced to be sure that we speak the same language. Consider it as a prerequisite, just check that all that stuff is clear in your mind and if not, have a look at the references.

- *Numbers* used in the book are natural integers $0, 1, 2, \cdots \in \mathbb{N}$ or signed integers $\cdots, -1, 0, 1, \cdots \in \mathbb{Z}$ mainly as indices. Physical quantities take their values in (sets of) real numbers $\mathbb{R}$ or complex numbers $\mathbb{C}$. Real and complex numbers are called *scalars*. A complex number z can be viewed as a pair (a, b) of real numbers denoted $z = a + ib$ where $i^2 = -1$. The real number $a = \Re e(z)$ is called the real part whereas the real number $b = \Im m(z)$ is called the imaginary part. The *complex conjugate* of z is $\overline{z} = a - ib$ and its *modulus* or absolute value is $\mid z \mid = \sqrt{z\overline{z}} = \sqrt{a^2 + b^2}$. The exponential representation $z = \mid z \mid e^{i\theta(z)}$ involves the modulus and the *argument* $\theta(z)$. The function $\theta(z)$ is usually chosen as being a continuous function on an open set of $\mathbb{C}$ called a continuous determination of the argument. Two continuous determinations on the same set differ by an integer multiple of 2π and the principal determination is defined by $\theta(z) : \mathbb{C} \backslash \{\Re e(z) \leq 0\} \to]-\pi, +\pi[$ and given by $\theta = 2\arctan(\frac{b}{a+\sqrt{a^2+b^2}})$.
- *Vector space*: given scalars $\mathbb{R}$ or $\mathbb{C}$, a vector space V is a set of elements, called vectors, with two operations: the addition of two vectors and the product of a vector by a scalar, both defined axiomatically in a way quite obvious for the common intuition and giving a new vector as the result. Given a set of vectors $\mathbf{v}^i$ of a vector space V and a set of scalars a_i, the combination of products by a scalar and vector additions is a new vector called a *linear combination* and denoted $a_i\mathbf{v}^i$. This expression uses the *Einstein summation convention* on repeated

indices: $a_i\mathbf{v}^i$ denotes $\sum_i a_i\mathbf{v}^i$. If things are written with care, it always involves an upper and a lower index.

The null scalar 0 and the null vector $\mathbf{0}$ are defined so that $0\mathbf{v} = \mathbf{0}$ and $\mathbf{0} + \mathbf{v} = \mathbf{v}$ for any vector. Given vectors are said to be *linearly independent* if the only linear combination which gives $\mathbf{0}$ is $\sum_i 0\mathbf{v}^i$ else they are said to be linearly dependent. The maximum number of vectors which you can find in a set of linearly independent vectors of V is called the *dimension* $\dim(V)$ of the vector space. A maximum set of linearly independent vectors is called a *basis* and any vector can be expressed as a linear combination of the basis vectors.

The canonical examples of vector spaces are the sets of n-tuples of scalars: $(b^1, \cdots, b^n)$ or $(c^1, \cdots, c^n)$ with $b^1, \cdots, b^n, c^1, \cdots, c^n \in \mathbb{C}$. Explicitly, the addition is given by $(b^1, \cdots, b^n) + (c^1, \cdots, c^n) = (b^1 + c^1, \cdots, b^n + c^n)$ and the product by a scalar is given by $a(b^1, \cdots, b^n) = (ab^1, \cdots, ab^n)$ with $a \in \mathbb{C}$.

Given a set of vectors $\mathbf{v}^1, \cdots, \mathbf{v}^p$, they *span* a vector space $V = \mathrm{span}(\mathbf{v}^1, \cdots, \mathbf{v}^p)$ which is the set of all the linear combinations of those vectors.

- *Multi-linear forms*: a p-multi-linear form is a map $f : \{\mathbf{v}^1, \mathbf{v}^2, \cdots, \mathbf{v}^p\} \in V \to f(\mathbf{v}^1, \mathbf{v}^2, \cdots, \mathbf{v}^p) \in \mathbb{C}$ from p vectors to a scalar (complex numbers $\mathbb{C}$ are taken as the default set of scalars) such that $f(\mathbf{v}^1, \cdots, a_i\mathbf{v}^i, \cdots, \mathbf{v}^p) = a_i f(\mathbf{v}^1, \cdots, \mathbf{v}^i, \cdots, \mathbf{v}^p)$, *i.e.* the map is linear in all entries. Particular cases are the *linear forms* which map linearly a single vector to a scalar and *bilinear forms* which map two vectors to a scalar.

  Note that the set of p-multi-linear forms for a given p are obviously a vector space themselves since a linear combinations of m p-forms $f_1, \cdots, f_m$ is defined as a new m-multi-linear form by $(a^i f_i)(\mathbf{v}^1, \cdots, \mathbf{v}^p) = a^i f_i(\mathbf{v}^1, \cdots, \mathbf{v}^p)$. In particular, the set of linear forms on a vector space V is called the *algebraic dual vector* space and is denoted V^*. In the finite dimension case, the dual space V^* is *isomorphic* to V, *i.e.* there is a one-to-one correspondence between the spaces preserving the various operations. For instance, if a vector $\mathbf{v}$ of V is a set of n numbers $(b^1, \cdots, b^n)$, a general linear form f is also given by a set of n numbers $(c_1, \cdots, c_n)$ so that the action of the linear form on the vector is $f(\mathbf{v}) = b^1 c_1 + \cdots + b^n c_n = b^i c_i$.

  Note the lower indices for the coefficients of forms.

  To stress on the duality, the action of f can be written as a *duality product* $f(\mathbf{v}) = < f, \mathbf{v} >$.

In the case of the canonical example, a basis is given by vectors $\mathbf{e}_i = (0, \cdots, 1, \cdots, 0)$ where 1 in the ith position is the only non zero component.

Given a basis $\mathbf{e}_i$ of a vector space V, the set of vectors $\mathbf{e}^i$ of V^* such that $< \mathbf{e}_i, \mathbf{e}^j > = \delta_i^j$, where δ_j^i is the *Kronecker delta symbol* equal to 1 when $i = j$ and else equal to 0, is a basis of V^* called the ***dual basis*** and $< b^i \mathbf{e}_i, c_j \mathbf{e}^j > = b^i c_i$.

- Functional analysis and function spaces: some sets of functions are vector spaces of particular interest called function spaces or functional vector spaces. The main characteristic of such spaces is that they are usually infinite dimensional and not isomorphic to their dual spaces. For instance, the set of square integrable functions on the $[0, 2\pi]$ interval is a functional vector space. A basis is the infinite set of trigonometric functions $\{1, \cos(x), \sin(x), \sin(2x), \cos(2x), \cdots\}$.
- Important examples of function spaces: let Ω be a bounded open set in $\mathbb{R}^n$ or $\mathbb{R}^n$ itself. We denote by $C^m(\Omega)$ the set of functions which are continuous in Ω together with all their first m derivatives.

  In a multi-variable context, the following notation is introduced: $\alpha = (\alpha_1 \cdots \alpha_n) \in \mathbb{N}^n$ is a multi-index with $|\alpha| = \sum_i \alpha_i$ and $D^\alpha = (-i)^{|\alpha|} \frac{\partial^{|\alpha|}}{\partial^{\alpha_1} x^1 \ldots \partial^{\alpha_n} x^n}$. The first m derivatives of a function f are the $D^\alpha f$ with $|\alpha| \leq m$.

  The subset consisting of the functions which have a compact support (*i.e.* which vanish outside a compact subset of Ω) is denoted by $C_0^m(\Omega)$. The corresponding space of infinitely differentiable functions (also called smooth functions) are denoted by $C^\infty(\Omega)$ and $C_0^\infty(\Omega)$.
- *Tensor product*: the tensor product is a fundamental operation of linear algebra which transforms multi-linear forms to linear ones. Given two vector spaces V and W, there exists a unique vector space (up to an isomorphism), denoted $V \otimes W$ and such that for any vector space U, the space of linear maps from $V \otimes W$ to U is isomorphic to the space of bilinear maps from $V \times W$ to U where $V \times W$ is the Cartesian product, *i.e.* the set of pairs $(v \in V, w \in W)$. The vector space $V \otimes W$ is called the tensor product.

  For example, in the case of functional spaces, the tensor product of two functions $f(x)$ and $g(x)$ of a single variable is the two variable function given by: $f \otimes g(x, y) = f(x)g(y)$.

  As another example, consider a vector space V of finite dimension n

and its dual space V^*. The tensor product $V^* \otimes V^*$ associates to a pair of linear forms on V, *i.e.* an element of $V^* \times V^*$ given by the coefficients a_i and b_i (a single index), a bilinear form on V given by the coefficients $(a \otimes b)_{i,j} = a_i b_j$ (two indices) such that the action on two vectors given by the coefficients v^i and w^i is the scalar $(a \otimes b)_{i,j} v^i w^j = a_i b_j v^i w^j$. Tensor product is an important tool to manipulate multi-linear maps as vectors spaces. For instance, if V is a vector space and V^* its dual, p-multi-linear maps form the $\otimes^p V^*$ vector space. A basis of this space may be built by taking the tensor products of p basis vectors of V^* and the dimension of this space is $\dim(\otimes^p V^*) = \dim(V)^p$.

- Direct sum: a subspace U of V is a subset of V such that any linear combination of elements of U is in U which is therefore itself a vector space. Suppose that a vector space V has two subspaces U and W such that any vector $\mathbf{v} \in V$ can be written in a unique way as the sum $\mathbf{u} = \mathbf{v} + \mathbf{w}$ where $\mathbf{u} \in U$ and $\mathbf{w} \in W$. The subspace V is in this case the *direct sum* of the vector spaces U and W which is denoted $V = U \oplus W$. One has $\dim(V) = \dim(U) + \dim(W)$.
- Scalar product and norm: new operations have often to be added to the bare structure of a vector space.
  The *scalar product* is a form $(\mathbf{v}, \mathbf{w}) \to \mathbb{C}$ with the following properties:

  - linear in the first variable, *i.e.* $(a\mathbf{u} + b\mathbf{v}, \mathbf{w}) = a(\mathbf{u}, \mathbf{w}) + b(\mathbf{v}, \mathbf{w})$.
  - $(\mathbf{v}, \mathbf{w}) = \overline{(\mathbf{w}, \mathbf{v})}$
  - positive, *i.e.* $(\mathbf{v}, \mathbf{v}) \geq 0$ and the equality arises only when $\mathbf{v} = \mathbf{0}$.

  In the case of a real vector space, this form is simply bilinear but in the complex case it is *sesqui-linear*, *i.e.* semi-linear in the second argument:$(\mathbf{v}, a\mathbf{w}) = \bar{a}(\mathbf{v}, \mathbf{w})$. A vector space with a scalar product is called a *pre-Hilbert space*.
  The *norm* $\|\mathbf{v}\|$ is the (positive) square root of the scalar product of a vector $\mathbf{v}$ with itself: $\|\mathbf{v}\|^2 = (\mathbf{v}, \mathbf{v})$.
  In the case of the canonical example, the canonical scalar product is given by $((b^1, \cdots, b^n), (c^1, \cdots, c^n)) = b^1\overline{c^1} + \cdots + b^n\overline{c^n}$.
  Two norms $\|.\|_1$ and $\|.\|_2$ are *equivalent* if there exist two real strictly positive constants c and C such that $c\|\mathbf{v}\|_1 \leq \|\mathbf{v}\|_2 \leq C\|\mathbf{v}\|_1$.
  A vector which has a norm equal to 1 is called a unit vector or a normalised vector. Two vectors are *orthogonal* if their scalar product is equal to 0.
  If V is a vector space with a norm $\|.\|_V$, this defines a norm on the dual space V^* by $\|f\|_{V^*} = \frac{\sup_{\mathbf{v} \in V} f(\mathbf{v})}{\|\mathbf{v}\|_V} = \sup_{\|\mathbf{v}\|_V = 1} f(\mathbf{v})$ for any linear form

$f \in V^*$.
We use the name of the concerned vector space as a subscript to indicate the vector product or the norm when this is necessary but it will be omitted if there is no ambiguity.

- The norm determines a *topology*, *i.e.* a system of subsets called open sets, on the vector space (which is therefore called a topological vector space) which is necessary to introduce the concepts of convergence and limit. Two equivalent norms determine the same topology.
The distance between two vectors is defined by $\rho(\mathbf{u}, \mathbf{v}) = \|\mathbf{u} - \mathbf{v}\|$ and the (open) ball of centre $\mathbf{u}$ and radius $\epsilon \in \mathbb{R}$ is the set of vectors $\mathbf{v} \in B(\mathbf{u}, \epsilon)$ so that $\rho(\mathbf{u}, \mathbf{v}) < \epsilon$. A set Ω is called *open* if for any $\mathbf{u} \in \Omega$ there exists an ϵ such that the ball $B(\mathbf{u}, \epsilon)$ is contained in Ω. A set E is *closed* if its complement (*i.e.* elements which are not in E) is open. The *closure* $\overline{\Omega}$ of a set is the smallest closed set which contains it and the *interior* $Int(\Omega)$ is the largest open set which is contained in it[1]. The *boundary* $\partial\Omega$ is the set of elements which are in the closure and not in the interior.
A vector space with a norm which is complete (*i.e.* every *Cauchy sequence* of the space $\{\mathbf{u}_n\}$, *i.e.* such that $\rho(\mathbf{u}_n, \mathbf{u}_m) \to 0$ as $n, m \to +\infty$, has its limit in the space) is called a *Banach space*.
- Quotient: if W is a vector subspace of the vector space V, equivalence classes in V can be defined in the following way: two elements of V are equivalent if they only differ by a vector of W. The set of equivalence classes is itself a vector space called the *quotient space* V/W and one has $V = W \oplus (V/W)$. If a norm $\|.\|_V$ is defined on the vector space V, a norm on the quotient space may be defined by $\|\mathbf{u}\|_{V/W} = \inf_{\mathbf{v} \in \mathbf{u}} \|\mathbf{v}\|_V$ for any equivalence class $\mathbf{u}$ of V/W.
- It is usual to build integration upon *Lebesgue measure theory* which is quite technical and mathematically demanding. We just recall here that a fundamental concept is the one of *null measure set*. For all practical purpose, null measure sets on $\mathbb{R}$ are finite and denumerable sets of points (although there are non-denumerable null measure sets), *e.g.* the set of rational numbers, and null measure sets on $\mathbb{R}^n$ are denumerable sets of subsets of $\mathbb{R}^p$ with $p < n$. Two functions are equal *almost everywhere (a.e.)* if they only differ on a null measure set.
For instance $\mathcal{L}^p(\Omega)$ is the set of functions f such that $\int_\Omega |f(\mathbf{x})|^p d\mathbf{x} < \infty$ where $d\mathbf{x}$ denotes the Lebesgue measure on Ω. The sets of functions

[1]Note that if Ω is an open set we have $Int(\Omega) = \Omega$.

which are equal a.e. are equivalent classes. The set $\mathcal{N}(\Omega)$ of functions equal to zero a.e. is a subspace of $\mathcal{L}^p(\Omega)$ and one defines $L^p(\Omega) = \mathcal{L}^p(\Omega)/\mathcal{N}(\Omega)$.

L^p spaces are Banach spaces with the norm $\|f\|_{L^p} = (\int_\Omega |f(\mathbf{x})|^p d\mathbf{x})^{1/p}$. $L^\infty(\Omega)$ is the Banach space of functions bounded a.e. on Ω which has been given the norm $\|f\|_{L^\infty} = \operatorname{ess\,sup}_\Omega |f(\mathbf{x})|$ where "ess" stands for "essentially", *i.e.* the norm is the smallest value M for which $|f(\mathbf{x})| \leq M$ almost everywhere.

If Ω is the whole $\mathbb{R}^n$, the integral may be defined only locally, *i.e.* on any compact subset. $L^p_{loc}(\mathbb{R}^n)$ is the space of functions f such that $f\varphi \in L^p(\mathbb{R}^n)$ for any $\varphi \in C_0^\infty(\mathbb{R}^n)$.

One of the main result of the integration theory is the *Lebesgue dominated convergence theorem*: if $\{f_n\}$ is a sequence of functions in $L^1(\Omega)$ and if there exists a function $g \in L^1(\Omega)$ so that $|f_n(\mathbf{x})| \leq g(\mathbf{x})$ a.e., the sequence is dominated. Then, if this sequence is convergent a.e. to a function f, this function is also integrable and $\int_\Omega f_n(\mathbf{x})d\mathbf{x} \to \int_\Omega f(\mathbf{x})d\mathbf{x}$. This result is sometimes called the *Lebesgue bounded convergence theorem* if one considers a constant $M > 0$ instead of the function g.

As an easy corollary, series and integral can be swapped with confidence provided the following holds: let us consider that for every integer N, the previous sequence $f_n \in L^1(\Omega)$ is such that $|\sum_{n=1}^N f_n(\mathbf{x})| \leq g(\mathbf{x})$ a.e. Then, if $\sum_{n=1}^\infty f_n(\mathbf{x})$ is convergent a.e. to a sum function $f(\mathbf{x})$, this function is also integrable, the series $\sum_{n=1}^\infty \int_\Omega f_n(\mathbf{x})d\mathbf{x}$ is convergent and $\int_\Omega f(\mathbf{x})d\mathbf{x} = \int_\Omega \sum_{n=1}^\infty f_n(\mathbf{x})d\mathbf{x} = \sum_{n=1}^\infty \int_\Omega f_n(\mathbf{x})d\mathbf{x}$.

- *Hilbert spaces*: important examples of infinite dimensional vector spaces are the Hilbert spaces. A Hilbert space is a complete space with a scalar product (*i.e.* is a pre-Hilbert Banach space). For instance, given a domain Ω of $\mathbb{R}^n$, the space of square integrable functions $L^2(\Omega)$ (introduced in the previous section) with the scalar product $(f, g) = \int_\Omega f(\mathbf{x})\overline{g(\mathbf{x})}d\mathbf{x}$ is a Hilbert space.

  One of the important properties of the Hilbert spaces is the *Riesz representation theorem*: given any linear form f on a Hilbert space H, there exists one and only one vector $\mathbf{u}$ such that for all vectors $\mathbf{v} \in H$, one has $f(\mathbf{v}) = (\mathbf{u}, \mathbf{v})$. Hilbert spaces are therefore *reflexive spaces*, *i.e.* they are isomorphic to their dual spaces. A common abuse of notation is to identify a Hilbert space with its dual and to write $(\mathbf{u}, \mathbf{v}) =< \mathbf{u}, \mathbf{v} >$, where $\mathbf{v}$ denotes both the element of the Hilbert space in the scalar product and its corresponding linear form by the Riez theorem in the duality product.

Via the scalar product, the notion of orthogonality is available in Hilbert spaces. Given a Hilbert space H and V a Hilbert subspace, *i.e.* a subspace which is also a Hilbert space, the set $V^{\perp}$ of elements of H orthogonal to all the elements of V is also a Hilbert subspace of H and one has $H = V \oplus V^{\perp}$.

- Operators and functionals: various operations may be defined on vector spaces. Mappings from a vector space onto a vector space are usually called *operators* while mappings from a vector space onto scalars are called *functionals*. Given two Banach spaces V and W, $\mathcal{L}(V, W)$ is the set of *linear operators* from V to W, *i.e.* operators which preserve linear combinations: if $L \in \mathcal{L}(V, W)$, for all $\mathbf{v}^1, \mathbf{v}^2 \in V$ and $a, b \in \mathbb{C}$ we have $L(a\mathbf{v}^1 + b\mathbf{v}^2) = aL(\mathbf{v}^1) + bL(\mathbf{v}^2)$. The set $\mathcal{L}(V, V)$ is written $\mathcal{L}(V)$. The space V is called the *domain* $\mathrm{dom}(L)$ of the operator L and the vector space spanned by the elements of W which can be obtained by the action of the operator on an element of V is called the *range*, *image* or *codomain* $\mathrm{cod}(L)$ of L. The *kernel* or *nullspace* $\ker(L)$ is the subspace $\ker(L) = \{\mathbf{v} \in V = \mathrm{dom}(L), L(\mathbf{v}) = \mathbf{0} \in W\}$.
  A linear operator L is *bounded* if there is a constant C such that $\|L(\mathbf{v})\|_W \leq C\|\mathbf{v}\|_V$ for all $\mathbf{v} \in V$. A linear operator is continuous (*i.e.* if $\mathbf{v}^n \to \mathbf{v}$ then $L(\mathbf{v}^n) \to L(\mathbf{v})$) if and only if it is bounded.
  The subset $\mathcal{B}(V, W)$ of bounded linear operators is a normed linear space with the norm $\|L\|_{\mathcal{B}} = \sup_{\|\mathbf{v}\|_V = 1} \|L(\mathbf{v})\|_W$.
  A bounded operator is *compact* if the image of every bounded sequence contains a convergent sub-sequence. In particular, the identity map I such that $I(\mathbf{v}) = \mathbf{v}$ is not compact on an infinite dimensional space.
  If the image of a linear operator has a finite dimension, the operator is a *finite rank operator*. All the finite rank operators are compact.
  Linear operators in $\mathcal{L}(V, W)$ can be combined linearly by just taking the linear combination of their action on the resulting vectors which gives a structure of vector space to $\mathcal{L}(V, W)$ itself. The *composite* of two operators L, M in $\mathcal{L}(V)$ is given by their successive applications: $L \circ M : \mathbf{v} \to L(M(\mathbf{v}))$. This can be viewed as a product which makes $\mathcal{L}(V)$ an algebra.
  In functional spaces, important linear operators are obtained by combinations of multiplications by functions and partial derivatives. Parenthesis are often omitted when expressing the action of an operator and one writes for instance $L\mathbf{v}$ instead of $L(\mathbf{v})$.
- A *matrix* is a linear operator A between finite dimensional vector spaces

which can be represented by a rectangular array of scalars $[a_{ij}]$ (called the elements of the matrix) so that $(A\mathbf{v})_i = \sum_j a_{ij}v_j$. Note that in the context of matrix algebra, one is often not very careful about upper and lower indices. The multiplication by a scalar and the addition of matrices are obvious from the vector space structure of linear operators. The composite of two matrices is given by their successive applications: $C\mathbf{v} = AB\mathbf{v}$ and the corresponding array for C is given by $[c_{ij}] = [\sum_k a_{ik}b_{kj}]$ called the *matrix product* which is not commutative. If the domain and image vector spaces have the same dimension, the matrix is a *square matrix*. A square matrix I such that $AI = IA = A$ for all the square matrices A of the same dimensions is a *unit matrix*. The *trace* $Tr(A)$ of a square matrix $A = [a_{ij}]$ is the sum of its diagonal elements: $Tr(A) = \sum_i a_{ii}$. If the columns of elements of a square matrix A are considered as a set of n vectors of dimension n (they are in fact the $A\mathbf{e}_i$ vectors), a scalar can be built from the matrix via a n-linear form. One considers the totally skew-symmetric n-linear form[2] which gives 1. when it is fed with the columns of the unit matrix, *i.e.* the $\mathbf{e}_i$. The scalar resulting from the action of this form on the columns of the matrix A is called the *determinant* $det(A)$. If the determinant of a matrix A is not equal to zero, the matrix is *regular* and there exists a matrix A^{-1} called the *inverse matrix* of A such that $A^{-1}A = AA^{-1} = I$ and else the matrix is *singular*.
A vector $\mathbf{v}$ can be considered as a single column matrix and the action of a matrix on such a vector is merely a matrix product.

- *Fourier transformation*: the Fourier transformation $\mathcal{F}$ is an extremely important linear operator which can be defined on $L^2(\mathbb{R}^n)$.
  Given a function $f \in L^2(\mathbb{R}^n) : \mathbf{x} \in \mathbb{R}^n \to \mathbb{R}$ and if the duality product on $\mathbb{R}^n$ is denoted $< \mathbf{k}, \mathbf{x} >= k_i x^i$, the Fourier transform $\mathcal{F}[f] = \hat{f}$ is given by $\hat{f}(\mathbf{k}) = \frac{1}{(2\pi)^{n/2}} \int_{\mathbb{R}^n} e^{-i<\mathbf{k},\mathbf{x}>} f(\mathbf{x}) d\mathbf{x}$.
  It is not obvious that the previous integrals exist for all functions in $L^2(\mathbb{R}^n)$. Technically, one has to start with a space where the existence is obvious, $L^1(\mathbb{R}^n)$ or $C_0^\infty(\mathbb{R}^n)$, and then one extends the operator to $L^2(\mathbb{R}^n)$.
  The inverse transform is given by:
  $f(\mathbf{x}) = \frac{1}{(2\pi)^{n/2}} \int_{\mathbb{R}^n} e^{i<\mathbf{k},\mathbf{x}>} \hat{f}(\mathbf{k}) d\mathbf{k}$.
  One of the fundamental properties of the Fourier transformation is that it makes differentiation algebraic: $\widehat{D^\alpha u}(\mathbf{k}) = \mathbf{k}^\alpha \widehat{u}(\mathbf{k})$ and $\widehat{\mathbf{x}^\alpha u}(\mathbf{k}) =$

[2]See the section below on n-covectors for further details.

$D^\alpha \widehat{u}(\mathbf{k})$ where $\mathbf{x}^\alpha = (x^1)^{\alpha_1} \cdots (x^n)^{\alpha_n}$ and $\mathbf{k}^\alpha = k_1^{\alpha_1} \cdots k_n^{\alpha_n}$.
A fundamental property is the *Parseval-Plancherel theorem*: the Fourier transform is an isometry in the L^2 norm, *i.e.* it conserves the scalar product and the norm: $(f, g)_{L^2} = (\mathcal{F}[f], \mathcal{F}[g])_{L^2}$

- *Convolution*: another useful operation is the convolution. Given two functions $f, g \in L^2(\mathbb{R}^n)$, the convolution product is defined by $f \star g(\mathbf{x}) = \int_{\mathbb{R}^n} f(\mathbf{y}) g(\mathbf{x} - \mathbf{y}) d\mathbf{y} =< f(\mathbf{y}), g(\mathbf{x} - \mathbf{y}) >_\mathbf{y}$ where the subscript $\mathbf{y}$ indicates that the integration involved in the duality product is performed along this variable.
  Surprisingly this operation is commutative and associative: $f \star g = g \star f$ and $(f \star g) \star h = f \star (g \star h) = f \star g \star h$.
  The fundamental result relating the Fourier transform and the convolution is the *Faltung theorem*: if f and g are two functions having Fourier transform and such that the convolution exists and is integrable: $\mathcal{F}[f \star g] = \mathcal{F}[f]\mathcal{F}[g]$.
- *Distribution*: the duality in infinite dimensional function space leads to a fundamental tools of mathematical physics: Distributions or generalized functions. The space of test functions $\mathcal{D}$ is defined as the set of functions infinitely differentiable on $\mathbb{R}^n$ and with a bounded support, *i.e.* $C_0^\infty(\mathbb{R}^n)$. Those functions are gentle enough to be integrated on their whole domain and differentiated everywhere as many times you like. The *dual topological space* $\mathcal{D}'$ of continuous (this requirement makes the difference with the algebraic dual), linear forms on $\mathcal{D}$ is the space of distributions. A fundamental example is the Dirac delta distribution (often improperly called Dirac delta function in physics) δ which associates to a test function $\phi \in \mathcal{D}$ its value at 0: $\delta(\phi) =< \delta, \phi >= \phi(0)$. The space of distributions is larger than the space of test functions and in a sense, contains it: on the one hand, with any test function ϕ is associated a distribution D_ϕ such that the action of this distribution on another test function χ is given by: $< D_\phi, \chi >= \int_{\mathbb{R}^n} \phi(\mathbf{x}) \chi(\mathbf{x})\, d\mathbf{x}$ and on the other hand, there is no test function associated with the Dirac distribution.
  Another important distribution is $\mathrm{vp}\{1/\|\mathbf{x}\|\}$ (where vp stands for the *Cauchy principal value*) defined by $< \mathrm{vp}\{1/\|\mathbf{x}\|\}, \varphi >= \lim_{\epsilon \to 0} \int_{\mathbb{R}^n - \{\|\mathbf{x}\| < \epsilon\}} \varphi(\mathbf{x})/\|\mathbf{x}\|\, d\mathbf{x}$.
  Nevertheless, there is a common and useful abuse of notation in physics which writes the action of the Dirac delta distribution and other singular distributions (those which do not correspond to a test function) using the integral symbol: $< \delta, \phi >= \int_{\mathbb{R}^n} \delta(\mathbf{x}) \phi(\mathbf{x}) d\mathbf{x} = \phi(0)$. In the

same spirit, one writes $< \delta_{\mathbf{y}}, \phi >= \int_{\mathbb{R}^n} \delta(\mathbf{x}-\mathbf{y})\phi(\mathbf{x})d\mathbf{x} = \phi(\mathbf{y})$.
We will often use the duality product notation to write the integral of the product of two functions f, g on their common domain of definition Ω: $< f, g >= \int_\Omega fg \, d\mathbf{x}$ in the real case and $< f, g >= \int_\Omega f\overline{g} \, d\mathbf{x}$ in the complex case.
Note that this corresponds also to the scalar product on L^2: $(f, g)_{L^2} =< f, g >$.[3]

- Operations such as derivation, Fourier transformation and convolution can be applied to distributions. As far as the Fourier transformation is concerned, not all the distributions have a transformation and one has to reduce the space of distributions by increasing the space of test functions (see here how the duality plays). The set of rapidly decreasing functions $\mathcal{S}$ is introduced, *i.e.* functions which decrease more rapidly to zero than any power of $\|\mathbf{x}\|$ when $\|\mathbf{x}\| \to \pm\infty$. The set of tempered distribution $\mathcal{S}'$ is the topological dual space of $\mathcal{S}$, *i.e.* the set of continuous linear functionals on $\mathcal{S}$ and any tempered distribution admit a Fourier transform. Operators on distributions are described by giving the resulting distribution but this one is itself described through its action on test functions φ. Therefore one has the following definitions,[4] given only in the case of test functions and distributions on $\mathbb{R}$ for the sake of simplicity:
  - Derivative of a distribution: $< \frac{dD}{dx}, \varphi >= - < D, \frac{d\varphi}{dx} >$.
    One often introduces the notation $\{dD(x)/dx\}$ to mean that the derivation is taken in the sense of the functions and not of the distributions. For instance, if a function $f : \mathbb{R} \to \mathbb{R}$ is C^∞ except in a single point $x = a$ where there is a discontinuity $\lim_{\epsilon\to 0^+} f(a+\epsilon) - \lim_{\epsilon\to 0^-} f(a+\epsilon) = disc_f(a)$ (also denoted $[f]_a$), one has $\frac{df}{dx} = \{\frac{df}{dx}\} + disc_f(a)\, \delta(x-a)$.
  - Multiplication of a distribution by a function: $< f\, D, \varphi >=< D, f\varphi >$. Yes, this seemingly trivial operation needs a definition! You can always multiply a distribution by a test function but you cannot for instance multiply a function with a discontinuity at the origin with the Dirac distribution. An important negative property of distributions is that the product of two distributions does not always exist.

[3] See the remark above on the abuse of notation due to the reflexive nature of the Hilbert spaces.

[4] which are chosen to match the definitions for functions when the distribution can be associated with a function.

- Fourier transform of a tempered distribution: $< \mathcal{F}[D], \varphi > = < D, \mathcal{F}[\varphi] >$.
  For instance, we have $\mathcal{F}[\delta(x)] = \chi_{\mathbb{R}}(k)$ (where χ_Ω is the characteristic function of the set Ω: $\chi_\Omega(x) = 1$ if $x \in \Omega$ and $\chi_\Omega(x) = 0$ if $x \notin \Omega$) and $\mathcal{F}[\mathrm{sgn(x)}] = \frac{i}{\pi}\,\mathrm{pv}\left(\frac{1}{k}\right)$ where $\mathrm{sgn}(x) = \frac{x}{|x|}$.
- Convolution of two distributions: $< D(x) \star E(x), \varphi(x) > = < D \otimes E(x,y), \varphi(x+y) >$, where the righthand duality product takes place in $\mathbb{R}^2$ with x and y variables.

Another interesting example of tempered distribution is the *Dirac comb* $\text{III}(x) = \sum_{n\in\mathbb{Z}} \delta(x-n)$ which is its own Fourier transform $\mathcal{F}[\text{III}(x)] = \text{III}(k) = \sum_{n\in\mathbb{Z}} \delta(k-n)$. Introducing this distribution in the Parseval-Plancherel theorem (extended to the duality product between $\mathcal{S}$ and $\mathcal{S}'$) together with a function $\varphi \in \mathcal{S}$ gives the *Poisson summation formula* $\sum_{n\in\mathbb{Z}} \varphi(n) = \sum_{n\in\mathbb{Z}} \widehat{\varphi}(n)$, a useful trick for convergence acceleration.

- *Sobolev spaces*: these are the Banach spaces defined for integers m, p by $W^{m,p}(\Omega) = \{u : u \in L^p(\Omega),\ D^\alpha u \in L^p(\Omega) \text{ for } |\alpha| \leq m\}$.
  The spaces $W^{m,2}(\Omega) = H^m(\Omega)$ are of particular interest because they are Hilbert spaces with the scalar product defined by $(u,v)_{H^m} = \sum_{|\alpha|\leq m} (D^\alpha u, D^\alpha v)_{L^2}$ and the corresponding norm $\|u\|^2_{H^m} = (u,u)_{H^m}$.
  It can be shown that $u \in H^m(\mathbb{R}^n)$ if and only if $(1+k^2)^{m/2}\widehat{u}(\mathbf{k}) \in L^2(\mathbb{R}^n)$ and that the norm $\|u\|'_{H^m} = (\int_{\mathbb{R}^n} (1+k^2)^{m/2}\widehat{u}(\mathbf{k})\, d\mathbf{k})^{1/2}$ with $k^2 = \sum_i (k_i)^2$ is equivalent to the first defined norm.
  The advantage of this definition is that it can be extended to any real value s and therefore allows the definition of negative and/or fractional index spaces H^s . Note that in the case of negative indices, the elements of H^s are not all in $L^2 = H^0$ and these spaces are rather distribution spaces than genuine function spaces.
  The *Rellich-Kondrachov theorem* states that we have the embeddings $H^s(\mathbb{R}^n) \subset H^t(\mathbb{R}^n)$ if $s > t$ and that the corresponding inclusion maps are compact.
  The *Sobolev lemma* states that $H^s(\mathbb{R}^n) \subset C^k(\mathbb{R}^n)$ if $s > k + n/2$.
  $H^{-s}(\mathbb{R}^n)$ is the topological dual space of $H^s(\mathbb{R}^n)$ according to the classical duality pairing corresponding to the L^2 scalar product.
  Given Ω, the space $H^s_0(\Omega)$ is the closure of $C^\infty_0(\Omega) = \mathcal{D}(\Omega)$ in $H^s(\Omega)$ (*i.e.* every element in $H^s_0(\Omega)$ is the limit according to the norm of $H^s(\Omega)$ of a sequence of functions in $C^\infty_0(\Omega)$). If $\partial\Omega$ is "regular enough", $H^s_0(\Omega)$ is the subspace of elements of $H^s(\Omega)$ equal to zero on $\partial\Omega$. In the case $\Omega = \mathbb{R}^n$, $H^s_0(\Omega)$ is the same as $H^s(\Omega)$ but it is a strict subset in the

other cases. $H^{-s}(\Omega)$ is the topological dual of $H_0^s(\Omega)$.
One has the following situation: $H_0^s(\Omega) \subset L^2(\Omega) \subset H^{-s}(\Omega)$ called a *Gelfand triplet* or a *rigged Hilbert space*. The space L^2 is called the *pivot space* and its scalar product provides the duality pairing. $H^{-s}(\Omega)$ can be without contradiction "larger" than $H_0^s(\Omega)$ (it is a distribution space) and "equal" since it is isomorphic according to the Riesz theorem. Another fundamental example of rigged Hilbert space is $\mathcal{D}(\Omega) \subset L^2(\Omega) \subset \mathcal{D}'(\Omega)$ where the "rigging" spaces are not Hilbert ones.

- *Trace theorems* answer the question of the regularity the restriction of functions (and distributions) on the boundary of a domain or another lower dimensional sub-domain. Consider Ω an open domain of $\mathbb{R}^n$ and a piecewise smooth hyper-surface Γ of co-dimension 1 (*i.e.* of dimension $n-1$) contained in $\overline{\Omega}$ (and in particular which may coincide with the boundary $\partial\Omega$). The *restriction operator* γ is introduced by $\gamma : C^\infty(\overline{\Omega}) \to C^\infty(\Gamma), \gamma(\mathbf{u}) = \mathbf{u}|_\Gamma$. This operator can be extended to a continuous operator on some Sobolev spaces: if the hyper-surface Γ is either compact or a portion of a hyper-plane, for $s > 1/2$ the operator γ can be extended to a continuous operator $\gamma : H^s(\Omega) \to H^{s-1/2}(\Gamma)$.
Considering the case $s = 1$ as an example, an alternative definition of the space $H^{1/2}(\Gamma)$ can be defined as the quotient space $H^1(\Omega)/H_0^1(\Omega)$ with the quotient norm $\|\mathbf{u}\|_{H^{1/2}(\Gamma)} = \inf_{\{\mathbf{v}\in H^1(\Omega), \gamma(\mathbf{v})=\mathbf{u}\}} \|\mathbf{v}\|_{H^1(\Omega)}$. Accordingly, one has $H^1(\Omega) = H_0^1(\Omega) \oplus H^{1/2}(\Gamma)$ and the trace inequality $\|\mathbf{u}|_\Gamma\|_{H^{1/2}(\Gamma)} \leq C\|\mathbf{u}\|_{H^1(\Omega)}$.
- Given a linear operator $L \in \mathcal{L}(V, V)$, the *adjoint operator* is $L^* \in \mathcal{L}(V^*, V^*)$ such that $< L\mathbf{u}, \mathbf{v} >=< \mathbf{u}, L^*\mathbf{v} > , \forall \mathbf{u} \in V, \forall \mathbf{v} \in V^*$.
In the functional case, given a differential operator $L : C^\infty(\Omega) \to C^\infty(\Omega)$ of order m defined by $L = \sum_{|\alpha|\leq m} a_\alpha(x)D^\alpha$, the *formal adjoint* L^* is such that $< L\varphi, \chi >=< \varphi, L^*\chi > , \forall\varphi, \forall\chi \in C_0^\infty(\Omega)$. Integration by parts shows that $L^*\varphi = \sum_{|\alpha|\leq m}(-1)^{|\alpha|}D^\alpha(a_\alpha\varphi)$.
For real matrices, the adjoint of a matrix $A = [a_{ij}]$ is called the *transposed matrix* $A^T = [a_{ji}]$ where lines are written as columns. For complex matrices, the adjoint is the complex conjugate transposed or Hermitian transposed $A^H = [\overline{a_{ji}}]$. A matrix A is *symmetric* if $A = A^T$ and it is (symmetric) *definite positive* if $(\mathbf{v}, A\mathbf{v}) > 0$, $\forall \mathbf{v} \neq \mathbf{0}$. A matrix A is *Hermitian* if $A = A^H$. The transposed $\mathbf{v}^T$ of a column vector $\mathbf{v}$ is a row vector, *i.e.* a single row matrix. The matrix product $\mathbf{v}^T\mathbf{w}$ is the canonical scalar product of real finite dimensional vectors. The matrix product $\mathbf{v}\mathbf{w}^T = [v_iw_j]$ resulting in a general matrix is called the *dyadic product* of the two vectors.

An operator L is *self-adjoint* if $L = L^*$. In the case of unbounded operators, physicists often confuse self-adjoint operators with merely Hermitian (in the complex case) or symmetric (in the real case) operators or even with formally self-adjoint operators and so we do in this book. An *extension* B of an operator A of domain dom(A) is an operator with a domain dom$(B) \supset$ dom(A) such that $B|_{\text{dom}(A)} = A$ and this situation is denoted $B \supset A$. An operator is self-adjoint if $A^* = A$ but the most common situation is $A^* \supset A$ which corresponds to symmetric and Hermitian operators. We will disregard this distinction except in the following simple example. Consider A_0 an operator acting on $H_0^1([a,b])$ (*i.e.* the set of square integrable functions φ defined on the interval $[a,b]$, with a square integrable derivative, and verifying the boundary conditions $\varphi(a) = \varphi(b) = 0$), defined by $A_0\varphi = i\frac{d\varphi}{dx}$. This operator is Hermitian: $< A_0\varphi, \chi >= \int_a^b i\frac{d\varphi}{dx}\overline{\chi}dx = \int_a^b \overline{i\frac{d\chi}{dx}}dx + i[\varphi\overline{\chi}]_a^b =< \varphi, A_0^*\chi >$ with $A_0^*\chi = \frac{d\chi}{dx}$ for any function $\chi \in H^1([a,b])$ which is obtained because of the boundary conditions on φ. There are no boundary conditions needed on $\chi \in H^1([a,b]) \supset H_0^1([a,b])$ and obviously $A_0^* \supset A_0$. Given $\theta \in [0, 2\pi[$, the operator $A_\theta\varphi = i\frac{d\varphi}{dx}$ with the domain dom$(A_\theta) = H_\theta^1([a,b]) = \{\varphi \in H^1([a,b]), \varphi(a) = e^{i\theta}\varphi(b)\}$ is now considered, the boundary term becomes $i(\varphi(b)\overline{\chi(b)} - \varphi(a)\overline{\chi(a)}) = i\varphi(b)(\overline{\chi(b)} - \overline{e^{-i\theta}\chi(a)})$. This term vanishes if and only if $\chi(a) = e^{i\theta}\chi(b)$, *i.e.* $\chi \in H_\theta^1([a,b])$ and $A_\theta^* = A_\theta$ is a self-adjoint operator. This is in fact a one-parameter family of self-adjoint operators with $A_0 \supset A_\theta \supset A_0^*$ but with $A_\theta \neq A_{\theta'}$ for $\theta \neq \theta'$.

Choosing $\chi \in C_0^\infty$ in the duality product avoids any question about the boundary condition and leads to the careless concept of formally self-adjoint operator A such that $< A\varphi, \chi >=< \varphi, A\chi >$ for any $\chi \in C_0^\infty$. One says that $Lf = g$ *weakly* for $f, g \in L_{loc}^1$ if $< g, \varphi >=< f, L^*\varphi >$ $, \forall\varphi \in C_0^\infty(\Omega)$.

- The *Green function* of an operator L with constant coefficients is a distribution $G(\mathbf{x}-\mathbf{y}) \in \mathcal{D}' \otimes \mathcal{D}'$ such that $L_{\mathbf{x}}^* G(\mathbf{x}-\mathbf{y}) = \delta(\mathbf{x}-\mathbf{y})$ where the fact that the operator acts on the $\mathbf{x}$ variables is emphasized. One also has to request that $G|_{\partial\Omega} = 0$ or to consider the free space Green function but we stay on a rather formal level here and let such considerations on side. The Green function is usually used to (formally) invert the differential operator. Consider once again $Lf = g$. By means of the fact that the Dirac distribution is the neutral element of the convolution, one has $f(\mathbf{x}) = f \star \delta =< f(\mathbf{y}), \delta(\mathbf{x}-\mathbf{y}) >_{\mathbf{y}} = < f(\mathbf{y}), L^*G(\mathbf{x}-\mathbf{y}) >_{\mathbf{y}}$

$=< Lf(\mathbf{y}), G(\mathbf{x}-\mathbf{y}) >_{\mathbf{y}} = < g(\mathbf{y}), G(\mathbf{x}-\mathbf{y}) >_{\mathbf{y}} = g \star G$. The formula $f = g \star G$ is only valid in free space, *i.e.* if the domain is $\mathbb{R}^n$. In the case of a bounded domain, boundary conditions have to be taken into account as it will be explained later.

- Up to now, the geometrical domains were open sets of $\mathbb{R}^n$. A serious treatment of the geometrical framework of physics requires the concept of *manifold*. A *manifold* M is a set of points which is locally homeomorphic to $\mathbb{R}^n$ in the sense that any neighbourhood of a point can be continuously mapped on an open set of $\mathbb{R}^n$ (n is the same for all the points of M and is called the *dimension* of the manifold). With such a mapping, the points in the neighbourhood can be distinguished by an ordered set of n real numbers called the *(local) co-ordinates*. Nevertheless we can not hope to be always able to find a set of co-ordinates which covers the whole manifold at once. Therefore, it is allowed to cover the manifold with several overlapping open sets each endowed with a particular co-ordinate system. The regularity of the manifold is given by the regularity of the so-called *transition functions*: considering two co-ordinate systems defined on overlapping opens sets U and $\tilde{U}$ by the mappings $\varphi : U \to \mathbb{R}^n$ and $\tilde{\varphi} : \tilde{U} \to \mathbb{R}^n$, the invertibility of the continuous co-ordinate mappings induces a transition mapping $\tilde{\varphi}\varphi^{-1} : \varphi(U \cap \tilde{U}) \subset \mathbb{R}^n \to \tilde{\varphi}(U \cap \tilde{U}) \subset \mathbb{R}^n$. the regularity of the manifold is the one of the transition functions: a differentiable C^m manifold is such that all the transition functions are C^m.
  The $\mathbb{R}^n$ are trivial examples of manifolds and in fact the only manifold used here is $\mathbb{R}^3$. In this case, why bother about manifolds? Because $\mathbb{R}^n$ can also be considered as a vector space, for instance. On the one hand, in "vector space $\mathbb{R}^n$", two elements can be added but not in "manifold $\mathbb{R}^n$" and on the other hand, changing all the n-uples $\{x^1, \cdots, x^n\}$ in "manifold $\mathbb{R}^n$" to n-uples $\{y^1 = (x^1)^3, \cdots, y^n = (x^n)^3\}$ is a valid global change of co-ordinates leaving the manifold unchanged but makes no sense in "vector space $\mathbb{R}^n$"!
- The position of a point in a manifold M of dimension n is given by an ordered set of n numbers $(x^1, \cdots, x^n)$ called the co-ordinates. Each co-ordinate can also be viewed as a function on the manifold.

A *curve* γ is an application from an interval of $\mathbb{R}$ on the manifold M : $\mathbf{r}(t) = (x^1(t), \cdots, x^n(t))$ where t is the parameter. If $f(x^1, \cdots, x^n)$ is a scalar function on the space, the composition of this function with the curve gives a function from $\mathbb{R}$ to $\mathbb{R}$: $f(x^1(t), \cdots, x^n(t))$.

The derivation with respect to t of this function gives, applying the chain rule: $\frac{d}{dt} f(x^1(t), \cdots, x^n(t)) = \frac{\partial f}{\partial x^1} \frac{dx^1}{dt} + \cdots + \frac{\partial f}{\partial x^n} \frac{dx^n}{dt}$. This expression can be viewed as the duality product $< df, \mathbf{v}_\gamma >$ of the covector $df = \frac{\partial f}{\partial x^1} dx^1 + \cdots + \frac{\partial f}{\partial x^n} dx^n = \frac{\partial f}{\partial x^i} dx^i$ with the vector $\mathbf{v}_\gamma = \frac{dx^1}{dt} \frac{\partial .}{\partial x^1} + \cdots + \frac{dx^n}{dt} \frac{\partial .}{\partial x^n} = \frac{dx^i}{dt} \frac{\partial .}{\partial x^i}$. The vector $\mathbf{v}_\gamma$ is the tangent to the curve given in the form of a first order linear differential operator and the covector df is the *differential* of the function. One takes the general definitions:

- A *vector* (in the geometric sense), at a point of the manifold, is a first order linear differential operator on the functions on the manifold. In a co-ordinate system $(x^1, \cdots, x^n)$, a vector $\mathbf{v}$ at a point of co-ordinates $(p^1, \cdots, p^n)$ is represented by a set of n numbers $(v^1, \cdots, v^n)$, the *components* of the vector, so that the result of the action of this vector on a function $f(x^1, \cdots, x^n)$ is the scalar: $\mathbf{v}(f) = v^1 \frac{\partial f}{\partial x^1} + \cdots + v^n \frac{\partial f}{\partial x^n}|_{(x^1=p^1, \cdots, x^n=p^n)}$.
- In a co-ordinate system, a basis for the vectors are the partial derivatives with respect to the co-ordinates so that a vector can be written $\mathbf{v} = v^1 \frac{\partial}{\partial x^1} + \cdots + v^n \frac{\partial}{\partial x^n}$.
- A *covector*, at a point of the space, is a linear form on the vectors at this point.
- In a co-ordinate system, a basis for the covectors are the differential of the co-ordinates so that a covector can be written $\alpha = \alpha_1 dx^1 + \cdots + \alpha_n dx^n$. They form a basis dual to the partial derivatives: $< dx^i, \frac{\partial .}{\partial x^j} >= \delta^i_j$ and therefore $< \mathbf{v}, \alpha >= \alpha_i v^i$.
- A *(co)vector field* is a set of (co)vectors so that with each point of the manifold is associated a (co)vector. In a co-ordinate system, it is represented by a set of n functions on the co-ordinates. The result of the action of a covector field on a vector field is a scalar function on the co-ordinates.
- A covector field is also called a *1-form*.
- The differential of a scalar function is a 1-form.
- In older terminology, vectors were called *contravariant vectors* and covectors were called *covariant vectors*.
- Vector fields and 1-forms have both n components, *i.e.* they can be represented (at least locally) by sets of n functions of the co-ordinates. There is a strong temptation to they say that the sets of vector fields and 1-forms are n-dimensional vector spaces but this forgets the fact that the components are functions that are

themselves elements of infinite dimensional functional spaces.

- Geometrical tensor spaces are generated by tensor products of vector and covectors. For example, $A = A^i_{jk} dx^i \otimes \frac{\partial \cdot}{\partial x^j} \otimes \frac{\partial \cdot}{\partial x^k}$ is a rank 3 tensor, once contravariant and twice covariant and the A^i_{jk} are its n^3 components in the x^i co-ordinate system. A tensor field is a set of tensors so that with each point of the manifold is associated a tensor. A rank k ($k \in \mathbb{N}$) tensor field has n^k components.
- Skew-symmetric tensors play a fundamental role in differential geometry. A *k-covector* ω is a totally skew-symmetric tensor of rank k ($k \in \mathbb{N}$), *i.e.* it is a k-linear form on vectors such that the swapping of two vectors changes the sign of the resulting scalar: $\omega(\cdots, \mathbf{v}^i, \cdots, \mathbf{v}^j, \cdots) = -\omega(\cdots, \mathbf{v}^j, \cdots, \mathbf{v}^i, \cdots)$. Note that k-covectors are identically zero if $k > n$.
  Given $\{1, \cdots, n\}$, the set of integers from 1 to n, a *permutation* is a bijection $\sigma \in \mathcal{P}_n : \{1, \cdots, n\} \to \{1, \cdots, n\}$. A *transposition* is a permutation which swaps two elements and leaves the others at the same place. Any permutation can be decomposed in a finite sequence of transpositions. The *signature* $\varepsilon(\sigma)$ of a permutation is the number equal to 1 if the number of transposition is even and to -1 if it is odd. It does not depend of course on the particular set of transpositions used to describe the permutation. The *Levi-Civita symbol* $\varepsilon_{i_1 i_2 \cdots i_n}$ is equal to $\varepsilon(\sigma)$ if the indices are a permutation σ of $\{1, \cdots, n\}$: $\{i_1 = \sigma(1), \cdots, i_n = \sigma(n)\}$ and equal to 0 if some indices are repeated. The general properties of k-covectors is $\omega(\mathbf{v}^1, \cdots, \mathbf{v}^k) = \varepsilon(\sigma)\omega(\mathbf{v}^{\sigma(1)}, \cdots, \mathbf{v}^{\sigma(k)})$.
  The dimension of the vector space formed by the k-covectors on a vector space of dimension n is $\binom{n}{k} = \frac{n!}{k!(n-k)!}$.
  A *k-form* is a field of k-covectors so that it is a map from the sets of k vector fields on M to the scalar functions on M. Note that k-forms are usually called *differential forms* or *exterior forms*. The vector space of k-forms on a manifold M is denoted $\bigwedge^k(M)$.
  The *exterior product* of $\alpha \in \bigwedge^k(M)$ and $\beta \in \bigwedge^j(M)$ is $\alpha \wedge \beta \in \bigwedge^{k+j}(M)$ such that

$$\alpha \wedge \beta(\mathbf{v}^1, \cdots, \mathbf{v}^{k+j}) =$$
$$\frac{1}{k!j!} \sum_{\sigma \in \mathcal{P}_{k+j}} \varepsilon(\sigma)\alpha(\mathbf{v}^{\sigma(1)}, \cdots, \mathbf{v}^{\sigma(k)})\beta(\mathbf{v}^{\sigma(k+1)}, \cdots, \mathbf{v}^{\sigma(k+j)}).$$

  The main properties of the exterior product of forms of are:

  $\alpha \wedge \beta = (-1)^{jk} \beta \wedge \alpha, \quad \forall \alpha \in \bigwedge^k(M), \beta \in \bigwedge^j(M)$.
  $\alpha \wedge (\beta + \gamma) = \alpha \wedge \beta + \alpha \wedge \gamma$, for all forms α, β, γ.

$(\alpha \wedge \beta) \wedge \gamma = \alpha \wedge (\beta \wedge \gamma) = \alpha \wedge \beta \wedge \gamma$, for all forms α, β, γ.

The exterior products of co-ordinate differentials form a basis for the forms so that any k-form α can be written $\alpha = \alpha(x^1, \cdots, x^n)_{i_1 \cdots i_k} dx^{i_1} \wedge \cdots \wedge dx^{i_k}$.

- n-forms have a single component and can be written in the form $f(x^1, \cdots, x^n) dx^1 \wedge \cdots \wedge dx^n$. If f is everywhere different from zero on the manifold, the n-form is called a *volume form*. Given a n-form ω, n vector fields $\mathbf{v}^i$ and a matrix A with constant elements acting on the vector fields as a linear operator, we have $\omega(A\mathbf{v}^1, \cdots, A\mathbf{v}^n) = det(A)\omega(\mathbf{v}^1, \cdots, \mathbf{v}^n)$ which gives the geometrical meaning of the determinant.
- Another fundamental operation is the *exterior derivative* of a form defined by:

$$d\alpha = d\alpha_{i_1 \cdots i_k} \wedge dx^{i_1} \wedge \cdots \wedge dx^{i_k} = \frac{\partial \alpha_{i_1 \cdots i_k}}{\partial x^i} dx^i \wedge dx^{i_1} \wedge \cdots \wedge dx^{i_k}.$$

In the righthand member, $d\alpha_{i_1 \cdots i_k}$ denotes the differential of the component $\alpha_{i_1 \cdots i_k}$ (these coefficients are functions of the co-ordinates) and the implicit summation on repeated indices is used. From this definition, it is clear that $d : \bigwedge^k(M) \to \bigwedge^{k+1}(M)$.
The exterior derivative of a function is its differential, the exterior derivative of a n-form is zero. The main properties of the exterior derivative of forms of are:

d is linear.
$d(\alpha \wedge \beta) = (d\alpha) \wedge \beta + (-1)^k \alpha \wedge d\beta, \;\; \forall \alpha \in \bigwedge^k(M)$ (*Leibnitz rule*).
$dd\alpha = 0$, for all forms α.

- Given an open set Σ of dimension $p \leq n$ in a manifold M of dimension n, sufficiently regular, and, for the simplicity of the presentation, which can be covered by a single local co-ordinate system $\{x^1, \cdots, x^n\}$ so that Σ can be parameterized by Σ: $\{x^1(\xi^1, \cdots, \xi^p), \cdots, x^n(\xi^1, \cdots, \xi^p), \forall\{\xi^1, \cdots, \xi^p\} \in \Omega \subset \mathbb{R}^p\}$. Depending on the context, Σ can be in fact a *submanifold* (*i.e.* a manifold of dimension p together with a regular map of this manifold into M), a hyper-surface or a chain which correspond to different objects from a formal point of view.[5] The *integral* $\int_\Sigma \alpha$ of a p-form α on Σ is defined

[5]The case $p = 1$ corresponds to a curve.

by:

$$\int_{\Sigma} \alpha = \int \cdots \int_{\Omega} \alpha_{i_1 \cdots i_p} \det \left(\frac{\partial(x^{i_1}, \cdots, x^{i_p})}{\partial(\xi^1, \cdots, \xi^p)} \right) d\xi^1 \cdots d\xi^p,$$

where the indices $i_1, \cdots, i_p$ are running between 1 and n (without repetition), $\det \left(\frac{\partial(x^{i_1}, \cdots, x^{i_p})}{\partial(\xi^1, \cdots, \xi^p)} \right)$ are the Jacobians (*i.e.* the determinants of the matrices whose elements are the partial derivatives of the x^i with respect to the ξ^j) and $d\xi^1 \cdots d\xi^p$ the Lebesgue measure on $\mathbb{R}^p$. This definition is in fact independent of the system of co-ordinates and of the parametrization used in the definition. The extension to the case where several local co-ordinate systems are necessary is purely technical and is based on the use of a partition of the unity.
The integration of a form is a linear operation. In order to emphasize the duality between the p-dimensional Σ and the p-forms, the abstract notation: $< \Sigma, \alpha >= \int_{\Sigma} \alpha$ can be used.
The main property is the *Stokes theorem*. Given a $(p-1)$-form α and an open set Σ of dimension p such that its boundary $\partial\Sigma$ (of dimension $p-1$) is regular enough, we have:

$$\int_{\Sigma} d\alpha = \int_{\partial\Sigma} \alpha.$$

- The marriage of differential forms and distribution theory leads to the concept of *de Rham current*. Firstly, a test p-form on a smooth manifold M (of dimension n) is a p-form the coefficients of which are $C_0^{\infty}(M)$ functions, *i.e.* infinitely differentiable and with a compact support. A p-current C is a continuous linear form on the test $(n-p)$-forms φ. The associated duality product is denoted $< C, \varphi >$. The set of currents is therefore the topological dual of the space of test forms.
  With a p-form α, is associated a p-current C_{α}, denoted α by abuse of notation, such that $< C_{\alpha}, \varphi >= \int_M \alpha \wedge \varphi$, for all test $(n-p)$-forms φ.
  The lower dimensional submanifolds or hyper-surfaces give currents similar to the singular distributions:[6] with a $(n-p)$-dimensional submanifold Σ is associated a p-current C_{Σ}, denoted Σ by abuse of notation, such that $< C_{\Sigma}, \varphi >= \int_{\Sigma} \varphi$, for all test $(n-p)$-forms φ.
  The product of a p-current C by a C^{∞} q-form α (not necessarily with a bounded support) is a $(p+q)$-current defined by: $< C \wedge \alpha, \varphi > = < C, \alpha \wedge \varphi >$, for all test $(n-p-q)$-form φ, and $C \wedge \alpha = (-1)^{pq} \alpha \wedge C$.

[6] A first technical step is to consider formal linear combinations of those geometrical objects in order to give them a vector space structure.

For instance, with a pair (Σ, α) where Σ is a $(n-p)$-dimensional submanifold and α is a $C^\infty(\Sigma)$ q-form (which needs only be defined on the support of Σ), is associated a $(p+q)$-current $C_{\Sigma\wedge\alpha}$, denoted $\Sigma \wedge \alpha$ by abuse of notation, such that $< C_{\Sigma\wedge\alpha}, \varphi >= \int_\Sigma \alpha \wedge \varphi$, for all test $(n-p-q)$-forms φ.
The exterior derivative of a p-current C is defined by $< dC, \varphi >= (-1)^{p-1} < C, d\varphi >$, for all test $(n-p-1)$-forms φ. If the current is associated with a form, the definition coincides with the former definition of the exterior derivative. In the case of a p-current associated with a $(n-p)$-dimensional manifold Σ, the previous definition via the Stokes theorem gives: $< d\Sigma, \varphi >= (-1)^{p-1} < \Sigma, d\varphi >= (-1)^{p-1} < \partial\Sigma, \varphi >$ hence $d\Sigma = (-1)^{p-1}\partial\Sigma$. For the p-current C and the C^∞ q-form α: $d(C \wedge \alpha) = dC \wedge \alpha + (-1)^p C \wedge d\alpha$.
If Σ is a $(n-1)$-dimensional submanifold (and its associated 1-current) and if ω is a p-form discontinuous on Σ, *i.e.* the components in any co-ordinate system are differentiable in the complement of Σ in M except across Σ where they suffer a jump $[\omega]_\Sigma$, then[7] $d\omega = \{d\omega\} + \Sigma \wedge [\omega]_\Sigma$.

- All those geometrical notions, the exterior product, the exterior derivative, and the integration of a form, do not rely on the definition of a scalar product or a norm and are therefore purely topological and differential but not metric.
- **Riemmanian spaces**: a scalar product on the tangent vectors of a manifold can be defined as a rank 2 totally covariant symmetric tensor (field) $\mathbf{g}$ called the *metric*. In a co-ordinate system, this tensor can be written $\mathbf{g} = g_{ij} dx^i \otimes dx^j$ where $g_{ij} = g_{ji}$ (the n^2 coefficients form a positive definite matrix with $n(n-1)/2$ independent values) and the scalar product of two vectors can be written $(\mathbf{v}, \mathbf{w}) = g_{ij} v^i w^j$
If the coefficients g_{ij} are considered to form a matrix, the coefficients of the inverse matrix are denoted g^{ij} (with $g^{ik} g_{kj} = \delta^i_j$) and define a rank 2 totally contravariant symmetric tensor $g^{ij} \frac{\partial}{\partial x^i} \otimes \frac{\partial}{\partial x^j}$.
In the context of differential forms, the metric is mostly involved in the *Hodge star operator* $* : \bigwedge^p(M) \to \bigwedge^{(n-p)}(M)$ which maps p-forms on $(n-p)$-forms. The p- and $(n-p)$-covectors have the same number of components and the map is linear, one-one and $** = (-1)^{p(n-p)}$. For any p-form α expressed in an arbitrary (covector) basis $\{\mathbf{e}^{i_1}, \cdots, , \mathbf{e}^{i_n}\}$

[7] See the derivation of a discontinuous function in the section on the distributions above for the notation $\{d\omega\}$.

by $\alpha_{j_1\cdots j_p}\mathbf{e}^{i_1}\wedge\cdots\wedge\mathbf{e}^{i_p}$, the action of the Hodge operator $*$ is given by:

$$*(\alpha_{j_1\cdots j_p}\mathbf{e}^{i_1}\wedge\cdots\wedge\mathbf{e}^{i_p}) =$$
$$\frac{1}{(n-p)!}\varepsilon_{i_1\cdots i_n}|\det[g_{ij}]|^{1/2}\alpha_{j_1\cdots j_p}g^{i_1j_1}\cdots g^{i_pj_p}\mathbf{e}^{i_{p+1}}\wedge\ \cdots\wedge\mathbf{e}^{i_n}.$$

The Hodge operator allows the definition of a scalar product on the vector spaces $\bigwedge^p(M)$ of p-forms which makes them Hilbert spaces[8] by setting:

$$(\alpha,\beta) = \int_M \alpha\wedge *\beta.$$

The *coderivative* $\delta = (-1)^{n(p+1)+1} * d*$ is the formal adjoint of the exterior derivative[9] since one has $(d\alpha,\beta) = (\alpha,\delta\beta)$ for forms with $C_0^\infty(M)$ components. One has also $\delta\delta = 0$.
The *Laplace-Beltrami operator* Δ (*Laplacian* for short) is defined by $\Delta = (d+\delta)(d+\delta) = \delta d + d\delta$ and is self-adjoint.

- Of course, many of the previous operations are simpler and indeed even trivial in the *three-dimensional Euclidean space* $\mathbb{E}^3$ that is $\mathbb{R}^3$. This one is considered as a manifold and is equipped with a special metric such that there exist global co-ordinates called *Cartesian co-ordinates* $\{x^1 = x, x^2 = y, x^3 = z\}$ where the metric has the form: $\mathbf{g} = dx\otimes dx + dy\otimes dy + dz\otimes dz$. In these co-ordinates, the Hodge operator has the following action:
$*dx = dy\wedge dz$, $*dy = dz\wedge dx$, $*dz = dx\wedge dy$,
$*(dx\wedge dy) = dz$, $*(dz\wedge dx) = dy$, $*(dy\wedge dz) = dx$,
$*1 = dx\wedge dy\wedge dz$, $*(dx\wedge dy\wedge dz) = 1$.
- There are in fact several mathematical structures which can be considered as "natural" descriptions of our three-dimensional perception of "space". $\mathbb{R}^3$ as a bare manifold obviously lacks structures but defining Cartesian co-ordinates is too arbitrary since it involves, for instance, the choice of a distinguished point, the origin O of co-ordinates $\{0,0,0\}$. Another candidate is the "vector space $\mathbb{R}^3$" which we note $\mathbb{V}^3$ to make the distinction with the manifold. The addition of two points is now a valid but meaningless operation. A sounder choice is to consider $\mathbb{A}^3$, the affine space associated with $\mathbb{V}^3$ (loosely speaking obtained by forgetting the origin). The physical points are elements of $\mathbb{A}^3$, they can

[8] The situation is in fact more subtle since it depends on the regularity of the components of the forms as functions of the co-ordinates but we consider here that they are all in $L^2(M)$.

[9] which could have then been denoted d^*.

not be added but their differences are vectors of $\mathbb{V}^3$ and the elements of $\mathbb{V}^3$ operate on points of $\mathbb{A}^3$ as displacements. The introduction of the Euclidean distance gives the affine Euclidean space which we still call $\mathbb{E}^3$ and where displacements (and not points) have a length. The distance between two points is then defined. Of course, in practice, an origin point and three mutually orthogonal unit vectors are chosen such that they define a particular Cartesian co-ordinate system.

- The peculiarities of $\mathbb{E}^3$ allow a simpler setting called *vector analysis* which takes advantage of the Cartesian co-ordinates but forget almost all the geometric relevance! We do not want advocate the giving up of vector analysis but we just want it considered as a computational trick rather than a genuine geometrical framework. In $\mathbb{E}^3$, 0-forms are scalar functions $v(x, y, z)$ and 3-forms are pseudo-scalars or *densities* $\rho(x, y, z)dx \wedge dy \wedge dz = \rho * 1$. Both fields have only one single component so that they are merged into the single concept of *scalar field* using the Hodge operator. Similarly, the 1-forms $\alpha_x dx + \alpha_y dy + \alpha_z dz$ are the *field intensities* and the 2-forms $\beta_x dy \wedge dz + \beta_y dz \wedge dx + \beta_z dx \wedge dy$ are the *flux densities* and both have three components so that they are merged in the concept of *vector field*. In this case, those vector fields are considered as *proxies* for 1-forms and 2-forms.
  A vector field is written $\mathbf{v} = v_x \mathbf{e}^x + v_y \mathbf{e}^y + v_y \mathbf{e}^y$ where the unit vectors $\mathbf{e}$ are as well unit 1-forms as unit 2-forms associated with Cartesian co-ordinates. This is safe as long as only Cartesian co-ordinates are used. The scalar product of two vectors is called the *dot product* and is defined by: $\mathbf{v} \cdot \mathbf{w} = v_x w_x + v_y w_y + v_z w_z$ and the associated norm is of course $|\mathbf{v}|^2 = \mathbf{v} \cdot \mathbf{v}$. Dot product can be traced back in differential geometry as the scalar product of two 1-forms or of two 2-forms but also to the metric free exterior product of a 1-form and a 2-form. This is usually enough to cloud the geometrical meaning of vector analysis computations! The *cross product* of two vectors is defined by: $\mathbf{v} \times \mathbf{w} = (v_y w_z - v_z w_y)\mathbf{e}^x + (v_z w_x - v_x w_z)\mathbf{e}^y + (v_x w_y - v_y w_x)\mathbf{e}^z$. It comes mostly from the exterior product of two 1-forms.[10]
  The exterior derivative gives rise to several operators.
  The exterior derivative of a 0-form $df = \frac{\partial f}{\partial x}dx + \frac{\partial f}{\partial y}dy + \frac{\partial f}{\partial z}dz$ corresponds to the *gradient* of a scalar field: $\mathbf{grad}\, f = \frac{\partial f}{\partial x}\mathbf{e}^x + \frac{\partial f}{\partial y}\mathbf{e}^y + \frac{\partial f}{\partial z}\mathbf{e}^z$.

[10] There are many other features in differential geometry not introduced here such as the Lie derivative and the inner product. In electromagnetism, the cross product of the velocity together with the magnetic flux density in the Lorentz force is in fact the inner product of a vector with a 2-form.

For a 1-form $\mathbf{v} = v_x dx + v_y dy + v_z dz$, the exterior derivative $d\mathbf{v} = (\frac{\partial v_z}{\partial y} - \frac{\partial v_y}{\partial z}) dy \wedge dz + (\frac{\partial v_x}{\partial z} - \frac{\partial v_z}{\partial x}) dz \wedge dx + (\frac{\partial v_y}{\partial x} - \frac{\partial v_x}{\partial y}) dx \wedge dy$ corresponds to the *curl* of a vector field: $\mathbf{curl}\,\mathbf{v} = (\frac{\partial v_z}{\partial y} - \frac{\partial v_y}{\partial z})\mathbf{e}^x + (\frac{\partial v_x}{\partial z} - \frac{\partial v_z}{\partial x})\mathbf{e}^y + (\frac{\partial v_y}{\partial x} - \frac{\partial v_x}{\partial y})\mathbf{e}^z$.
For a 2-form $\mathbf{w} = w_x dy \wedge dz + w_y dz \wedge dx + w_z dx \wedge dy$, the exterior derivative $d\mathbf{w} = (\frac{\partial w_x}{\partial w} + \frac{\partial w_y}{\partial y} + \frac{\partial w_z}{\partial z}) dx \wedge dy \wedge dz$ corresponds to the *divergence* of a vector field: $\text{div}\ \mathbf{w} = \frac{\partial w_x}{\partial w} + \frac{\partial w_y}{\partial y} + \frac{\partial w_z}{\partial z}$.
Alternate notations for these operations use the nabla operator ∇: $\mathbf{grad}\, f = \nabla f$, $\mathbf{curl}\,\mathbf{v} = \nabla \times \mathbf{v}$, and $\text{div}\ \mathbf{v} = \nabla \cdot \mathbf{v}$.
The Leibniz rule is expressed in the classical formulae:

$\mathbf{grad}(fg) = f\,\mathbf{grad}\, g + g\ \mathbf{grad}\, f.$
$\mathbf{curl}(f\mathbf{v}) = f\,\mathbf{curl}\,\mathbf{v} - \mathbf{v} \times \mathbf{grad}\, f.$
$\text{div}(f\mathbf{v}) = f\,\text{div}\,\mathbf{v} + \mathbf{v} \cdot \mathbf{grad}\, f.$
$\text{div}(\mathbf{v} \times \mathbf{w}) = \mathbf{w} \cdot \mathbf{curl}\,\mathbf{v} - \mathbf{v} \cdot \mathbf{curl}\,\mathbf{w}.$

and of course $dd = 0$ is nothing else than $\mathbf{curl}\ \mathbf{grad} = \mathbf{0}$ and $\text{div}\ \mathbf{curl} = 0$.
The Laplacian of a scalar field is $\Delta f = \text{div}\ \mathbf{grad}\, f = \frac{\partial^2 f}{\partial x^2} + \frac{\partial^2 f}{\partial y^2} + \frac{\partial^2 f}{\partial z^2}$.
The Laplacian of a vector field is $\Delta\mathbf{v} = \mathbf{grad}\ \text{div}\,\mathbf{v} - \mathbf{curl}\,\mathbf{curl}\,\mathbf{v} = (\Delta v_x)\mathbf{e}^x + (\Delta v_y)\mathbf{e}^y + (\Delta v_z)\mathbf{e}^z$.
Note that Laplacians are self-adjoint operators.
In vector analysis, a (rank 2) tensor (field) is a linear operator transforming a vector (field) into a vector (field). In a particular system of co-ordinates, it is usually given in the form of an array of 9 coefficients hence the common confusion with a square matrix. A "vector analysis" tensor is denoted $\underline{\underline{\alpha}}$ and its action on a vector $\mathbf{v}$ is simply denoted $\underline{\underline{\alpha}}\mathbf{v}$.
The Stokes theorem corresponds to various integral equalities.
If Γ is a curve with initial point $\mathbf{a}$ and end point $\mathbf{b}$, for any scalar field whose gradient exists: $\int_\Gamma \mathbf{grad}\, f \cdot d\mathbf{l} = f(\mathbf{b}) - f(\mathbf{a})$ where $d\mathbf{l}$ is the line element.
If Σ is a surface with boundary $\partial\Sigma$ (with an orientation inherited from Σ), for any vector field $\mathbf{v}$ whose curl exists : $\iint_\Sigma \mathbf{curl}\,\mathbf{v} \cdot \mathbf{n}\, ds = \int_{\partial\Sigma} \mathbf{v} \cdot d\mathbf{l}$ where ds is the surface element and $\mathbf{n}$ the normal vector on Σ.
If V is a volume with boundary ∂V, for any vector field $\mathbf{v}$ whose divergence exists : $\iiint_V \text{div}\ \mathbf{v}\, d\mathbf{v} = \iint_{\partial V} \mathbf{v} \cdot \mathbf{n}\, ds$ where $d\mathbf{v}$ is the volume element and $\mathbf{n}$ the outer normal vector on ∂V.
Often, multiple integrals are simply denoted by a single $\int$ just as in differential geometry.
The simplest way to give natural definitions of all the integrals involved

here above is to go back to the definition of the integration of differential forms. *Moreover, the traditional notations of vector analysis rely on metric concepts such as the normal vector (involving both orthogonality and unit length) and the scalar product while the definition of these integrals do not require any metric.*
Given a unit vector $\mathbf{e}$, the *directional derivative* $\frac{\partial \cdot}{\partial \mathbf{e}}$ is defined by $\frac{\partial f}{\partial \mathbf{e}} = \mathbf{grad}\ f \cdot \mathbf{e}$. Another fundamental integral identity is the *Green formula*:

$$\iiint_V (f\Delta g - g\Delta f) d\mathbf{v} = \iint_{\partial V} (f\frac{\partial g}{\partial \mathbf{n}} - g\frac{\partial f}{\partial \mathbf{n}})\, ds\,.$$

- Using the Leibnitz rule to integrate by parts, it is easy to find the formal adjoint to the vector analysis operators: $\mathbf{grad}^* = -\operatorname{div}$, $\mathbf{curl}^* = \mathbf{curl}$, and $\operatorname{div}^* = -\mathbf{grad}$.
- The identity $d\omega = \{d\omega\} + \Sigma \wedge [\omega]_\Sigma$ for de Rham currents leads to the following identities in vector analysis in the case of a function f and a vector field $\mathbf{v}$ differentiable in the complement of a surface Σ but undergoing the jumps $[f]_\Sigma$ and $[\mathbf{v}]_\Sigma$ across Σ respectively (the direction chosen to cross the surface is given by $\mathbf{n}$, the normal vector to the surface Σ):

  $\mathbf{grad}\, f = \{\mathbf{grad}\, f\} + \mathbf{n}[f]_\Sigma \delta_\Sigma$.
  $\mathbf{curl}\, \mathbf{v} = \{\mathbf{curl}\, \mathbf{v}\} + \mathbf{n} \times [\mathbf{v}]_\Sigma \delta_\Sigma$.
  $\operatorname{div}\ \mathbf{v} = \{\operatorname{div}\ \mathbf{v}\} + \mathbf{n} \cdot [\mathbf{v}]_\Sigma \delta_\Sigma$.

  where it is necessary to denote explicitly the singular distribution δ_Σ associated with the surface Σ (by definition $\int_{\mathbb{R}} \delta_\Sigma \varphi(\mathbf{x}) d\mathbf{x} = \varphi|_\Sigma$). Note that metric concepts creep again in a place where they are not necessary.
- $L^2(\mathbb{R}^3, \mathbb{C}^3) = [L^2(\mathbb{R}^3)]^3 = \mathbb{L}^2(\mathbb{R}^3)$ denotes the space of square integrable functions on $\mathbb{R}^3$ with values in $\mathbb{C}^3$ where the scalar product is defined by: $(\mathbf{v}, \mathbf{w}) = \int_{\mathbb{R}^3} \mathbf{v}(\mathbf{x}) \cdot \overline{\mathbf{w}(\mathbf{x})} d\mathbf{x}$.
- The Green function G of three-dimensional scalar Laplacian (in free space) is such that $\Delta_\mathbf{p} G(\mathbf{p} - \mathbf{q}) = \delta(\mathbf{p} - \mathbf{q})$ where $\mathbf{p}, \mathbf{q} \in \mathbb{E}^3$ and $\Delta_\mathbf{p}$ indicates that the derivatives in the Laplacian are taken with respect to the co-ordinates of $\mathbf{p}$ (but this is seldom explicitly indicated when there is no ambiguity). The expression of this Green function is $G(\mathbf{p} - \mathbf{q}) = \frac{-1}{4\pi|\mathbf{p}-\mathbf{q}|}$. The displacement vector $\mathbf{p} - \mathbf{q}$ is traditionally called $\mathbf{r}$ and its norm r and one can be written $G = \frac{-1}{4\pi r}$.
  In the two-dimensional scalar Laplacian case, $G = \frac{1}{2\pi} \ln(r)$.
  For the scalar Helmholtz equation, one has $(\Delta + k^2)H = \delta$ and $H = -\frac{e^{ikr}}{4\pi r}$.

In the two-dimensional scalar Helmholtz equation case, $H = \frac{1}{4i} H_0^{(1)}(kr)$ where $H_0^{(1)}$ denotes a Hankel function.
The Green function is physically interpreted as the field (or potential) generated by a (monopolar) point source. The introduction of Green functions in Green formula gives identities which are the basis of the *boundary element method*.

- The differential geometry on $\mathbb{R}^2$ provides a nice framework to introduce *complex analysis* on $\mathbb{C}$. Given two Cartesian co-ordinates x and y and the Euclidean metric $\mathbf{g} = dx \otimes dx + dy \otimes dy$, the associated Hodge star operator has the following action on 1-forms: $*dx = dy$ and $*dy = -dx$. This action is a $\pi/2$ rotation counterclockwise equivalent to the action of a multiplication by i in the complex plane. The knowledge of this Hodge star on the plane does not determine the Euclidean geometry but only the conformal geometry, *i.e.* the metric up to a scalar factor so that only angles but not lengths are relevant. In the complex plane, the complex variable $z = x + iy$ and its complex conjugate $\overline{z} = x - iy$ play the role of independent variables. Their differentials $dz = dx + idy$ and $d\overline{z} = dx - idy$ are a basis for the 1-forms. The dual basis for the vectors are the differential operators $\frac{\partial}{\partial z} = \frac{1}{2}(\frac{\partial}{\partial x} - i\frac{\partial}{\partial y})$ and $\frac{\partial}{\partial \overline{z}} = \frac{1}{2}(\frac{\partial}{\partial x} + i\frac{\partial}{\partial y})$. Note that $\frac{\partial}{\partial \overline{z}}\frac{\partial}{\partial z} = \frac{\partial}{\partial z}\frac{\partial}{\partial \overline{z}} = \frac{1}{4}\Delta$ and $dz \wedge d\overline{z} = -2i(dx \wedge dy)$.
A complex function of the complex variable $f : \mathbb{C} \to \mathbb{C}$ is associated with a pair (p, q) of real functions on $\mathbb{R}^2$ so that $f(z, \overline{z}) = p(x, y) + iq(x, y)$. Those complex functions are usually denoted $f(z)$ which is fully justified in the case of holomorphic functions as explained below.
Given two complex functions f and g, the exterior derivative of the 0-form f is $df = \frac{\partial f}{\partial z}dz + \frac{\partial f}{\partial \overline{z}}d\overline{z} = (\frac{\partial p}{\partial x} + i\frac{\partial q}{\partial x})dx + (\frac{\partial p}{\partial y} + i\frac{\partial q}{\partial y})dy$ and the exterior derivative of the 1-form $f\,dz + g\,d\overline{z}$ is $d(f\,dz + g\,d\overline{z}) = (\frac{\partial g}{\partial z} - \frac{\partial f}{\partial \overline{z}})dz \wedge d\overline{z}$.
Let γ be a closed curve in $\mathbb{C}$ (and therefore also in $\mathbb{R}^2$), *i.e.* a map $\gamma : t \in [a, b] \subset \mathbb{R} \to z(t) \in \mathbb{C}$ such that $\gamma(a) = \gamma(b)$, and such that there are no other multiple points. The line integral of the complex 1-form fdz associated with f (note that it is not a general 1-form since there is no term in $d\overline{z}$) is $\int_\gamma f dz = \int_\gamma (p\,dx - q\,dy) + i\int_\gamma (p\,dy + q\,dx)$. The exterior derivative is $d(fdz) = -\frac{\partial f}{\partial \overline{z}}\, dz \wedge d\overline{z}$.
If D is a bounded domain of $\mathbb{C}$ (and therefore also of $\mathbb{R}^2$), the domain integral of a complex 2-form $f\ dz \wedge d\overline{z}$ is $\int_D f\ dz \wedge d\overline{z} = -2\int_D p\,dx \wedge dy + 2i\int_D q\,dx \wedge dy$. In this case, the Stokes theorem becomes $\int_D d(fdz) = \int_{\partial D} fdz$. As for a 1-form $f\,d\overline{z}$, one has $\int_D d(fd\overline{z}) = \int_{\partial D} fd\overline{z}$ where

$\int_\gamma f d\bar{z} = \int_\gamma (p\,dx + q\,dy) + i\int_\gamma (q\,dx - p\,dy)$.
A function $f(z)$ is *holomorphic* or *analytic* on a on a topologically simple domain (contractible to a point) $D \subset \mathbb{C}$ if one of the following five equivalent conditions are satisfied:

(1) If at each point of D, f is $\mathbb{C}$-differentiable, *i.e.* if the limit $f'(z) = \lim_{|h|\to 0} \frac{f(z+h)-f(z)}{h}$ exists and is independent of the direction of $h \in \mathbb{C}$ (Fréchet derivative). In this case $f'(z) = \frac{\partial f(z)}{\partial z} = \frac{1}{2}(\frac{\partial p}{\partial x} + \frac{\partial q}{\partial y}) - \frac{i}{2}(\frac{\partial p}{\partial y} - \frac{\partial q}{\partial x})$.
(2) If the 1-form $f dz$ is closed, *i.e.* its exterior derivative vanishes, on D: $d(f\,dz) = -\frac{\partial f}{\partial \bar{z}}\,dz = 0$ at each point of D, *i.e.* if the *Cauchy-Riemann conditions* $\frac{\partial p}{\partial x} - \frac{\partial q}{\partial y} = 0$ and $\frac{\partial p}{\partial y} + \frac{\partial q}{\partial x} = 0$ are verified. This condition means that $f(z)$ behaves as if it were independent of $\bar{z}$. The operator $\frac{\partial}{\partial \bar{z}}$ is called the Cauchy-Riemann operator and, since it is a factor of the Laplacian, a holomorphic function f is also a harmonic function, *i.e.* $\Delta f = 0$.
(3) If $df = g\,dz$, *i.e.* if the exterior derivative of f has no $d\bar{z}$ component.
(4) If the line integrals $\int_\gamma f\,dz = 0$ for any closed curve γ contained together with its interior in D. As any closed curve can be defined as a boundary $\partial\Omega$ with $\Omega \subset D$, by the Stokes theorem $\int_{\partial\Omega} f dz = \int_\Omega d(f dz) = 0$.
(5) If f can be represented by an infinite power series $f(z) = \sum_{n=0}^{\infty} a_n (z - z_0)^n$ in a neighbourhood of any point of z_0 of D.

In the case of the function $1/z$ on a domain Ω including the origin $z = 0$, this function is holomorphic everywhere on Ω except at the origin where there is a singularity called a *pole* of order 1. Such a function which is holomorphic except for a discrete set of points is called *meromorphic*. The fact that the domain in which the function is holomorphic is no more topologically simple has important consequences: consider for instance a disc $D(r, 0)$ of radius r and centre $z = 0$. Using the polar representation $z = re^{i\theta}$, $dz = ire^{i\theta}d\theta$ on $\partial D(r, 0)$ and one has, for any $n \in \mathbb{Z}$, $\int_{\partial D(r,0)} z^n dz = \int_0^{2\pi} e^{in\theta} r^n ire^{i\theta} d\theta = ir^{n+1} \int_0^{2\pi} e^{i(n+1)\theta} d\theta = 2i\pi$ if $n = -1$ and 0 if $n \neq -1$. As $1/z$ is holomorphic on any domain not including the origin, its domain integral is null and this fact can be used to easily prove that $\int_{\partial\Omega} \frac{1}{z} dz = 2i\pi$ for any Ω including the pole $z = 0$. A holomorphic function f on Ω including the origin $z = 0$ can be written as an infinite power series $f(z) = \sum_{n=0}^{\infty} a_n z^n$ where the z^n are all holomorphic functions and where $a_0 = f(0)$. Therefore

$\int_{\partial\Omega} \frac{f(z)}{z} dz = \int_{\partial\Omega} \sum_{n=0}^{\infty} a_n z^{n-1} dz = 2i\pi f(0)$. It is easy to shift the pole at any ζ inside a domain Ω where f is holomorphic to obtain the *Cauchy's integral formula* or *Cauchy theorem*:

$$f(\zeta) = \frac{1}{2i\pi} \int_{\partial\Omega} \frac{f(z)dz}{z-\zeta}.$$

If the function f is meromorphic, it can be represented by a *Laurent series* $f(z) = \sum_{n=-p}^{\infty} a_n (z-z_0)^n$ (involving negative powers of z) in a neighbourhood D of a pole z_0 of order p. The coefficient a_{-1} denoted $R(z_0; f)$ is called the *residue* and we have: $R(z_0; f) = \lim_{z \to z_0} (z - z_0) f(z) = \frac{1}{2i\pi} \int_\gamma f(z)dz$ for a curve γ such that it is contained in D, and that z_0 is the only pole contained in its interior.
Given two complex functions f and g on D, a duality product $< f, g >= \int_D f\overline{g}\; dz \wedge d\overline{z}$ can be introduced so that $-\frac{\partial}{\partial \overline{z}}$ is the formal adjoint of $\frac{\partial}{\partial z}$ since $< \frac{\partial f}{\partial z}, g >= - < f, \frac{\partial g}{\partial \overline{z}} > + \int_{\partial D} f\overline{g}\, d\overline{z}$
The Green function of $\frac{\partial}{\partial z}$ is given by $\frac{\partial}{\partial \overline{z}} \frac{1}{\pi z} = \delta$, where δ is the Dirac distribution on $\mathbb{R}^2$, and it can be justified by $\int_{\partial D} f \frac{1}{2i\pi z} dz = f(0) = \int_D f \delta \frac{dz \wedge d\overline{z}}{-2i} = \int_D f \frac{\partial g}{\partial \overline{z}} \frac{dz \wedge d\overline{z}}{-2i} = \frac{1}{2i} \int_D d(fg\; dz) = \frac{1}{2i} \int_{\partial D} fg\; dz$. The theory of hyperfunctions is a distribution theory on $\mathbb{C}$ where test functions are holomorphic.

- *Landau notation*: given x_0 and given two functions f and g defined in a neighbourhood of $x_0 \in \overline{\mathbb{R}}$ (*i.e.* real numbers including the cases $x_0 \to \pm\infty$), $f = O(g)$ if and only if $\frac{f(x)}{g(x)} = O(1)$ which means that $|\frac{f(x)}{g(x)}|$ stays bounded in a neighbourhood of x_0. Moreover, $f = o(g)$ if and only if $\frac{f(x)}{g(x)} = o(1)$ which means that $\lim_{x \to x_0} \frac{f(x)}{g(x)} = 0$. It is legitimate to write, say, $2x = O(x) = o(x^2)$ for $x \to \infty$ with the understanding that we are using the equality sign in an unsymmetrical (and informal) way, in that we do not have, for example, $o(x^2) = O(x)$.
Landau notation is a practical tool to give asymptotic behaviours, *e.g.* for $\nu \in \mathbb{R}$, $J_\nu(x) = \sqrt{\frac{2}{\pi x}} \cos(x - (2\nu + 1)\frac{\pi}{4}) + O(\frac{1}{x\sqrt{x}})$ as $x \to +\infty$.
Landau notation is also handy in computer science, *e.g.* in describing the efficiency of an algorithm. It is common to say that an algorithm requires $O(n^3)$ steps, for example, without needing to specify exactly what is a step; for if $f = O(n^3)$, then $f = O(An^3)$ for any positive constant A.
- Perfect notations probably do not exist since absolutely rigorously ones should be intractable! It is difficult to avoid ambiguities and collisions. For instance using $\star$ for the convolution, $*$ for the Hodge star operator,

and the upper index * for algebraic duals may not be appreciated by presbyopic readers. Adopting non ambiguous but non standard notations may be a solution to avoid collisions. The danger is to lose the reader in a cumbersome formal deciphering game. Therefore we prefer to leave some ambiguities which may be removed by understanding.
Moreover we leave on side with regret some important issues which are not explicitly used here such as for instance the orientation of manifolds and twisted forms, the Hodge orthogonal decomposition theorem for forms which generalizes the Helmholtz decomposition theorem for vector fields.
There is of course a huge number of books on the mathematical tools for physics but almost everything in this appendix can be found in a more detailed and more rigorous version in the formidable Ref.[CBWMDB82]. The classical reference for functional analysis is Ref. [Yos80] but a more readable book for the physicist interested in the functional analysis for partial differential equations is Ref. [Fol95] and a concise introduction to basic analysis and functional analysis is Ref. [Fri82]. One of the best reference for distribution theory including de Rham currents is still Ref. [Sch66]. There are many good books on geometrical methods in physics such as the very pedagogical Ref. [BS91] or the quite comprehensive Ref. [Nak90] but a very concise presentation aimed at electromagnetism is Ref. [Bos91].

Bibliography

B. J. Ainslie and C. R. Day. A review of single-mode fibres with modified dispersion characterictics. *J. Lightwave Technol.*, 4:967, 1986.

G. P. Agrawal. *Nonlinear fiber optics*. Academic Press, 3rd edition, 2001.

A. Argyros. Guided modes and loss in Bragg fibres. *Opt. Express*, 10(24):1411–1417, 2002.

D. N. Arnold. Differential complexes and numerical stability. *ICM*, I:137–157, 2002.

M. Abramowitz and I. A. Stegun. *Handbook of mathematical functions*. Dover Publications, New York, 9th edition, 1965.

I. M. Bassett and A. Argyros. Elimination of polarization degeneracy in round waveguides. *Opt. Express*, 10(23):1342–1346, 2002.

S. E. Barkou, J. Broeng, and A. Bjarklev. Silica-air photonic crystal fiber design that permits waveguiding by a true photonic bandgap effect. *Optics Letters*, 24(1):46–48, 1999.

A. Bjarklev, J. Broeng, and A.S. Bjarklev. *Photonic Crystal Fibres*. Kluwer Academic Publishers, 2003.

P. M. Blanchard, J. G. Burnett, G. R. G. Erry, A. H. Greenaway, P. Harrison, B. Mangan, J. C. Knight, P. S. Russell, M. J. Gander, R. McBride, and J. D. C. Jones. Two-dimensional bend sensing with a single, multi-core optical fibre. *Smart Mater. Struct.*, 9(2):132–140, 2000.

A. Blanco, E. Chomski, S. Grabtchak, M. Ibisate, S. John, S. W. Leonard, C. Lopez, F. Meseguer, H. Miguez, J. P. Mondia, G. A. Ozin, O. Toader, and H. M. van Driel. Large-scale synthesis of a silicon photonic crystal with a complete tree-dimensional bandgap near 15 micrometres. *Nature*, 405:437–440, 2000.

L. C. Botten, M. S. Craig, and R. C. McPhedran. Complex Zeros of Analytic Functions. *Comp. Phys. Comm.*, 29:245–249, 1983.

J. Berenger. A perfectly matched layer for the absorption of electromagnetic waves. *J. Comput. Phys.*, (114):185–200, 1994.

F. Brezzi and M. Fortin. *Mixed and Hybrid Finite Element Methods*. Springer-Verlag, New-York, 1991.

D. Boffi, P. Fernandes, L. Gastaldi, and I. Periugia. Computational models of electromagnetic resonators: analysis of edge elements approximation. *SIAM J. Numer. Anal.*, 36:1264–1290, 1999.

D. Boffi, L. Gastaldi, and G. Naldi. Application of Maxwell equations. Proceedings of SIMAI 2002, to appear.

A. Bossavit and L. Kettunen. Yee-like schemes on a tetrahedral mesh with diagonal lumping. *Int. J. Numer. Modell*, pages 12:129–142, 1999.

T. A. Birks, J. C. Knight, and P. J. Russel. Endlessly single-mode photonic crystal fiber. *Opt. Lett.*, 22(13):961–963, 1997.

F. Benabid, J. C. Knight, and P. S. Russell. Particle levitation and guidance in hollow-core photonic crystal fiber. *Opt. Express*, 10(21):1195–1203, 2002.

F. Brezzi and D. Marini. A survey on mixed finite element approximations. *IEEE Trans. on Magnetics*, 30:3547–3551, 1994.

F. Brechet, J. Marcou, D. Pagnoux, and P. Roy. Complete analysis of the characteristics of propagation into photonic crystal fibers by the finite element method. *Opt. Fiber Technol.*, 6(2):181–191, 2000.

D. Boffi. A note on the de Rham complex and a discrete compactness property. *Appl. Math. Letters*, 14:33–38, 2001.

A. S. Bonnet. *Analyse mathématique de la propagation de modes guidés dans les fibres optiques.* PhD thesis, Université Paris VI, 1998.

A. Bossavit. Mixed finite elements and the complex of Whitney forms. The Mathematics of Finite Elements and Applications VI:74–79, 1988.

A. Bossavit. A rationale for 'edge-elements' in 3-D fields computations. *IEEE Trans. On Magnetics*, 24(1):74–79, 1988.

A. Bossavit. Un nouveau point de vue sur les éléments mixtes. *Matapli (Bull. Soc. Math. Appl. Industr.)*, 20:23–35, 1989.

A. Bossavit. Solving Maxwell equations in a closed cavity, and the question of spurious modes. *IEEE Trans. on Magnetics*, 26(2):702–705, 1990.

A. Bossavit. *Notions de géométrie différentielle pour l'étude des courants de Foucault et des méthodes numériques en électromagnétisme*, volume Méthodes numériques en électromagnétisme (A. Bossavit, C. Emson, I. Mayergoyz). Eyrolles, Paris, 1991. see also the English translation http://www.lgep.supelec.fr/mse/perso/ab/DGSNME.pdf.

A. Bossavit. *Électromagnétisme, en vue de la modélisation, Mathématiques et Applications 14.* Springer-Verlag, Paris, 1993.

A. Bossavit. *Computational Electromagnetism, Variational formulations, Edge elements, Complementarity.* Academic Press, San Diego, 1998.

A. Bossavit. "Generalized finite differences" in computational electromagnetics. *Progress in Electromagnetics Research, PIER 32 (Special Volume on Geometrical Methods for Comp. Electromagnetics)*, pages 45–64, 2001. http://cetaweb.mit.edu/pier/pier32/02.bossavit.pdf.

A. Bossavit. Chapitre 1: Géométrie de l'électromagnétisme et éléments finis. In G. Meunier, editor, *Champs et équations en électromagnétisme: Électromagnétisme et éléments finis, Vol. 1*, Paris, 2003. Hermès-Lavoisier.

D. Braess. *Finite Elements, Theory, Fast Solvers, and Applications in Solid Mechanics.* Cambridge University Press, Cambridge, 1997.

C. G. Broyden. A class of methods for solving nonlinear simultaneous equations.

Mathematics of Computation, 19:577–593, 1965.

P. Bamberg and S. Sternberg. *A Course in Mathematics for Students of Physics: volume 1 and volume 2*. Cambridge University Press, Cambridge, 1991.

J. Broeng, T. Sondergaard, S. E. Barkou, P. M. Barbeito, and A. Bjarklev. Waveguidance by the photonic bandgap effect in optical fibres. *Journal of Optics A*, 1:477–482, 1999.

Y. Choquet-Bruhat, C. De Witt-Morette, and M. Dillard-Bleick. *Analysis, Manifolds and Physics*. Elsevier, Amsterdam, revised edition, 1982.

C-S Chang and H-C Chang. Theory of the circular harmonics expansion method for multiple-optical-fiber system. *J. Lightwave Technology*, 12:415–417, 1994.

S. Coen, A. Hing Lun Chau, R. Leonhardt, J. D. Harvey, J. C. Knight, W. J. Wadsworth, and P. S. Russell. White-light supercontinuum generation with 60-ps pump pulses in a photonic crystal fiber. *Optics Letters*, 26:1356–1358, 2001.

A. R. Chraplyvy. Limitations on lightwave communications imposed by optical-fibre nonlinearities. *J. Lightwave Technol.*, 8:1548, 1990.

R. F. Cregan, B. J. Mangan, J. C. Knight, T. A. Birks, P. S. Russell, P. J. Roberts, and D. C. Allan. Single-mode photonic band gap guidance of light in air. *Science*, 285:1537–1539, 1999.

C. Conca, J. Planchard, and M. Vanninathan. *Fluids and Periodic Structures*. Wiley-Masson, Paris, 1995.

P. Curie. Sur la symétrie des phénomènes physiques, symétrie d'un champ électrique et d'un champ magnétique. *Journal de Physique*, page 393, 1894.

M. Clemens and T. Weiland. Discrete electromagnetism with the finite integration technique. *Progress in Electromagnetics Research, PIER 32 (Special Volume on Geometrical Methods for Comp. Electromagnetics)*, pages 65–87, 2001. http://cetaweb.mit.edu/pier/pier32/03.clemens.pdf.

E. Desurvire. *Erbium-Doped Fiber Amplifiers, Principles and Applications*. John Wiley and Sons, New-York, 2002.

P. Dular, C. Geuzaine, F. Henrotte, and W. Legros. A general environment for the treatment of discrete problems and its application to the finite element method. *IEEE Trans. on Magnetics*, 34(5):3395–3398, 1998. See also the Internet address http://www.geuz.org/getdp/.

R. de L. Kronig and W. G. Penney. Quantum mechanics of electrons in crystal lattices. *Proc. Roy. Soc. Lond. A*, 130:499–513, 1931.

H. Van der Lem, A. Tip, and A. Moroz. Band structure of absorptive two-dimensional photonic crystals. *J. Opt. Soc. Am. B.*, 20(6):1334–1341, 2003.

J. M. Dudley, L. Provino, N. Grossard, H. Maillotte, R. Windeler, B. Eggleton, and S. Coen. Supercontinuum generation in air-silica microstructured fibers with nanosecond and femtosecond pulse pumping. *J. Opt. Soc. Am. B*, 19(4):765–771, 2002.

B. Dillon and J. Webb. A comparison of formulations for the vector finite element analysis of waveguides. *IEEE Trans. Microw. Theory Tech.*, 42(2):308–316, 1994.

B. J. Eggleton, C. Kerbage, P. S. Westbrook, R. S. Windeler, and A. Hale. Microstructured optical fiber devices. *Opt. Express*, 9(13):698–713, 2001.

B. J. Eggleton, P. S. Westbrook, C. A. White, C. Kerbage, R. S. Windeler, and G. L. Burdge. Cladding-Mode-Resonnances in Air-Silica Microstructure Optical Fibers. *Journal of Lightwave Technology*, 18(8):1084–1099, 2000.

J. G. Fleming and S.-Y. Lin. Three-dimensional photonic crystal with a stop band from 1.35 to 1.95 μm. *Optics Letters*, 24(1):49–51, 1999.

J. W. Fleming. Material dispersion in lightguide glasses. *Electron. Lett.*, 14(11):326–328, 1978.

J. R. Folkenberg, N. A. Mortensen, K. P. Hansen, T. P. Hansen, H. R. Simonsen, and C. Jacobsen. Experimental investigation of cut-off phenomena in nonlinear photonic crystal fibers. *Optics Letters*, 28(20):1882–1884, 2003.

G.B. Folland. *Introduction to Partial Differential Equations*. Princeton University Press, Princeton, 1995.

A. Friedman. *Foundation of Modern Analysis*. Dover, New York, 1982.

A. Ferrando, E. Silvestre, P. Andrés, J.-J. Miret, and M.V. Andrés. Designing the properties of dispersion-flattened photonic crystal fibers. *Optics Express*, 9(13):687–697, 2001.

A. Ferrando, E. Silvestre, J.-J. Miret, P. Andrés, and M. V. Andrés. Donor and acceptor guided modes in photonic crystal fibers. *Optics Letters*, 25(18):1328–1330, 2000.

A. Ferrando, E. Silvestre, J.-J. Miret, P. Andrés, and M. V. Andrés. Vector description of higher-order modes in photonic crystal fibers. *J. Opt. Soc. Am. A*, 17(7):1333–1340, 2000.

A. Ferrando, E. Silvestre, J.-J. Miret, and P. Andrés. Nearly zero ultraflattened dispersion in photonic crystal fibers. *Opt. Lett.*, 25(11):790–792, 2000.

D. Felbacq, G. Tayeb, and D. Maystre. Scattering by a random set of parallel cylinders. *J. Opt. Soc. Am. A*, 9:2526–2538, 1994.

P. L. François and C. Vassallo. Finite cladding effects in W-fibers: a new interpretation of leaky modes. *Applied Optics*, 22(19):3109–3120, 1983.

Y. Fink, J. N. Winn, S. Fan, C. Chen, J. Michel, J. D. Joannopoulos, and E. L. Thomas. A dielectric omnidirectional reflector. *Science*, 282:1679–1682, 1998.

D. Felbacq and F. Zolla. *Scattering Theory of Photonic Crystals, Introduction to Complex Mediums for Optics and Electromagnetics*. SPIE Press, 2003.

A. Ferrando, M. Zacarés, P. Fernandez de Cordoba, D. Binosi, and J. A. Monsoriu. Spatial soliton formation in photonic crystal fibers. *Optics Express*, 11(5):452–459, 2003.

S. Guenneau, C.G. Poulton A.B. Movchan, and A. Nicolet. Coupling between electromagnetic and mechanic vibrations of thin-walled and composite structures. *Quat. Jour. Mech. Appl. Math.*, to appear.

B. Gralak, M. de Dood, G. Tayeb, S. Enoch, and D. Maystre. Theoretical study of photonic bandgap in woodpile crystals. *Phys. Rev. E*, 67:066601–1–18, 2003.

C. Geuzaine. *High-Order Hybrid Finite Element Schemes for Maxwell's Equations Taking Thin Structures and Global Quantities into Account*. PhD thesis, Université de Liège, Liège, 2002.

S. Guenneau, C. Geuzaine, A. Nicolet, A.B. Movchan, and F. Zolla. Electromagnetic waves in periodic structures. In *Proceedings of the Eleventh International Symposium on Applied Electromagnetics and Mechanics*, pages 116–117, 2003.

P.W. Gross and P.R. Kotiuga. *Electromagnetic Theory and Computation: A Topological Approach.* Cambridge University Press, Cambridge, 2004.

S. Guenneau, S. Lasquellec, A. Nicolet, and F. Zolla. Design of photonic crystal fibers using finite elements. *International Journal for Computation and Mathematics in Electrical and Electronic Engineering/ COMPEL*, 21(4):534–539, 2002.

S. Guenneau and A. B. Movchan. Analysis of elastic band structures for oblique incidence. *Arch. Rat. Mech. Anal.*, 171(1):129–150, 2004.

S. Guenneau, A. Nicolet, C. Geuzaine, F. Zolla, and A.B. Movchan. Comparisons of finite element and Rayleigh methods for the study of conical Bloch waves in arrays of metallic cylinders. *International Journal for Computation and Mathematics in Electrical and Electronic Engineering/ COMPEL*, to appear.

S. Guenneau, A. Nicolet, F. Zolla, C. Geuzaine, and B. Meys. A finite element method for spectral problem in optical fibers. *COMPEL, The International Journal for Computation and Mathematics in Electrical and Electronic Engineering*, 20(1):120–131, 2001.

S. Guenneau, A. Nicolet, F. Zolla, and S. Lasquellec. Modelling of photonic crystal optical fibers with finite elements. *IEEE Trans. on Magnetics*, 38(2):1261–1264, march 2002.

S. Guenneau, A. Nicolet, F. Zolla, and S. Lasquellec. Theoretical and numerical study of photonic crystal fibers. *Progress In Electromagnetic Research, PIER*, pages 271–305, 2003.

S. Guenneau, C. Poulton, and A. B. Movchan. Oblique propagation of electromagnetic and elastic waves for an array of cylindrical fibres. *Proc. Roy. Soc. Lond. A*, 459:2215–2163, 2003.

S. Guenneau, C.G. Poulton, and A.B. Movchan. Conical propagation of electromagnetic waves through an array of cylindrical inclusions. *Physica B*, 338:149–152, 2003.

S. Guenneau. *Homogenization of Quasi-Crystals and Analysis of Modes in Photonic Crystal Fibres.* PhD thesis, University of Aix-Marseille I, april 2001.

A. B. Yakovlev G. W. Hanson. *Operator Theory for Electromagnetics, An Introduction.* Springer-Verlag, New York, Heidelberg, Berlin, 2002.

S. Guenneau and F. Zolla. Homogenization of three-dimensional finite photonic crystals. *PIER*, 27:91–127, 2000.

J. Herrmann, U. Griebner, N. Zhavoronkov, A. Husakou, D. Nickel, J. C. Knight, W. J. Wadsworth, P. S. Russell, and G. Korn. Experimental evidence for supercontinuum generation by fission of higher-order solitons in photonic fibers. *Phys. Rev. Lett.*, 88(17):173901, 2002.

A. V. Husakou and J. Herrmann. Supercontinuum generation of higher-order solitons by fission in photonic crystal fibers. *Phys. Rev. Lett.*, 87(20):203901, 2001.

R. Hiptmair. Higher order Whitney forms. *Progress in Electromagnetics Research, Volume on Geometrical Methods for Comp. Electromagnetics)*, pages 271–299, 2001. http://cetaweb.mit.edu/pier/pier32/11.hiptmair.1.pdf.

K. C. Ho, P. T. Leung, A. M. van den Brink, and K. Young. Second quantization of open sytems using quasinormal modes. *Phys. Rev. E*, 3:1409–1417, 1999.

F. Henrotte, B. Meys, H. Hedia, P. Dular, and W. Legros. Finite element modelling with transformation techniques. *IEEE Trans. on Magnetics*, 35(3):1434–1437, 1999.

P. Helluy, S. Maire, and P. Ravel. Intégration numérique d'ordre élevé de fonctions régulières ou singulières sur un intervalle. *C.R. Acad. Sci. Paris*, 327(2):843–848, 1998.

T. Itoh, G. Pelosi, and P. Silvester, editors. *Finite Element Software for Microwave Engineering.* Wiley, New York, 1996.

P. R. Mc Isaac. Symmetry-induced modal characteristics of uniform waveguides-I: Summary of results. *IEEE Trans. on Microwave Theory Tech.*, 23(5):421–429, 1975.

P. R. Mc Isaac. Symmetry-induced modal characteristics of uniform waveguides-II: Theory. *IEEE Trans. on Microwave Theory Tech.*, 23(5):429–433, 1975.

K. Jänich. *Vector Analysis.* Springer Verlag, New-York, 2001.

J. D. Jackson. *Classical Electrodynamics.* John Wiley & Sons, INC, New York, third edition, 1999.

J. Jin. *The Finite Element Method in Electromagnetics.* Wiley, New York, second edition, 2002.

J. D. Joannopoulos, R. Meade, and J. N. Winn. *Photonic Crystals Molding the Flow of Light.* Princeton University Press, 1995.

S. John. Strong localization of photons in certain disordered dielectric superlattices. *Phys. Rev. Let.*, 58:2486–2489, 1987.

S. John. Electromagnetic absorption in a disordered medium near a photon mobility edge. *Phys Rev Lett*, 53:2169–2172, 1994.

P. Kaiser and H. W. Astle. Low-loss single-material fibers made from pure fused silica. *Bell Syst. Tech. J.*, 53(6):1021–1039, 1974.

J. C. Knight, J. Arriaga, T. A. Birks, A. Ortigosa-Blanch, W. J. Wadsworth, and P. S. Russell. Anomalous dispersion in photonic crystal fibers. *IEEE Photon. Tech.. Lett.*, 12(7):807–809, 2000.

T. Kato. *Perturbation Theory for Linear Operators.* Springer-Verlag, Berlin, 1995.

P. Kravanja and M. Van Barel. *Computing the Zeros of Analytic Functions.* Springer, Berlin, 2000.

J.C. Knight, J. Broeng, T.A. Birks, and P. S. J. Russel. Photonic Band Gap Guidance in Optical Fibers. *Science*, 282:1476–1478, 1998.

J. C. Knight, T. A. Birks, and S. Russell. Properties of photonic crystal fiber and the effective index model. *J. Opt. Soc. Am. A*, 15(4):746–750, 1998.

J. C. Knight, T. A. Birks, P. S. J. Russel, and D. M. Atkin. All silica single-mode optical fiber with photonic crystal cladding. *Opt. Lett.*, 21(19):1547–1549, 1996.

B. Kuhlmey, C. Martijn de Sterke, R. McPhedran, P. Robinson, G. Renversez, and D. Maystre. Microstructured optical fibers: where's the edge. *Optics Express*, 10(22):1285–1290, 2002.

Ch. Kittel. *Introduction to Solid State Physics.* Wiley Text Books, 1995.

B. Kuhlmey, R. C. McPhedran, and C. M. de Sterke. Modal 'cutoff' in microstructured optical fibers. *Optics Letters*, 27(19):1684–1686, 2002.

D. B. Keck, R. D. Maurer, and P. C. Schultz. Ultimate lower limit of attenuation in glass optical waveguides. *Appl. Phys. Lett.*, 22:307–309, 1973.

B. Kuhlmey, G. Renversez, and D. Maystre. Chromatic dispersion and losses of microstructured optical fiber. *Applied Optics OT*, 42(4):634–639, 2003.

B. Kuhlmey. *Theoretical and Numerical Investigation of the Physics of Microstructured Optical Fibers.* PhD thesis, Université Aix-Marseille III and University of Sydney, 2003.

B. Kuhlmey, T. P. White, G. Renversez, D. Maystre, L.C. Botten, C. Martijn de Sterke, and R. C. McPhedran. Multipole method for microstructured optical fibers II: implementation and results. *J. Opt. Soc. Am. B*, 10(19):2331–2340, 2002.

J.M. Lourtioz, H. Benisty, V. Berger, J.-M. Gérard, D. Maystre, and A. Tchelnokov. *Les cristaux photoniques.* Hermes, Paris, 2003.

Ph. Langlet, A.-C. Hladky-Hennion, and J.-N. Decarpigny. Analysis of the propagation of plane acoustic waves in passive periodic materials using the finite element method. *J. Acoust. Soc. Am.*, 98(5):2792–2800, 1995.

K. M. Lee, P. T. Leung, and K. M. Pang. Dyadic formulation of morphology-dependent resonances. I. Completeness relation. *J. Opt. Soc. Am. B*, 16(9):1409–1417, 1999.

K. M. Lee, P. T. Leung, and K. M. Pang. Dyadic formulation of morphology-dependent resonances. II. Perturbation theory. *J. Opt. Soc. Am. B*, 16(9):1418–1430, 1999.

M. Lassas, J. Liukkonen, and E. Somersalo. Complex Riemannian metric and absorbing boundary condition. *J. Math. Pures Appl.*, 80(9):739–768, 2001.

K. M. Lo, R. C. McPhedran, I. M. Bassett, and G. W. Milton. An electromagnetic theory of dielectric waveguides with multiple embedded cylinders. *J. Lightwave Technology*, 12:396–410, 1994.

P. T. Leung and K. M. Pang. Completeness and time-independent perturbation of morphology-dependent resonances in dielectric spheres. *J. Opt. Soc. Am. B*, 13(5):805–817, 1996.

J.-F. Lee, D.-K. Sun, and Z. Cendes. Full-wave analysis of dielectric waveguides using tangential vector finite elements. *IEEE Trans. Microw. Theory Tech.*, 39(8):1262–1271, 1991.

P. T. Leung, W. M. Suen, C. P. Sun, and K. Young. Waves in open systems via a biorthogonal basis. *Phys. Rev. E*, 57(5):6101–6104, 1998.

R. B. Lehoucq, D. C. Sorensen, and C. Yang. *ARPACK Users' Guide: Solution of Large-Scale Eigenvalue Problems with Implicitly Restarted Arnoldi Methods.* SIAM, Philadelphia, 1998.

P. T. Leung, S. S. Tong, and K. Young. Two-component eigenfunction expansion for open systems described by the wave equation I: completeness of expansion. *J. Phys. A: Math. Gen.*, 30:2139–2151, 1997.

P. T. Leung, S. S. Tong, and K. Young. Two-component eigenfunction expansion for open systems described by the wave equation II: linear space structure. *J. Phys. A: Math. Gen.*, 30:2153–2162, 1997.

Dietrich Marcuse. *Theory of Dielectric Optical Waveguides.* Academic Press, San Diego, 2nd edition, 1991.

T. M. Monro, P. J. Bennett, N. G. R. Broderick, and D. J. Richardson. Holey fibers with random cladding distributions. *Opt. Lett.*, 25(4):206–208, 2000.

D. Mogilevstev, T. A. Birks, and P. S. Russell. Group-velocity dispersion in photonic crystal fibers. *Opt. Lett.*, 23(21):1662–1664, 1998.

R.C. McPhedran and D.H. Dawes. Lattice sums for a dynamic scattering problem. *Jour. Electromagn. Waves Appl.*, 6:1327–1340, 1992.

R. B. Melrose. *Geometric scattering theory.* Cambridge University Press, 1995.

B. Meys. *Ph. D. Thesis, Modélisation des champs électromagnétiques aux hyperfréquences par la méthode des éléments finis.* PhD thesis, Université de Liège, Liège, 1999.

N. A. Mortensen, J. R. Folkenberg, M. D. Nielsen, and K. P. Hansen. Modal cutoff and the V parameter in photonic crystal fibers. *Optics Letters*, 28(20):1879–1881, 2003.

A.B. Movchan and S. Guenneau. Localised modes in split ring resonators. *Physical Review B,* to appear.

A. Mitchell and D. Griffiths. *The Finite Differences Method in Partial Differential Equations.* Wiley, Chichester, 1980.

W. N. MacPherson, M. J. Gander, R. McBride, J. D. C. Jones, P. M. Blanchard, J. G. Burnett, A. H. Greenaway, B. Mangan, T. A. Birks, J. C. Knight, and P. S. Russell. Remotely addressed optical fibre curvature sensor using multicore photonic crystal fibre. *Optics Commun.*, 193:97–104, 2001.

A. A. Maradudin and A. R. McGurn. Out of plane propagation of electromagnetic waves in a two-dimensional periodic dielectric medium. *Journal of Modern Optics*, 41(2):275–284, 1994.

A.B. Movchan, N.V. Movchan, and C.G. Poulton. *Asymptotic Models of Fields in Dilute and Densely Packed Composites.* Imperial College Press, London, 2002.

R. C. McPhedran, N. A. Nicorovici, L. C. Botten, and Bao Ke-Da. Green's functions, lattices sums, and Rayleigh's identity for a dynamic scattering problem. In G. Papanicolaou, editor, *IMA Volumes in Mathematics and its Applications.* Springer, New York, 1997.

N. A. Mortensen. Effective area of photonic crystal fibers. *Optics Express*, 10(7):341–348, 2002.

R. J. Mears, L. Reekie, I. M. Jauncey, and D. N. Payne. Low-noise Erbium-doped fiber amplifier operating at 1.54 μm. *Electron. Lett.*, 23(17):884–885, 1987.

M. Midrio, M. P. Singh, and C. G. Someda. The space filling mode of holey fibers: An analytical vectorial solution. *Journal of Lightwave Technology*, 18(7):1031–1037, 2000.

D. Maystre and P. Vincent. Diffraction d'une onde électromagnétique plane par un objet cylindrique non-infiniment conducteur de section arbitraire. *Optics Communications*, 5(5):327–330, 1972.

M. Maeda and S. Yamada. Leaky modes on W-fibers: mode structure and attenuation. *Applied Optics*, 16(8):2198–2203, 1977.

J.-C. Nédélec. Mixed finite elements in $\mathbb{R}^3$. *Numer. Math.*, 35:314–341, 1980.

J.-C. Nédélec. A new family of mixed finite elements in $\mathbb{R}^3$. *Numer. Math.*, 50:57–81, 1986.

J.-C. Nédélec. *Notions sur les techniques d'éléments finis.* Ellipses, Mathématiques & Applications n°7, Paris, 1991.

M. Nakahara. *Geometry, Topology and Physics.* IOP, Bristol, 1990.

M. Nevière. The Homogeneous Problem. In *Electromagnetic Theory of Gratings*, volume 22 of *Topics in Current Physics*, chapter 5. Springer-Verlag, 1980.

A. Nicolet, S. Guenneau, C. Geuzaine, and F. Zolla. Modeling of electromagnetic waves in periodic media with finite elements. *Journal of Computational and Applied Mathematics,* to appear.

A. Nicolet, S. Guenneau, F. Zolla, C. Gueuzaine, B. Kuhlmey, and G. Renversez. Numerical investigation of photonic crystal fibers by spectral and multipole methods. In A.B. Movchan, editor, *Proceedings of IUTAM 2002, Asymptotics, Singularities and Homogenisation in Problems of Mechanics*, pages 23–31, Liverpool, July 2002. Kluwer Academic Press.

S. W. Ng, P. T. Leung, and K. M. Lee. Dyadic formulation of morphology-dependent resonances. III. degenerate perturbation theory. *J. Opt. Soc. Am. B*, 19(1):154–164, 2002.

N.A. Nicorovici, R.C. McPhedran, and L.C. Botten. Photonic band gaps for arrays of perfectly conducting spheres. *Phys. Rev. E*, 52:1135–1145, 1995.

M. Nevière and E. Popov. *Light propagation in periodic media*, chapter 14. Marcel Dekker, 2003.

A. Nicolet, F. Zolla, and S. Guenneau. Modelling of twisted optical waveguides with edge elements. *Eur. Phys. Jour. Appl. Phys.*, to appear.

A. Ortigosa-Blanch, J. C. Knight, W. J. Wadsworth, J. Arriaga, B. J. Mangan, T. A. Birks, and P. S. Russell. Highly birefringent photonic crystal fibers. *Optics Letters*, 25(18):1325–1327, 2000.

T. Okoshi. *Optical Fibers.* Academic Press, New York, 1982.

J. M. Pottage, D. Bird, T. D. Hedley, J. C. Knight, T. A. Birks, P. S. Russell, and P. J. Roberts. Robust photonic band gaps for hollow core guidance in PCF made from high index glass. *Optics Express*, 11(22):2854–2861, 2003.

J. H. Price, W. Belardi, T. M. Monro, A. Malinowski, A. Piper, and D. J. Richardson. Soliton transmission and supercontinuum generation in holey fiber, using a diode pumped ytterbium fiber source. *Opt. Express*, 10(8):382–387, 2002.

W.H. Press, B.P.Flannery, S.A. Teukolsky, and W.T. Vetterling. *Numerical Recipes.* Cambridge University Press, 1986.

J.B. Pendry. Low energy electron diffraction. *Academic Press, London*, 1974.

J.B. Pendry. Photonic band structures. *Jour. Mod. Opt.*, 41(2):209–229, 1994.

C.G. Poulton, S. Guenneau, and A.B. Movchan. Non-commuting limits and effective properties for oblique propagation of electromagnetic waves through an array of aligned fibres. *Physical Review B,* to appear.

C.G. Poulton, S. Guenneau, A.B. Movchan, and A. Nicolet. Transverse propagating waves in perturbed periodic structures. In A.B. Movchan, editor, *Asymptotics, singularities and homogenisation in problems of mechanics, Proceedings of the IUTAM Symposium*, pages 147–157. Kluwer Academic Publishers, 2003.

M. Plihal and A.A. Maradudin. Photonic band structure of two-dimensional systems: the triangular lattice. *Phys. Rev. B*, 44:8565–8571, 1991.

C.G. Poulton, A.B. Movchan, R.C. McPhedran, N.A. Nicorovici, and Y.A. Antipov. Eigenvalue problems for doubly periodic elastic structures and phononic band gaps. *Proc. Roy. Soc. Lond. A*, 456:2543–2559, 2000.

C. Poirier. *Guides d'Ondes Électromagnétiques: Analyse Mathématique et Numérique*. PhD thesis, Université de Nantes, Nantes, 1994.

C. G. Poulton. *Asymptotics and Wave Propagation in Cylindrical Geometries*. PhD thesis, University of Sydney, 1999.

C. D. Poole and R. E. Wagner. Phenomenological approach to polarization dispersion in long single-mode fibers. *Electron. Lett.*, 22(17):1029–1030, 1986.

Lord Rayleigh. On the influence of obstacles arranged in rectangular order upon the properties of a medium. *Phil. Mag.*, 34:481–502, 1892.

B. Daya Reddy. *Introductory Functional Analysis with Applications to Boundary Value Problems and Finite Elements*. Springer, New York, 1998.

R. D. Richtmyer. *Principles of Advanced Mathematical Physics*, volume 1. Springer-Verlag, New York, Heidelberg, Berlin, 1978.

G. Renversez, B. Kuhlmey, and R. McPhedran. Dispersion management with microstructured optical fibers: Ultra-flattened chromatic dispersion with low losses. *Optics Letters*, 28(12):989–991, 2003.

W. H. Reeves, J. C. Knight, and P. S. Russell. Demonstration of ultra-flattened dispersion in photonic crystal fibers. *Optics Express*, 10(14):609–613, 2002.

P. St. Russell, E. Marin, A. Diez, S. Guenneau, and A.B. Movchan. Enhancing sound-light interaction in periodically microstructured photo-sonic crystals. *Optics Express*, 11(20):2555–2560, 2003.

M. Reed and B. Simon. *Methods of Modern Mathematical Physics*. Academic Press, New York, 1978.

P. Raviart and J.M. Thomas. A mixed finite element method for second order elliptic problems. In A. Dold and B. Eckmann, editors, *Mathematical Aspects of Finite Element Methods*, New-York, 1977. Springer Verlag.

P. Raviart and J.M. Thomas. *Introduction à l'analyse numérique des équations aux dérivées partielles*. Masson, Paris, 1983.

S. C. Rasleigh and R. Ulrich. Polarisation mode dispersion in single-mode fibres. *Optics Letters*, 3:60–62, 1978.

W. Rudin. *Real and Complex Analysis*. McGraw-Hill Science, 1986.

P. Russell. Photonic crystal fibers. *Science*, 299(5605):358–362, 2003.

J. K. Ranka, R. S. Windeler, and A. J. Stentz. Visible continuum generation in air-silica microstructure optical fibers with anomalous dispersion at 800 nm. *Opt. Lett.*, 25(1):25–27, 2000.

Y. Saad. Sparskit: a basic tool-kit for sparse matrix computations. 1999. http://www-users.cs.umn.edu/ saad/software/SPARSKIT/sparskit.html.

K. Sakoda. *Optical Properties of Photonic Crystals*, volume 80. Springer Series in Optical Sciences, 2001.

L. Schwartz. *Méthodes mathématiques pour les sciences physiques.* Hermann, Paris, 1987 edition, 1965.

L. Schwartz. *Théorie des Distributions.* Hermann, Paris, third edition, 1966.

M. Schechter. *Operator Methods in Quantum Mechanics.* North Holland, New York, Oxford, 1980.

G. Scharf. *From Electrostatics to Optics.* Springer Verlag, Berlin Heidelberg, 1994.

H. S. Sözüer and J. P. Dowling. Photonic band calculations for woodpile structures. *Journal of Modern Optics*, 41(2):231–239, 1994.

R. H. Stolen and E. P. Ippen. Raman gain in glass optical waveguides. *Appl. Phys. Lett.*, 22(6):276–278, 1973.

K. Suzuki, H. Kubota, S. Kawanishi, M. Tanaka, and M. Fujita. Optical properties of a low-loss polarization-maintaining photonic crystal fiber. *Opt. Express*, 9(13):676–680, 2001.

A. W. Snyder and J. D. Love. *Optical Waveguide Theory.* Chapman & Hall, 1983.

M. J. Steel and R. M. Osgood. Elliptical-hole photonic crystal fibers. *Opt. Lett.*, 26(4):229–231, 2001.

S. J. Savory and F. P. Payne. Pulse propagation in fibers with polarization-mode dispersion. *J. Lightwave Technology*, 19(3):350–357, 2001.

M. J. Steel, T. P. White, C. Martijn de Sterke, R. C. McPhedran, and L. C. Botten. Symmetry and degeneracy in microstructured optical fibers. *Opt. Lett.*, 26(8):488–490, 2001.

A. W. Snyder and W.R. Young. Modes of optical waveguides. *J. Opt. Soc. Am.*, 68:297–309, 1978.

A. Taflove. *Computational Electrodynamics: The Finite-Differences Time-Domain Method.* Artech House, Boston, 1995.

F. L. Teixeira and W. C. Chew. Unified analysis of perfectly matched layers using differential forms. *Microw. Opt. Technol. Lett.*, 20(2):124–126, 1999.

F. L. Teixeira. On aspects of the physical realizability of perfectly matched absorbers for electromagnetic waves. *Radio Sci.*, 38(2):8014, 2003.

B. Temelkuran, S. D. Hart, G. Benoit, J. D. Joannopoulos, and Y. Fink. Wavelength-scalable hollow optical fibres with large photonic bandgaps for CO_2 laser transmission. *Nature*, 420:650–653, 2002.

L. N. Trefethen and D. Bau III. *Numerical Linear Algebra.* SIAM, Philadelphia, 1995.

E. Tonti. On the geometrical structure of the electromagnetism. *Gravitation, Electromagnetism and Geometrical Structures, for the 80th birthday of A. Lichnerowicz*, pages 281–308, 1995.

E. Tonti. A direct discrete formulation of field laws: The cell method. *CMES - Computer Modeling in Engineering & Sciences*, 2(2):237–258, 2001.

E. Tonti. Finite formulation of the electromagnetic field. *Progress in Electromagnetics Research, PIER 32 (Special Volume on Geometrical Methods for Comp. Electromagnetics)*, pages 1–44, 2001. http://cetaweb.mit.edu/pier/pier32/01.tonti.pdf.

G. A. Thomas, B. I. Shraiman, P. F. Glodis, and M. J. Stephen. Towards the clarity limit in optical fibre. *Nature*, 404:262–264, 2000.

R. Ulrich. Far-infrared properties of metallic mesh and its complementary structure. *Infrared Phys.*, 7:37–55, 1967.

R. Ulrich. Interference filters for the far infrared. *Appl. Optics*, 7:1987–1996, 1968.

R. Vallée. *Fibre Optic Communication Devices*, chapter 2. Photonics. Springer, Berlin, 2001.

Y. A. Vlasov, X.-Z. Bo, J. C. Sturm, and D. J. Norris. On-chip natural assembly of silicon photonic bandgap crystals. *Nature*, 414:289–293, 2001.

L. Vardapetyan and L. Demcowicz. Full-wave analysis of dielectric waveguides at a given frequency. *Mathematics of Computation*, (72):105–129, 2003.

L. Vardapetyan, L. Demcowicz, and D. Neikirk. *hp*-vector finite element method for eigenmode analysis of waveguides. *Comput. Methods Appl. Mech. Engrg.*, (192):185–201, 2003.

M. A. van Eijekelenborg, M. C. J. Large, A. Argyros, J. Zagari, S. Manos, N. A. Issa, S. Fleming, R. C. McPhedran, C. M. de Sterke, L. C. Botten, and M. J. Steel. Microstructured polymer optical fibre. *Optics Express*, 9(7):319–327, 2001.

G.N. Watson. *A Treatise on the Theory of Bessel Functions*. Cambridge University Press, 1944.

H. Whitney. *Geometric Integration Theory*. Princeton University Press, Princeton, 1957.

W. Wijngaard. Guided normal modes of two parallel circular dielectric rods. *J. Opt. Soc. Amer.*, 63:944–949, 1973.

T. P. White, B. Kuhlmey, R. C. McPhedran, D. Maystre, G. Renversez, C. Martijn de Sterke, and L.C. Botten. Multipole method for microstructured optical fibers I: formulation. *J. Opt. Soc. Am. B*, 10(19):2322–2330, 2002.

W. J. Wadsworth, J. C. Knight, A. Ortigosa-Blanch, J. Arriaga, E. Silvestre, and P. S. Russell. Soliton effects in photonic crystal fibres at 850 nm. *Electron. Lett.*, 36(1):53–55, 2000.

T.P. White, R.C. McPhedran, L.C. Botten, G.H. Smith, and C.M. de Sterke. Calculations of air-guided modes in photonic crystal fibers using the multipole method. *Optics Express*, 9(13):721–732, 2001.

T. P. White, R. C. McPhedran, C. M. de Sterke, L. C. Botten, and M. J. Steel. Confinement losses in microstructured optical fibers. *Opt. Lett.*, 26(21):1660–1662, 2001.

B. R. Washburn, S. E. Ralph, and R. S. Windeler. Ultrashort pulse propagation in air-silica microstructure fiber. *Opt. Express*, 10(13):575–580, 2002.

E. Yablonovitch. Inhibited spontaneous emission in solid state physics and electronics. *Phys. Rev. Lett.*, 58(20), 1987.

K. Yee. Numerical solution of initial boundary value problems involving maxwell's equations in isotropic media. *IEEE Trans. Antennas and Propagation*, (14):302–307, 1966.

E. Yablonovitch, T. J. Gmitter, and K. M. Leung. Photonic band structures: the face-centered cubic case employing non-spherical atoms. *Phys. Rev. Lett.*, 67:2295, 1991.

Y.G. Yardley, R.C. McPhedran, N.A. Nicorovici, and L.C. Botten. Addition formulae and the Rayleigh identity for arrays of elliptical cylinders. *Phys. Rev. E*, 60:6068, 1999.

K. Yosida. *Functional Analysis*. Springer, Berlin, 6th edition, 1980.

P. Yeh, A. Yariv, and E. Marom. Theory of Bragg fiber. *Journal of the Optical Society of America*, 68(9):1196–1201, 1978.

F. Zolla and S. Guenneau. Artificial ferro-magnetic anisotropy: homogenization of 3D finite photonic crystals. In A. B. Movchan, editor, *Proceedings of IUTAM 2002, Asymptotics, Singularities and Homogenization in Problems of Mechanics*, pages 375–384, Liverpool, July 2002. IUTAM, Kluwer Academic Press.

Index